Richard Owen

Richard Owen

Biology without Darwin, a Revised Edition

NICOLAAS A. RUPKE

The University of Chicago Press
Chicago and London

Nicolaas Rupke is professor of the history of science at Göttingen University and director of the Institut für Wissenschaftsgeschichte. His previous books include *The Great Chain of History* and *Alexander von Humboldt: A Metabiography*, the latter of which has been reprinted by the University of Chicago Press.

The University of Chicago Press, Chicago 60637
The University of Chicago Press, Ltd., London

Printed in the United States of America

18 17 16 15 14 13 12 11 10 09   1 2 3 4 5

ISBN-13: 978-0-226-73177-3 (paper)
ISBN-10: 0-226-73177-4 (paper)

Library of Congress Cataloging-in-Publication Data
Rupke, Nicolaas A.
Richard Owen : biology without Darwin / Nicolaas A. Rupke. —Rev. ed.
p. cm.
Includes bibliographical references and index.
ISBN-13: 978-0-226-73177-3 (pbk. : alk. paper)
ISBN-10: 0-226-73177-4 (pbk. : alk. paper) 1. Owen, Richard, 1804–1892.
2. Naturalists—Great Britain—Biography. I. Title.
QH31.094R86 2009
508.092—dc22
[B] 2008037786

⊗ The paper used in this publication meets the minimum requirements of the American National Standard for Information Sciences—Permanence of Paper for Printed Library Materials, ANSI Z39.48-1992.

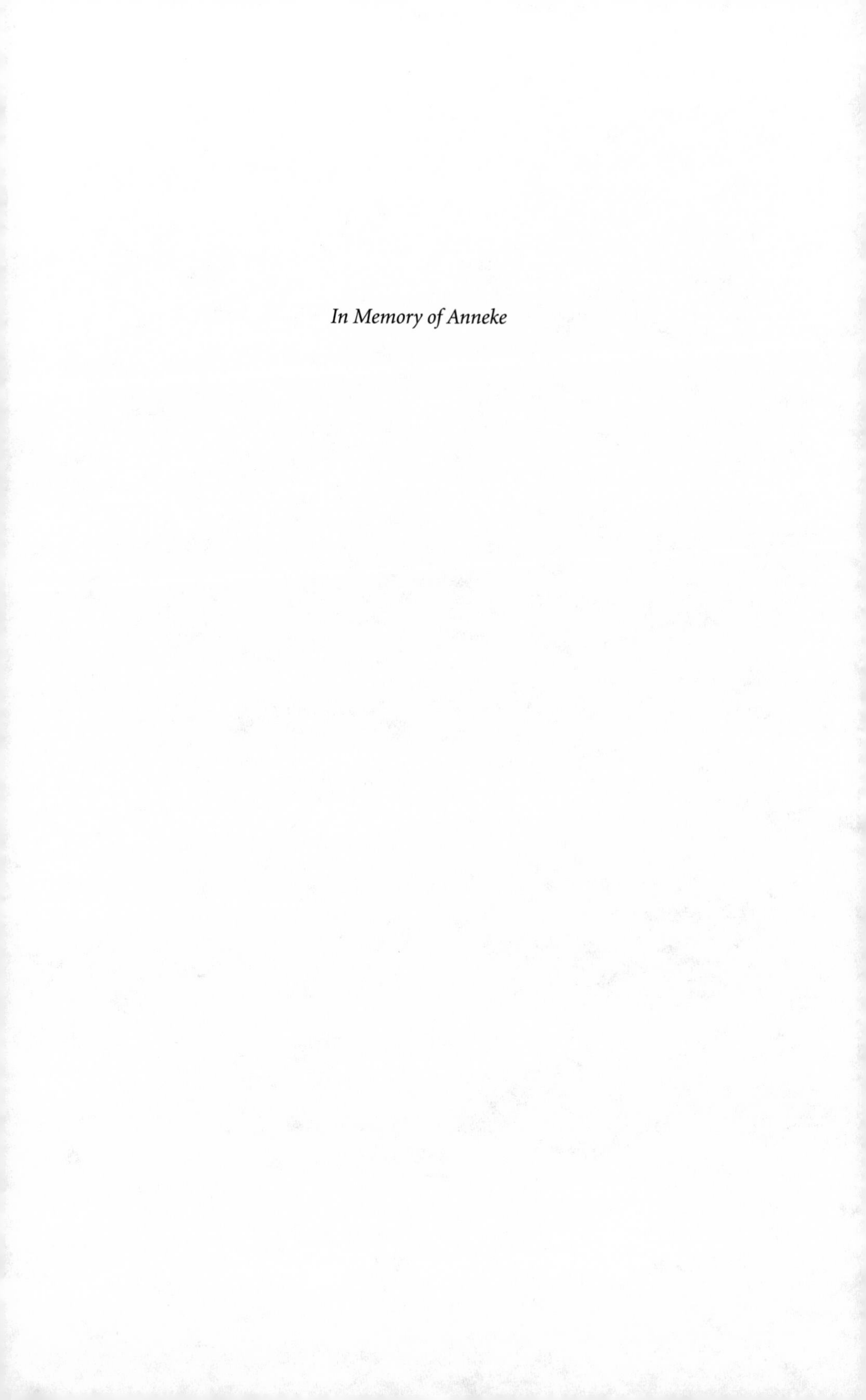

In Memory of Anneke

Contents

Illustrations

Figures

Tables

Preface

The appearance of this book in 1994 was greeted in the reviewing press as a fitting rehabilitation of Richard Owen. No longer maltreated as Charles Darwin's creationist whipping boy, he was put forward as Britain's leading biologist of the mid-nineteenth century, who well before the *Origin of Species* advocated an evolutionary theory of the origin of species, albeit a non-Darwinian one. Alluding to Owen's fame as an osteologist, Janet Browne quipped that "Old Bones" was "the skeleton in the cupboard of evolutionary science," whose image had now been restored to that of an "endlessly fascinating scientific thinker," in spite of—or possibly also because of—his having been at odds with Darwin.[1] Similarly, a majority of the critics (who in gratifyingly large numbers discussed the book) highlighted that Owen had been the towering representative of a non-Darwinian tradition in Victorian biology and paleontology. It is perhaps fitting that I acknowledge the collective thrust of my reviewers' responses by giving the new edition of *Richard Owen* a different subtitle, replacing the original *Victorian Naturalist* with the more specific *Biology without Darwin.*

Owen represented "biology without Darwin" in more than one sense. The two great naturalists differed above all about the nature of organic evolution. They held opposing views on the origin of life from lifeless matter (Owen postulated multiple spontaneous "emergences"; Darwin kept the issue at arm's length), on the mechanism of species development (Owen stressed an inner, "genetic" cause, Darwin external, natural selection), on the pattern of evolution through geologic time (Owen saw in it a structural logic, Darwin the haphazardness of contingency), and on "man's place in nature" (Owen stressed the unity of humanity and its distance from the apes; the Darwinians constructed close racialist links between "lower humans" and "higher apes"). In addition,

Owen tried, much more than Darwin, to bring processes of morphogenesis to bear on the origin of species, and as such he was an early representative of what today we refer to as evo-devo, the field of evolutionary biology that integrates the study of how individual organisms develop with the development of species. Owen and Darwin also occupied quite different spaces in Victorian science: Owen as museum curator, his thinking circumscribed by osteological collections, Darwin as scientific voyager and man of means whose interests lay in living populations.

To bring out these and other characteristic features of Owen's scientific career—to look at Owen on his own terms—I joined those who in recent years have stressed the importance of places and practices in the production as well as reception of scientific knowledge,[2] experimenting with a narrative structure that directed attention to the institutional coordinates of Owen's biological thought. In putting the successive chapters together, I abandoned the conventional practice of following the time line of Owen's life, using instead a narrative that moves from context to content. Several critics apprehended the intent of this construction,[3] but not all.[4] I therefore want to make clear more explicitly than in the 1994 edition that the narrative structure of *Richard Owen* serves an argument, namely that Owen's scientific work is best understood by examining it in the context of his museum career and as integral to his institution-building agenda. In the words of one reviewer, the book depicts "Owen as strategist."[5]

My story is thus crafted as a historical nest of boxes,[6] starting with a broad conspectus of Owen's positions at the Hunterian Museum and the British Museum (Natural History) and of his participation in the metropolitan museum movement in general. This encapsulates a synopsis of his duties and opportunities as a museum curator and expositor of the collections—his acquisitions policy, cataloguing efforts, and teaching of and lecturing to students, colleagues, and popular audiences. Further enclosed, then, I offer a close-up of Owen's actual research accomplishments, which in turn embeds the story of Owen's clash with Darwin and his followers over the origin of species and related issues.

Trevor Levere put it well: "In this account, Owen exists in and through his work, and we discover his personality principally through that medium."[7] It should go without saying that I do not consider my representation of Owen in any sense definitive. Other portraits have been painted and more will surely appear. One such, as suggested by Jane Camerini and other reviewers, could usefully dwell on Owen-the-man, his character, private life, family circumstances, and the like, more than I have done in my introductory chapter "Personality Matters."[8] My purpose, however, has been less to present a psychological portrait than to demonstrate the formative importance for Owen's scientific oeuvre of his institutional location as well as of the patronage politics associated with his museum ambitions.

The present edition of *Richard Owen* is an updated abridgment of the 1994 original. It omits most of the long quotations, a few illustrations, and several chapter sections that to a student readership may seem overly detailed, technical, or tangential. Nevertheless, additions have been made, for example, references to recent research into the "third way" in mid-nineteenth-century thinking about the origin of species—the theory of the autogenous and heterogenous origin of species[9]—which help clarify Owen's disagreements with both Lamarck and Darwin over how one species may give rise to another. These changes, together with a variety of small improvements throughout, enhance the book's accessibility to a twenty-first-century audience, strengthening the argument and assisting comprehension by means of a more streamlined narrative. Further references to recent Owen-related writings have been added to the text and bibliography. The new publications for the most part relate to one of three aspects of Owen's activities and impact: the hippocampus, or "brain," controversy with T. H. Huxley; the founding of London's Natural History Museum; and "the survival of Richard Owen,"[10] that is, his enduring legacy of idealist biology through the late nineteenth century and indeed till today.[11]

Given our strong continuing interest in issues of evolutionary biology and its history, this new, more reader-friendly edition of *Richard Owen* should prove timely. The original edition is out of print, with secondhand copies offered at many times the cover price. I am much indebted to the University of Chicago Press and in particular to Christie Henry, executive editor for life sciences and geography, for issuing this edition in paperback on the occasion of the Darwin bicentenary.

Acknowledgments

This study was supported by research fellowships at the Wellcome Institute for the History of Medicine in London (Wellcome Trust Fellowship), the Netherlands Institute for Advanced Study in Wassenaar, the University of Tübingen (Alexander von Humboldt Fellowship), the National Humanities Center in North Carolina (Walter Hines Page Fellowship), and the Institute of Advanced Studies at the Australian National University in Canberra (Senior Fellowship). For their support, I owe a debt of gratitude to my colleagues at these institutions, and in particular to W. F. Bynum, Wolf von Engelhardt, and the late Eugene Kamenka. It is a pleasure, also, to acknowledge stimulating conversations with, and specific items of information as well as moral encouragement from, many friends and colleagues, among whom are Rosemary Ashton, Michael Bassett, Edith Gilchrist, Jacob Gruber, Raina Haig, Renato Mazzolini, Alec Panchen, Peter Prange, Phillip Sloan, Francis Warner, Paul van Woerkom, and especially Karen Wonders. Furthermore, I am much indebted to Trevor Levere and David Livingstone for their insightful comments on the manuscript of the new, abbreviated edition, and to Pamela Bruton for her skillful copyediting. Last but not least, the generous assistance of the staff at the archives and libraries mentioned in the list of manuscript sources in the bibliography is gratefully acknowledged.

Chronology

1804:	Born in Lancaster on July 20, the second of six children, the younger of two sons. His father, Richard Owen Senior, was a West India merchant; his mother, born Catherine Parrin, was of Huguenot descent.
1807:	Geological Society founded.
1809:	Richard's father dies.
1813:	Hunterian Museum opens.
1810–20:	Attends local grammar school; is reportedly "lazy and impudent."
1820–24:	Apprenticed to local surgeons/apothecaries, thus gaining access to postmortem dissections at the county jail.
1824:	Matriculates at Edinburgh University, where he attends the extramural lectures on anatomy by John Barclay; stays for only about half a year (Oct. 1824 to spring 1825).
1825:	Moves to St. Bartholomew's Hospital, London; appointed prosector to the lectures by John Abernethy.
1826:	Becomes a member of the Royal College of Surgeons (Aug. 18); sets up in medical practice. Zoological Society founded.
1827:	At the age of twenty-two is made assistant curator to William Home Clift at the Hunterian Museum on Abernethy's urging to assist with the preparation of the catalogues of Hunter's collections (Mar. 6).
1827–56:	Employed at the Royal College of Surgeons.
1828:	Appointed lecturer in comparative anatomy, St. Bartholomew's Hospital; start of his career as a lecturer. Begins cooperation with the Zoological Society and dissections of animals that died at the Regent's Park Zoo.
1830:	Georges Cuvier visits the Hunterian Museum in August.
1831:	Makes return visit to Cuvier in Paris during July–August. Back in London, writes "Report on the Museum d'anatomie comparée." British Association for the Advancement of Science founded. Natural History

Department in the new building of the British Museum in Bloomsbury opens to the public.

1832: Publishes *Memoir on the Pearly Nautilus.* Begins work on marsupials and monotremes, summarized in Todd's *Cyclopaedia of Anatomy and Physiology*, vol. 3. William Home Clift dies.

1833: Founds the short-lived *Zoological Magazine.*

1833–40: Five volumes of his *Descriptive and Illustrated Catalogue of the Physiological Series* are published.

1834: Appointed to the newly established chair of comparative anatomy, St. Bartholomew's Hospital. Elected fellow of the Royal Society (Dec. 18) at the age of thirty for his work on marsupials and monotremes.

1834–37: Rebuilding of the Royal College of Surgeons, including the Hunterian Museum (Apr. 9, 1834–Feb. 14, 1837).

1835: Marries Caroline Clift (July 20) after an engagement of nearly eight years. Describes *Trichina spiralis*, the parasite that causes trichinosis. Starts work on T. L. Mitchell's Australian fossils.

1835–36: Becomes member of Select Committee on "the conditions, management and affairs of the British Museum."

1835–52: Occupies apartments in the building of the Royal College of Surgeons.

1836: Appointed Hunterian professor on the rotating scheme. The enlarged Hunterian Museum is completed; Owen exhibits the fossil collection for the first time. Starts cooperation with Charles Darwin, studying mammalian fossils brought back from South America (*Toxodon*, etc.). Elected fellow of the Linnean Society.

1837: Becomes first permanent Hunterian professor, at the age of thirty-two. Elected Fullerian professor but is obliged to decline. Elected fellow of the Geological Society for his *Toxodon* work-in-progress. His only child, William, is born (Oct. 6).

1838: Receives Wollaston Medal of the Geological Society for his work-in-progress on Darwin's fossils. Attends the Gesellschaft deutscher Naturforscher und Ärzte in Freiburg, Germany, where he is treated as a celebrity. Visits Dutch and German museums on the way. His mother dies (Nov. 25).

1839: Begins "Contributions to the Natural History of the Anthropoid Apes." Delivers papers on the Stonesfield opossum to the Geological Society; BAAS "Report on British Fossil Reptiles" (pt. 1); and paper on the moa to the Zoological Society (Nov. 12).

1840: Victoria marries Albert (Feb. 10). Elected a member of the Athenaeum Club under Rule II (Feb. 25). Elected corresponding member of the Institute of France. Publishes *The Zoology of the Voyage of H.M.S. Beagle*, part 1, *Fossil Mammalia.* Carlyle's lectures "On Heroes, Hero-Worship and the Heroic in History."

1840–41: Is first president of the Microscopical Society.

1840–45: Publishes *Odontography.*

1841: Begins work on the osteological catalogue. Kew Gardens, under William Hooker, opens to the public.

1842: Publication in April of the revised BAAS "Report on British Fossil Reptiles" (pt. 2); introduces the order *Dinosauria.* Owen and Thomas Carlyle become acquainted. Chartist riots. Delivers BAAS "Report on the British Fossil Mammalia" (pt. 1). Public dinner in Lancaster in honor of Owen and William Whewell. Robert Peel offers Owen a civil list pension of £200 (Nov. 1); strong support from William Buckland. Appointed joint conservator with William Clift.

1842–56: Conservator at the Hunterian Museum.

1843: Confirmation of moa reconstruction arrives (Jan. 19). Starts new series of Hunterian lectures, arranged by taxonomic grouping rather than by organs. Meets Charles Dickens (May 12). Delivers BAAS "Report on the British Fossil Mammalia" (pt. 2). Publishes *Lectures on the Comparative Anatomy and Physiology of the Invertebrate Animals* (2nd ed., 1855). New charter of the Royal College of Surgeons is adopted, introducing moderate reform.

1843–46: Serves as a member of the Commission of Inquiry into the Health of Towns.

1844: Elected a member of the Literary Society Club. Robert Chambers's *Vestiges of the Natural History of Creation* published. Is offered and declines the presidency of the Geological Society (Nov. 6). First visit to Peel. Meets Carl Gustav Carus. Peel commissions portrait of Owen for Drayton Manor gallery to complement Cuvier's. Ray Society founded.

1845: Elected to "The Club" at the age of forty (May 20). Declines Peel's offer of a knighthood (July 24). Trip to Naples; has copy made of Cromwell's portrait in Pitti Palace, Florence. Begins formal move for a national museum of comparative anatomy, to be formed by uniting the British Museum fossils with the osteological collection of the Hunterian Museum.

1846: Delivers BAAS "Report on the Archetype and Homologies of the Vertebrate Skeleton." Publishes *Lectures on the Comparative Anatomy and Physiology of the Vertebrate Animals*, part 1, *Fishes.* Receives Royal Medal of the Royal Society for his work on belemnites (Nov. 30).

1847: Darwin visits while Owen works on *On the Archetype and Homologies of the Vertebrate Skeleton.* Has Oken's *Lehrbuch der Naturphilosophie* translated; uproar in the Ray Society. Palaeontographical Society founded. New charter of the Royal Society is introduced, representing moderate reform. First memorial of the scientists to the prime minister for adequate representation of naturalists on the Board of Trustees of the British Museum.

1847–48: Serves as a member of the Metropolitan Sanitary Commission.

1847–49: Royal Commission on "the constitution and government of the British Museum"; Owen gives evidence on May 23, 1848 and Feb. 15, 1849.

1848: Revolutions for popular government throughout Europe. Chartist unrest. Publishes *On the Archetype and Homologies of the Vertebrate Skeleton.* William Buckland becomes the first scientific trustee by election of the British Museum.

1849: Clift and his wife, Caroline, die. Publishes *On the Nature of Limbs* and *On Parthenogenesis.* Is turned down as council member of the Royal College of Surgeons.

1849–50: Serves as a member of the Commission on the Smithfield Market and the Meat Supply of London. Advises against intramural slaughterhouses.

1849–84: Publishes *History of British Fossil Reptiles.*

1850: Peel dies following a riding accident (July 2).

1851: Owen's *annus mirabilis.* Delivers Croonian lecture on the megatherium (May 8). The Great Exhibition; appointed chairman of Jury IV (on raw materials, alimentary substances, etc.). Offered knighthood in the Order of Merit of the King of Prussia via Alexander von Humboldt to fill the vacancy created by the death of Hans Christian Oersted. Presented with the former residence of the King of Hanover, in Kew. Is encouraged to apply for Charles König's position as curator at the British Museum. Awarded the Copley Medal of the Royal Society for his oeuvre (Dec. 1). Writes review of Lyell's work on progressive development.

1852: Moves to grace-and-favour Sheen Lodge in Richmond Park. For second time offered and declines presidency of the Geological Society. Awarded honorary doctor of civil law degree from Oxford (June 23). A third exhibition room is added to the Hunterian Museum.

1852–53: Two essays on Owen's oeuvre appear in *Quarterly Review.*

1852–92: Lives at Sheen Lodge, Richmond Park.

1853: Supervises the construction of prehistoric "monsters" for the Crystal Palace Garden, Sydenham. Presides at the dinner in the iguanodon (Dec. 31).

1853–56: Sharp rise in tension with his superiors at the College of Surgeons.

1854: Is offered and declines the chair of natural history at Edinburgh, vacant upon the death of Edward Forbes. "Plurality of worlds" debate flares up.

1855: Exhibition universelle, Paris; elected chairman of Jury XI (prepared and preserved alimentary substances). Created Knight of the Legion of Honour. Publishes *Principes d'ostéologie comparée.*

1855–60: Oxford University Museum erected.

1856: Resigns the curatorship of the Hunterian Museum (May 26). Antonio Panizzi appointed principal librarian at the British Museum. Appointed the first superintendent of the natural history collections at the British Museum (Aug. 14) at a salary of £800 per year. Buckland dies (Aug. 14).

1856–83: Employed at the British Museum.

1857: Awarded the Prix Cuvier (Feb. 2). John Edward Gray and Owen, in annual reports for 1856, emphasize the need for space for the natural history collections at the British Museum.

1857–61: Gives lectures on paleontology at the Royal School of Mines. Surging hostility from T. H. Huxley.

1858: Second memorial of the scientists to the government to prevent removal of natural history collections from Bloomsbury (July 6). Is president of the BAAS, Leeds; starts a campaign to separate natural history from the British Museum in Bloomsbury. Memorial by "Darwinians" advocating dispersal of natural history collections (Nov. 19).

1859: Writes "Report with Plan for a Museum of Natural History" (Feb. 10). Elected one of eight foreign associates of the Institute of France (Apr. 25). Delivers the first of the revamped Rede lectures, "On the Classification and Geographical Distribution of the Mammalia." Awarded honorary doctor of laws degree from Cambridge University. The first 1,250 copies of Darwin's *Origin of Species* roll off the press (Nov. 24). William Broderip dies.

1859–61: Is Fullerian professor at the Royal Institution.

1860: Gives four lectures to the royal household at Buckingham Palace (Apr.). The Gregory Committee "On the British Museum" appointed. Writes hostile essay on Darwin's *Origin of Species* for *Edinburgh Review. Palaeontology*, a compilation of his lectures at the Royal School of Mines, is published. *Essays and Reviews* by leading Anglicans creates a furor among the orthodox.

1860–63: Parliamentary debates about the museum question.

1861: Befriends Paul B. Du Chaillu (Feb.). Gives lecture at the Royal Institution, "On a National Museum of Natural History" (Apr. 26). British Museum trustees vote in favor of removal of natural history from Bloomsbury. Prince Albert dies (Dec. 14).

1862: Publishes *On the Extent and Aims of a National Museum of Natural History*. J. W. Colenso's *Pentateuch* adds to turmoil among orthodox Anglicans.

1863: Describes the *Archaeopteryx*. Publishes *Monograph on the Aye-aye*. Delivers lecture to YMCA at Exeter Hall (Nov. 17) supporting Old Testament criticism.

1864: Visits south of France to inspect Bruniquel Cavern (Jan. and Feb.). Gives lectures to the royal family at Windsor Castle (Mar. and Apr.).

1865: Publishes *Memoir on the Gorilla*.

1866: William Whewell dies following a riding accident (Mar. 6). Publishes *Memoir on the Dodo*. Fourth memorial by the scientists, urging administrative independence for natural history (May 14).

1866–68: Publishes *On the Anatomy of Vertebrates* (3 vols.).

1869: Pays first visit to Egypt, with the Prince of Wales (Jan.–Feb.).

1870–71: Pays second visit to Egypt, for health reasons (Nov.–Jan.).

1870–74: Royal Commission on "scientific instruction and the advancement of science."

1871: Third offer of the presidency of the Geological Society; again declined. Alfred Waterhouse's design for the Natural History Museum approved by the British Museum trustees.

1872: Writes "On Longevity," further attacking biblical literalism.

1873–81: Natural History Museum erected.

1873: Pays third visit to Egypt (Jan. 2–Apr. 3). Caroline Owen dies (May 7).

1874: Pays fourth and last visit to Egypt (Jan.). Gladstone offers a choice of honors; Owen takes the "inexpensive" Companion of the Bath (May 15).

1875: Refers to himself as "a poor broken reed" (Sept. 7); suffers from bronchitis.

1877: Elected member of the Royal Commission for the International Exhibition at Melbourne, 1880. Publishes *Researches on the Fossil Remains of the Extinct Mammals of Australia.*

1879: Publishes *Memoirs on the Extinct Wingless Birds of New Zealand.* Fifth memorial of the scientists (Mar. 25).

1880–83: Supervises the removal of the natural history collections from Bloomsbury to Kensington.

1881: William Holman-Hunt paints Owen's portrait (Jan.). On Easter Monday the British Museum (Natural History) opens its doors. International Medical Congress; Owen unveils statue of William Harvey in Folkestone (Aug. 6); is involved in the vivisection controversy.

1883: Retires from the British Museum (Natural History) (Dec. 31).

1884: Upon retirement from public service, is created a Knight Commander of the Bath by Gladstone.

1885: Corresponds with Gladstone about "Genesis and geology." Statue of Darwin unveiled on the grand staircase of the Natural History Museum (June) (Owen's statue unveiled in 1897).

1886: Owen's only child, William, commits suicide.

1892: Owen suffers from deafness and stomatitis; dies "of old age" (Dec. 18).

A Note on Citation

In the text the date of delivery of lectures is indicated; in the bibliography the date of publication is given. Publication dates of books in the bibliography refer to the particular editions used in this study; these do not always tally with the dates in the text, where, unless otherwise stated, the years of first editions are given. The notes contain abbreviated reference material; full documentation is given in the bibliography. Minor items from periodicals, however, are not separately listed in the bibliography but are cited in full in the notes. Mention of the Geological, Linnean, Zoological, and Royal societies pertains to those of London. The British Association for the Advancement of Science is abbreviated to BAAS. Familiar paleontological names are given in roman type, while less familiar ones have been italicized in accordance with taxonomic practice. The German *Umlaut*, when written as a diaeresis, has not been taken into account in the alphabetical ordering of the bibliography and the index. The German "sharp *s*" has invariably been transliterated as *ss*. The collective noun "man," as in "man's place in nature," to denote people in general has been retained only when the use of gender-inclusive alternatives would have meant putting words into the mouths of Victorians anachronistically.

1

Personality Matters

Blackened Bronze

"Than Professor Owen there is not a more distinguished man of science in the country," the *Times* declared on Friday, January 25, 1856;[1] and indeed, by the middle of the 1850s, when Richard Owen left his curatorial position at the Hunterian Museum to take charge of the natural history departments at the British Museum, he had become the single most visible scientist of the British empire. His name was mentioned in one breath with Isaac Newton's, and he was idolized as Britain's answer to France's Georges Cuvier and Germany's Alexander von Humboldt. Although such comparisons served political purposes and inflated his reputation to outsized proportions, it is fair to say that among Britain's Victorian naturalists Owen came second in importance only to Charles Darwin.

In the course of his active career of nearly six decades (1827–83), Owen, more than any of his colleagues, came to represent "natural history" in the public mind. His company was sought by royalty, prime ministers, and literati. Virtually every honor that could be bestowed upon a scientific figure came his way: the Wollaston, Royal, and Copley medals, the Prix Cuvier, a civil list pension, a knighthood, honorary doctorates from Oxford, Cambridge, Dublin, and Edinburgh, and many, many more. The list of honorary distinctions attached to *The Life of Richard Owen* (written by his grandson) contains a staggering one hundred items.[2]

The range and volume of Owen's scientific oeuvre are vast and unequaled by the majority of naturalists. He published over six hundred papers, and more than a dozen book-length studies flowed from his pen.[3] The subjects of his best-known papers ranged from the Australian marsupials and monotremes to the gorillas of Central Africa. Even more sensational was his work on extinct vertebrates, among them such popular showpieces of paleontology as the dinosaurs, the dinornis, and the *Archaeopteryx*. In naming and describing these fossil finds, he accumulated an inordinate number of "firsts."

When at the end of 1883 he retired from the British Museum (Natural History), a commemorative dinner card pictorially summarized the breadth and depth of his many and innovative papers and memoirs. It showed invertebrates and vertebrates, indigenous and colonial specimens, present-day and extinct species, and marine and terrestrial dwellers; and on an inset were the symbols of the two different approaches to the interpretation of organic form that he perfected: functionalism, of which his work on the New Zealand moa was emblematic, and transcendentalism, represented by his theory of the ideal vertebra (see the appendix).

When his long life, which spanned most of the nineteenth century (1804–92), came to an end, the Royal Society set up an executive and a general committee for the purpose of honoring Owen with a "suitable Memorial." The general committee was made up of no fewer than 164 people and included the Prince of Wales, who served as chairman of the executive committee, the Duke of Teck, the archbishop of Canterbury, the Lord High Chancellor, the presidents of the Royal, Linnean, Geological, Zoological, and several other British and foreign societies, and leading representatives from the worlds of museums and universities.

As many as 332 donations were received, amounting to £1,103 8s. 11d.—substantially more than Owen's annual salary—and a bronze statue of Owen was commissioned and later placed in the main hall of the Natural History Museum in South Kensington; the trustees of the British Museum provided a pedestal of Numidian marble. The donors included many of the best-known names of science and medicine, not just from Britain but from the European continent, the United States, and all major British colonies. Friends and former foes alike contributed, among whom were Henry Acland, Alexander Agassiz, George Baden-Powell, J. D. Dana, Michael Foster, Francis Galton, Carl Gegenbaur, Archibald Geikie, Oscar Hertwig, J. D. Hooker, T. H. Huxley, Lord Kelvin, Lord Lister, O. C. Marsh, Lord Playfair, Ferdinand von Richthofen, J. Burdon Sanderson, Herbert Spencer, John Tyndall, August Weismann, and, in addition to the grand old man of Liberal politics—William E. Gladstone—an assortment of dukes, earls, admirals, and generals.[4]

Three-quarters of a century later, however, Owen's name had long faded from the firmament, and Charles C. Gillispie bypassed Owen in his *Edge of Objectivity* (1960) as a "now forgotten" naturalist.[5] Outside a restricted circle of specialists, Owen was no longer remembered or, equally ignominiously, was confused with the social reformer Robert Owen. How did such a sharp reversal of renown come about? What was responsible for the extreme difference in the assessment of Owen's importance between the mid–nineteenth and the mid–twentieth century? The answer is simple: Owen was systematically written out of Victorian history by Darwin and his followers; to the extent that his memory was kept alive, it was for his critique of Darwin's theory of evolution by natural selection. His institutional power, his scientific eminence, his competitive treatment of rivals, his theistic theory of evolution, which clashed with Darwin's, his strategy of covert promotion of evolutionary thinking—all these and more made Owen a controversial figure, both feared and hated by a substantial number of his fellow naturalists.

The Darwinians managed to elbow Owen out of the history books in a peculiarly Victorian way, namely by making his personality the issue. Blackening the behavior of their adversary, they undercut his scientific credibility and succeeded in turning his stance vis-à-vis Darwinism into the touchstone of his historical worth. So successful were Darwin and his disciples in this that Owen's own grandson, Richard Starton Owen, in his *Life of Richard Owen* of 1894, omitted mention of Owen's attack on *The Origin of Species*, insinuated that someone else might have perpetrated it, and missed no opportunity to relate instances of cooperation between his grandfather and Darwin.[6] Moreover, in a bizarre instance of bowing down to the victors, the grandson, on the suggestion of C. Davies Sherborn, invited Owen's worst enemy, T. H. Huxley, to add to *The Life of Richard Owen* an assessment of "Owen's position in the history of anatomical science."[7]

In the Darwinian literature of the century 1860–1960, Owen has been made to serve the purpose of black countershading, to enhance the shiny white of Darwin and the Darwinians. This historiographical tradition culminated in the literature spawned by the centenary commemoration of Darwin's magnum opus, a representative example of which was Gavin de Beer's *Charles Darwin.* In the Natural History Museum there are, apart from Owen's statue, the statues of Darwin and Huxley, the white of their marble contrasting with Owen's blackened bronze. Nothing could be more poignantly symbolic of the roles in which these men have been cast than this black-white color contrast; nor could the place be more ironic, given that the founding of the Natural History Museum was Owen's work, accomplished in the face of Darwinian opposition.

One might argue that the issue of his personality has no bearing on the significance of his scientific contributions and should be ignored. After all, what would be left of Newton's worth as a mathematician and a physicist if that were made a function of his personality? Yet it may be apt to start a biography, albeit a scientific one, with a brief sketch of the dramatis persona, in the form of several contemporary sketches of his appearance and personality.

Portraits and Caricatures

Owen's six-foot-tall frame made a striking impression and provided an easy target for cartoonists. He was gaunt looking, with "an immense head. His face was very large, with a prodigious forehead, and very large eyes, which seemed highly speculative and pondering in thought, as well as watchful of all external things."[8] He had high cheekbones and a wide, thin-lipped mouth (fig. 1). With age, his raw-boned looks became more pronounced, and as if by sympathetic magic he began to resemble his vertebrate fossils. *Vanity Fair* depicted him as "Old Bones" and described him as "a simple-minded creature, although a bit of a dandy"[9] (fig. 2), while the *Period* showed Owen astride "his favourite hobby," his head resembling the birdlike skull of a megatherioid monster. A phrenologist cited its peculiarities of shape, together with those of Humphry Davy's head, to illustrate "two scientists with strongly developed imagination and a love of poetry."[10] Remarkable, too, were his large, supple hands, delicately depicted in Holman-Hunt's celebrated painting of Owen (fig. 20) and lovingly described by the Duke of Argyll:

> Owen had fingers and a thumb which seemed to me to be "opposable" through arcs of movement larger and wider than were attainable by other men. His fingers seemed to lap round the objects he handled, so as completely to invest them with a soft but universal application. It was beautiful to see him holding some delicate articulation of bones, so round and round that the whole of it was embraced, and yet so tenderly that not the most delicate portions could be crushed.[11]

By all contemporary accounts, Owen was a sociable person with an active interest in the fine arts of his day. His main form of relaxation was listening to music, and he was no mean vocalist himself; in addition, he played the cello and the flute. He regularly visited both concert hall and theater. In 1826, K. M. von Weber's opera *Oberon: or the Elf-King's Oath* was performed at Covent Garden; this first production consisted of thirty-one performances, and reputedly, Owen went to see them all, revealing something of the obsessive side to his character.[12] Among his friends in the performing arts was Jenny Lind, the Swedish opera star who made England her home for the second half of her life.

Owen delighted in contemporary literature, and among his favorite novels were George Eliot's *The Mill on the Floss* and Thackeray's *Vanity Fair*. Because of a similarity in handwriting with George Eliot, he was suspected of being the author of her *Scenes of Clerical Life*. Owen looked forward with keen anticipation to the monthly numbers of Dickens's works, and in *Our*

FIGURE 1. Richard Owen in 1850 at the start of his most successful decade. Lithograph portrait by T. H. Maguire, 1850 (courtesy Wellcome Institute Library, London; used by permission of The Wellcome Trust).

FIGURE 2. Owen caricatured as "Old Bones" in 1873, the year in which the construction of his greatest accomplishment, the Natural History Museum, was begun. ("Men of the Day, No. 57," *Vanity Fair*, Mar. 1, 1873.)

Mutual Friend, the character Mrs. Podsnap is described as a "fine woman for Professor Owen, quantity of bone, neck and nostrils like a rocking-horse, hard features, majestic head-dress in which Podsnap has hung golden offerings.[13] Owen's literary and artistic acquaintances included—in addition to Dickens—Charles Kingsley, Alfred Tennyson, and the painter William Turner. Kingsley once complimented Owen on a popular talk by stating: "I have long felt that you were a poet as well as a philosopher."[14] Another of Owen's pastimes was chess, and he was good enough to have played Howard Staunton, who for thirty years was the best chess player in England and for nearly a decade world champion.

In *The Life of Richard Owen* and in Owen's unpublished papers there are many confirmations of warm family relations and of "his charm of manner, his genial courtesy, and his kindness of heart."[15] When away from home, he would write long, entertaining, and loving letters to his wife, Caroline, and to his only son, William, sharing his prides and vanities. His correspondence to his mother and sisters was equally characterized by respectful affection.[16] Yet the fact that his son, at the age of forty-eight, committed suicide by jumping into the Thames, leaving his hat with purse, watch, and address card inside it on the bank,[17] may indicate that the father's authoritarian and ambitious personality had cast an undeclared shadow over the family.

The Life of Richard Owen, like so many Victorian "Life and Letters," was something of an *oratio pro domo* and presents a sanitized version of the man Owen. Not a glimpse is given of the dark, autocratic, and jealous side of his character, so glaringly exposed by his Darwinian detractors. Yet for every negative portrait of Owen, a positive one can be found; there are contrasting descriptions in diaries, contrasting comments in private correspondence, and contrasting autobiographical reminiscences about him. Let us look at just a few of these. As early as 1851, Huxley, just back from his voyage to Australia, wrote to one of Sydney's best-known resident naturalists, W. S. Macleay: "It is astonishing with what an intense feeling of hatred Owen is regarded by the majority of his contemporaries, with Mantell as arch-hater. The truth is, he is the superior of most, and does not conceal that he knows it, and it must be confessed that he does some very ill-natured tricks now and then."[18]

Gideon Algernon Mantell was a surgeon and a paleontologist whose interests considerably overlapped with Owen's. Mantell kept a journal covering the years 1818–52, and many entries from 1845 onward record conflict with Owen and a strong dislike of him. After a disagreement at the Geological Society about bones from the Cretaceous Wealden beds, which Mantell thought might have belonged to birds but which Owen attributed—correctly—to pterodactyls, Mantell confided to his diary: "It is deeply to be deplored that this eminent and highly gifted man, can never act with candour or liberality." Later, Mantell enjoined himself: "I must avoid this man in future: but it is very sad thus to be compelled to become as reserved and selfish as the characters I despise." This self-admonition notwithstanding, Mantell continued to cross swords with his adversary, and at the Royal Society the two men clashed over the award of the Royal Medal; Mantell regarded Owen's paper on belemnites unworthy of the honor, and Owen schemed to block the award of the Royal Medal to Mantell. Then the two clashed over moa bones, with Owen disparaging the quality of Mantell's specimens. "Poor envious man!" Mantell commented. Similar territorial skirmishes followed, leading Mantell to accuse Owen of perpetrating "atrocious falsehood[s]" and of acting with "lamentable turpitude."[19]

Another diary, however, covering the period 1835–71, gives an entirely different picture of Owen. It was kept by Caroline Fox, daughter of the Quaker Cornishman Robert Were Fox, known among other things for his work on the geothermal gradient.[20] Caroline kept a record of the comings and goings of the members of her father's circle, which included Thomas Carlyle, John Stuart Mill, John Stirling, and various other Coleridgeans and also, during the 1840s, "Professor Owen." Visits by and to Owen are mentioned, and he comes across as a thoroughly pleasant social companion. Talk is of phrenology, "which he considers the most remarkable chimera which has taken possession of rational heads for a long time"; of Carlyle's *French Revolution*, the style of which reminded Owen of Milton's; of Whewell's *Philosophy of the Inductive Sciences*, which Owen regarded in the same league as Bacon's *Novum Organum*; of Cromwell and Shakespeare; and of Goethe and Schiller. In 1841, not long after she had met Owen for the first time, Caroline Fox wrote: "He is a very interesting person, his face full of energetic thought and quiet strength. His eye has in it a fixedness of purpose, and enthusiasm for that purpose, seldom surpassed." After another visit by Owen, she enthused: "Owen was very delightful; he is such a natural creature, never affecting the stilted 'philosophe,' and never ashamed of the science which he so ardently loves. He is passionately fond of scenery; indeed, all that the Infinite Mind has impressed on matter has a charm and a voice for him. A truly Catholic soul!"[21]

Jane Carlyle, by contrast, reacted to Owen's social manner with apparent dislike. Huxley informed Tyndall once: "You know Mrs Carlyle said that Owen's sweetness reminded her of sugar of lead"; among the Darwinians, "sugar of lead" became a nickname for Owen.[22] Thomas Carlyle himself, however, although his most immediate impression of Owen was unfavorable, thought highly of him. Carlyle gave an amusing account of his first meeting with Owen, which occurred on August 24, 1842. Early that evening, he had fallen asleep on the sofa; a loud knock at the door woke him, and a man in white trousers entered. The maid announced the guest. Carlyle, however, heard nothing but a mumble. At first he mistook the unexpected guest for someone else, yet a "clammy, irresponsive" hand made him realize his mistake. The visitor announced that he had come "impowered to call" by Caroline Fox:

> he got seated, disclosed himself as a man of huge coarse head, with projecting brow and chin (like a cheese in the *last* quarter), with a pair of large protrusive glittering eyes, which he did not direct to me or to anybody, but sat staring with, into the blue vague! There sat he, and talked, in a copious, but altogether vague way; like a man lecturing, like a man hurried, embarrassed and not knowing well what to do. I thought with myself, Good Heavens can this be some vagrant Yankee; Lion-hunting Insipidity,—biped perhaps escaped from Bedlam; coming in on me by stealth? He talked a minute longer; he proved to be Owen the Geological Anatomist, a man of real faculty, whom I had wished

> to see; my recognition of him issued in peals of laughter, and I got two hours of excellent talk out of him: a man of real ability, who could tell me innumerable things![23]

At the time when Carlyle got to know Owen, Darwin held the "Geological Anatomist" in similarly high esteem; but nearly two decades later a radical change occurred when the two men parted company, not just over the issue of the origin of species but more comprehensively in a struggle for factional hegemony. Darwin's destructive descriptions of Owen's character in letters to loyal Darwinians have become part of the canon of the history of science, and only a mere reminder of these is needed here. Darwin plastered Owen's review of *The Origin of Species* with such adjectives as "spiteful," "unfair," "ungenerous," and "extremely malignant." Other adjectives included "false," "rude," "unjust," "illiberal," and "disingenuous," and Owen was accused of gross misrepresentation, abuse, bitter sneers, and mean conduct. In his *Autobiography* Darwin reminisced as follows:

> I often saw Owen, whilst living in London, and admired him greatly, but was never able to understand his character and never became intimate with him. After the publication of the *Origin of Species* he became my bitter enemy, not owing to any quarrel between us, but as far as I could judge out of jealousy at its success. Poor dear Falconer, who was a charming man, had a very bad opinion of him, being convinced that he was not only ambitious, very envious and arrogant, but untruthful and dishonest. His power of hatred was certainly unsurpassed. When in former days I used to defend Owen, Falconer often said, "You will find him out some day" and so it has passed.[24]

Both the positive and the negative reactions to Owen may well have been wholly justified. Nor is there a problem in reconciling such very contrasting opinions of Owen's character. A simple explanation was offered by the Duke of Argyll, an admirer of Owen, yet a man of independent statesman-like stature. He observed in his *Autobiography* that among colleagues Owen had the reputation of jealousy but that to outsiders, such as Argyll, "he was always most charming and instructive."[25] And indeed, as long as Owen was acknowledged as the supreme expert, by the nonexpert, by junior naturalists, by admiring disciples, and by distant colleagues, he was a model of kindness and generosity; but when interacting with near equals, whether Robert Grant at the Zoological Society, Gideon Mantell at the Geological Society and the Royal Society, or the Darwinians at meetings of the BAAS, Owen turned petty and failed to act with the magnanimity that, given his position and fame, he could well afford.

There are numerous instances of supportive testimonials written by Owen on behalf of colleagues who represented no threat. Young Acland, young Hooker, young Huxley, the anatomist John Goodsir in Edinburgh, William Turner in succession to Goodsir, the botanist George Allman in Dublin, the zoologist George Dickie in Belfast, and various foreign colleagues, all turned to Owen for support and nearly always with favorable results.[26] Owen engineered the election of the ornithologist John Gould, the son of a gardener and a man without advantages of education or position, into the prestigious Athenaeum Club, under Rule II (which allowed for the admission of distinguished individuals without balloting by the membership).[27] As a referee for the *Philosophical Transactions*, Owen wrote favorable reports on papers submitted for publication by the embryologist-surgeon Martin Barry, the anatomist-obstetrician Arthur Farre, John Lubbock, at the time working on insects, and several others; the few rejections he recommended were argued in conscientious detail.[28] Admirers and followers wrote touching responses; William White Cooper, one of Owen's pupils and an ophthalmic surgeon, wrote in 1839:

> Words are quite inadequate to describe the feelings of regard, of affection I may say, I entertain for you—You have filled up that void in my heart which the death of my father occasioned. You have been to me a parent, and so long as I live—in whatever clime I may be—in adversity or prosperity—my heart will turn towards you with feelings such as no language can express, but which have, and will cheer me in many an hour of sadness, for the anticipation of your kind looks and cordial greetings is in itself a cordial and a true source of comfort to my heart.[29]

Similarly strong admiration and gratitude were expressed by others who mentioned that in their student days Owen treated them "with all the high bred courtesy and unaffected and sympathetic kindness which were inborn in his nature."[30] The Scottish brachiopod expert Thomas Davidson, after many years of contact with Owen, wrote: "I shall *never* as long as I live cease to feel *grateful* to you for the truly generous manner with which you encouraged and helped me when a young man and when beginning my work on the Brachiopods."[31] Tokens of esteem in the form of book dedications came by and large from people distant to Owen, either in level of accomplishment or in field of interest. Books dedicated to Owen included *General Outline of the Animal Kingdom* (1838–41) by Owen's disciple Thomas Rymer Jones, *Zoological Recreations* (1847) by the amateur naturalist W. J. Broderip, a translation of Homer's *Iliad* (1865) by the linguist Edwin W. Simcox, and *Human Longevity* (1873) by the deputy librarian to the House of Lords W. J. Thoms.

Furthermore, the only substantive and sympathetic obituary assessment of Owen's scientific significance was written by another peripheral figure, St. George Jackson Mivart (1827–1900), onetime faculty member for biology and physiology at the short-lived Catholic University College at Kensington.[32]

The Rehabilitation of Owen

For a full seven decades after Mivart's essay of 1895, only a single article of substance on Owen's oeuvre appeared, in 1935, written in German by a German historian.[33] Not until 1965 did Owen begin to lose his Darwinian whipping-boy function, when a gradual reappraisal was begun with the appearance of Roy MacLeod's article "Evolutionism and Richard Owen." In the wake of this paper, the realization set in that Owen's work in fact formed part of the foundations of Darwin's theory. This new emphasis on Darwin's debt to Owen found its widely read expression in Dov Ospovat's study *The Development of Darwin's Theory* (1981) and, more recently, was significantly expanded upon by Robert J. Richards.[34] Following on from Ospovat, a further revision of Owen's place in "Darwin's century" was made, in particular by Adrian Desmond and Evelleen Richards, who have put the spotlight on Owen's own, non-Darwinian theory of evolution and his involvement, during the three decades directly preceding *The Origin of Species*, in what Desmond calls the "politics of evolution." Moreover, in a detailed study of Owen's Hunterian lectures of 1837, Phillip R. Sloan has depicted Owen, at the start of his career as a professor, standing "[o]n the edge of evolution."[35] In the meantime, Jacob W. Gruber and John C. Thackray have been engaged in a comprehensive compilation of Owen's correspondence and other unpublished manuscripts,[36] while Gruber and others have published detailed studies of specific items on Owen's list of accomplishments.

This gradual rehabilitation of Owen has been facilitated by a simultaneous reevaluation of Darwinian theory. For example, the Lyellian gradualism—to which Owen did not subscribe—implicit in Darwin's theory of natural selection, and the related reliance on the imperfection of the geological record, have increasingly been displaced by a neocatastrophist model of the history of life on earth, most famously in the case of the end of the reign of the dinosaurs at the Cretaceous-Tertiary boundary, widely presumed to have been caused by an asteroid impact. Moreover, Adolf Seilacher, Stephen Jay Gould, and others have reemphasized the importance of the architecture of form in explaining organic diversity and the structural constraints this imposes on Darwinian adaptation. Admittedly, they additionally have championed Darwinian contingency to elucidate the patterns and processes of evolution; yet others, Simon Conway Morris prominently among them, counter by arguing that a certain inevitability characterizes the course of organic evolution and thus are restoring for present-day use a central plank of Owen's epistemological platform.[37] The same is done by proponents of evo-devo, whose approach represents a revival of nineteenth-century structuralist biology and who, Ron Amundson urges, ought to recognize Owen as a precursor.[38]

Yet with the continued employment of evolutionary theory as the primary context for studying Owen, he is still left tied down to the procrustean bed of Darwinian historiography. Its implicit and incorrect assumption is that the focal concern of Owen's work, like Darwin's, was to solve the problem of the origin of species. In this study, in which for the first time Owen's scientific career is looked at in its entirety, the issue of evolution will be subordinated to that of Owen's museum-building agenda. In the course of his lifetime, the natural sciences in England went through a revolutionary process of emancipation marked by the reform of the Royal Society and of the medical corporations, the establishment of the Geological, Zoological, and many similar societies, the founding of natural history museums and of university science departments, and the emergence of a class of scientists occupying positions of acknowledged authority.

In this process, Owen played a major and, at times, a central role. I argue that crucial to the understanding of Owen's scientific oeuvre is his drive for the creation of a national museum of natural history, an effort in which he was the *primus inter pares*. To a significant extent he made his scientific work, his choice of subject matter, his theoretical approach, and the manner of its presentation serve the politics of institutionalization. The networks of patronage in particular, essential for the fulfillment of his museum ambitions, provide an overarching framework of explanation for his choices of theory.

Admittedly, to appreciate Owen's scientific accomplishments adequately, we must consider them at different levels of interaction with formative influences, including cognitive ones. At the most general of these, Owen's work was shaped by an international discourse about the meaning of organic form; less generally, his endeavors were molded by national debates about the significance of form and function; and in a yet more specific context, his thinking was driven simply by the logic of an adopted method or a program of research. In the next several chapters, these various intellectual influences will be considered, but before doing so, we need to step back and look at a set of overarching conditions: institutional ones.

2

Museum Politics

Unlike Darwin, who was a man of independent means and worked in the relative isolation of his country home, Owen had to earn a living, in the bustle of metropolitan London. His background was modest. Owen's father, a West India merchant, died when Richard was five years old. At the age of six he was sent to the grammar school in his native town of Lancaster. Apart from about half a year at Edinburgh University, from the autumn of 1824 till the spring of 1825, Owen enjoyed no university education, let alone an Oxford or Cambridge one, but worked his way up through a succession of apprenticeships. On the recommendation of John Barclay, his extramural anatomy teacher at Edinburgh, he was appointed prosector at St. Bartholomew's Hospital, London, working under its leading surgeon, John Abernethy. In 1826 Abernethy supported Owen for membership in the Royal College of Surgeons and, in 1827, helped him gain a curatorial position at the college's Hunterian Museum, in Lincoln's Inn Fields, London. Membership in the college made it possible for Owen to set up in medical practice. Around 1837, however, he let the practice lapse because, rather than establish himself as a surgeon, he decided to pursue a scientific career at the Hunterian Museum.[1] Yet Owen's scientific work was no less a means of making his way in the world than a medical practice would have been, and it should be looked at in this light.

The Museum Movement

In choosing museum work Owen did not move into a ready-made institutional niche for scientific study. On the contrary: both the concept and the architectural reality of museums as institutions of research, though at the time already well established in Paris,[2] were still being developed in Britain. In fact,

Owen himself was one of the main driving forces of the Victorian movement to provide accommodation for museum collections, to expand such collections, and to turn these to educational and research purposes. Owen was a museum man par excellence. Nearly his entire professional life was spent within the walls of two museums, while these in turn were significantly shaped by him. Owen's main career objective was to establish a separate, national museum of natural history. Writing to a friend toward the end of his employment, he confided: "As my strength fails, and I feel the term of my labours drawing nigh, how I long to see the conclusion of their main aim!—the exposition of our national treasures of natural history in a manner worthy of the greatest commercial and colonial empire of the world."[3] This aim was realized when in 1881 the Natural History Museum in South Kensington was completed and its galleries were filled with the collections of botany, geology, mineralogy, and zoology from the British Museum building in Bloomsbury, where they had been kept in overcrowded conditions together with the antiquities and other collections related to "intellectual man." It is my contention that to understand the dynamics of the museum movement is to understand Owen.

For a period of twenty-nine years, from 1827 till 1856, Owen was employed at the Hunterian Museum, first at the age of twenty-two as assistant conservator, then in 1842 at the age of thirty-eight as conservator. In 1856, shortly after having turned fifty-two, Owen became superintendent of the natural history collections at the British Museum.[4] When he retired at the end of 1883, aged seventy-nine, having seen his long-nurtured dream of a separate, national museum of natural history come true, he asked the trustees of the British Museum, as part of their submission to the Treasury of a record of Owen's museum service, to add to his time at the British Museum the length of his service at the Hunterian. To combine the two periods was to appreciate that his career could be seen as a continuous span of fifty-seven years of devotion to museum collections of national import. In this retrospective, Owen defined his accomplishments in terms of the care he had taken of the collections and in particular of the new space he had provided for their display.[5]

The path of Owen's professional life can be marked by milestones of museum construction—an activity that required a constant preoccupation with the issue of expansion, which in turn required the never-ending lobbying of bodies of trustees and the continual cultivation of friendships with many individuals in positions of power. The drive to expand existing museums and establish new ones was all the more engrossing in that it was not an isolated process of institutionalization but part of a much wider movement, in London, in the provinces, and abroad, and pertained not just to museums of natural history but also to museums of antiquities, fine arts, ethnography, and technology. Owen's career coincided with what one could call the age of museums. Nearly all of Britain's great museums were built during Owen's lifetime. In addition to the British Museum in Bloomsbury and the British Museum (Natural History) in South Kensington, these included the Victoria and Albert Museum and the National Art Gallery. In fact, some two hundred metropolitan, provincial, and university museums were founded in Britain during the decades of Owen's scientific career.[6] A majority of the great museums on the European mainland, too, date from this period. Simultaneous with the founding of natural history museums were other forms of institutionalization of the natural sciences, in particular scientific societies but also university chairs and science departments.[7] Architecturally most visible, however, were the museums.

Yet museums were by no means the only or even the main focus of architectural innovation of the Victorian era. Museum construction was part of a wider building drive that transformed cities, especially London. The architectural energy of the period was colossal and unparalleled. Many of the monuments for which the British capital is known today were placed on the streetscape in the course of Owen's life. Impressive examples in just about every category of building, in addition to museums and art galleries, can be cited—bridges, churches, clubs, university colleges, concert halls, hotels, libraries, markets, memorials, prisons, railway stations, theaters, town halls, government buildings, and also residential areas and parks. The importance of these construction projects for Owen must have been considerable. Wherever he went, whichever way he turned, he was surrounded by architectural change and innovation on a major scale, seeing monumental buildings replace old edifices or rise out of freshly broken ground. Many of these had a direct bearing on his daily routine and on the particulars of his professional and social position: his own museums—of course—but also the Athenaeum Club, the Museum of Practical Geology, and the 1862 International Exhibition buildings.

Let us consider in some detail the issues and debates that marked Owen's career path, both at the Royal College of Surgeons and at the British Museum, and examine how his institutional setting determined the scope of his work and his choice of research topics.

Attempted Hijack of the Hunterian Museum

A well-known aspect of Owen's tenure at the Hunterian Museum is the disagreement that developed between him and his employers, the Council of the Royal College of Surgeons, culminating in a bitter resignation row. The main cause of the problems has been sought in Owen's difficult personality and his reluctance to carry out menial curatorial tasks.[8] This interpretation, while not wrong in itself, fails to engage the larger issues of which the various points of conflict were the symbols. Tensions developed over the museum catalogues, salary matters, Owen's place of residence, his candidacy for membership in the governing body of the college,[9] and, most importantly, the interpretation of the Hunterian legacy.

In 1799 the Company of Surgeons, which would become the Royal College of Surgeons in the following year, was awarded custody of John Hunter's collections, purchased by Parliament for £15,000. The resulting Hunterian Museum was entrusted to the surgeons under a number of conditions, which included that a "Catalogue of the Preparations" be made available. Owen was appointed to help fulfill this condition.[10] As an employee of the Royal College of Surgeons, he had to report to the Board of Curators, later renamed the Museum Committee, which was composed of seven surgeons who were members of the governing body of the college, that is, the council. This consisted of twenty-one surgeons, a number enlarged, as a consequence of the new charter of 1843, to twenty-four. The councillors tended to appoint each other in order of seniority to the Court of Examiners, which was the most powerful college body, in that its members controlled admission to the surgical profession and determined who was qualified to teach surgery.[11]

Separate from this administrative hierarchy within the Royal College of Surgeons were the trustees of the Hunterian Museum, a board set up when Hunter's collections had been bought with public money. It was the board's nominal duty to oversee the affairs of the museum. Taken together, the Board of Trustees of the Hunterian Museum was a more distinguished body than the Council of the Royal College of Surgeons, and its membership was similar to that of the Board of Trustees of the British Museum. The Hunterian Board of Trustees was made up of sixteen *ex officio* trustees (six top government figures; the presidents of the Royal Society and the College of Physicians; the four censors of the College of Physicians; the regius professor of medicine and Lee's reader in anatomy at Oxford; and the regius professor of physic and the professor of anatomy at Cambridge), plus fourteen others, elected by the *ex officio* trustees, who tended to choose not only representatives from the world of medicine but also aristocrats and clergymen. A majority of the trustees attended meetings only rarely, some never, and the running of the Hunterian Museum fell largely to the Museum Committee.[12]

Over the years, a fundamental disagreement surfaced about the overarching issue of the scope and purpose of the Hunterian Museum itself. Almost from the start Owen wanted to expand the Hunterian Museum into something bigger than a medical museum. He had long nurtured the ambition of turning Hunter's legacy into a national museum of comparative anatomy to rival Continental establishments of a similar kind. The Museum Committee, by contrast, while pleased with the prestige that Owen's scientific reputation bestowed upon the surgical profession, in the end gave precedence to the narrow interest of the education of surgeons and refused to go along with Owen's wider vision of a national museum to cater to the needs not only of surgeons but also of scientists interested in the study of animal morphology broadly conceived.

Already in the spring of 1831 Owen expressed the view—writing to the then grand old man of the college, William Blizard—that priority should be given to quality rather than speed in the preparation of the museum catalogues (table 1), because a good set of catalogues would help legitimize the museum's claims that its contents were comparable to national collections of comparative anatomy on the European mainland, specifically those of Berlin, Paris, and Vienna.[13] A few months later, in the summer of 1831, Owen was given a historic chance to compare his institution with the internationally most famous one when he visited the Muséum national d'histoire naturelle (part of the Jardin des plantes, Paris), and especially its department of comparative anatomy. His visit took place at the invitation of none other than Georges Cuvier, who the year before had been entertained at the Hunterian Museum. At this time, intellectual Paris was still buzzing with excitement about the public clash in 1830 at the Institut de France, when Cuvier had locked horns with his archrival Étienne Geoffroy Saint-Hilaire. Later in life Owen referred to his presence in Paris at the tail end of this clash as the most memorable part of his visit,[14] yet when he wrote an account of his Parisian sojourn upon his return to London, it was a detailed report on the organization of Cuvier's department, "in which it is considered with respect to its origin and extent, the nature of its contents, the modes of preserving and displaying them, and the objects aimed at in their arrangement; also its government, its accessibility, and the provision for its present maintenance and future increase."[15]

Two of its features in particular were to influence him in the pursuit of his own museum. First, there was the composition of Cuvier's department, which included a rich osteological collection, far richer than its Hunterian counterpart. Hand in hand with this went a crucial difference in display principles of the skeletal material. Whereas the Hunterian setup was ruled by physiology, the Cuvierian was taxonomic: "the one illustrates the scheme of the animal, the other the scheme of nature—the one throws light on physiology, the other on natural history."[16] Second, there was the administrative organization, which differed fundamentally from the Hunterian's in that each department came under the direct control of a professor; Cuvier, for example,

TABLE 1. Hunterian Museum Catalogues Prepared by Richard Owen

1830	*Catalogue of the Hunterian Collection in the Museum of the Royal College of Surgeons in London*, part 4 (1), *Comprehending the First Division of the Preparations of Natural History in Spirit*
1832	*Catalogue of Preparations Illustrative of Human and Comparative Anatomy, Presented by Sir William Blizard to the Royal College of Surgeons in London*
	Descriptive and Illustrated Catalogue of the Physiological Series of Comparative Anatomy Contained in the Museum of the Royal College of Surgeons in London
1833	Vol. 1, *Organs of Motion and Digestion* (2nd ed., 1852)
1834	Vol. 2, *Absorbent, Circulating, Respiratory and Urinary Systems*
1835	Vol. 3 (1), *Nervous System and Organs of Sense*
1836	Vol. 3 (2), *Connective and Tegumentary Systems and Peculiarities*
1838	Vol. 4, *Organs of Generation*
1840	Vol. 5, *Products of Generation*
1845	*Descriptive and Illustrated Catalogue of the Fossil Organic Remains of Mammalia and Aves Contained in the Museum of the Royal College of Surgeons of England*
1845	*Synopsis of the Arrangement of the Preparations in the Museum* (2nd ed., 1850)
1853	*Descriptive Catalogue of the Osteological Series Contained in the Museum of the Royal College of Surgeons of England*, vol. 1, *Pisces, Reptilia, Aves, Marsupialia*; vol. 2, *Mammalia Placentalia*
1854	*Descriptive Catalogue of the Fossil Organic Remains of Reptilia and Pisces* (assisted by J. T. Quekett for Reptilia)
1856	*Descriptive Catalogue of the Fossil Organic Remains of Invertebrata* (cephalopods by Owen; bulk by John Morris)

The "Bibliography of Richard Owen" (Rev. Owen, *Life*, vol. 2, 333–34) attributes to Owen all six parts of the *Catalogue of the Hunterian Collection in the Museum of the Royal College of Surgeons in London*, which was prepared under William Clift and appeared in 1830–31. Only one part, however, was produced by Owen; Clift was responsible for the parts on the pathological series, and his son, William Home Clift, for those on the osteological collection and the "monsters and malformed parts" (Negus, *Hunterian Collection*, 43). Owen may have assisted Clift Sr. in the preparation of Clift's three parts.

was in charge of comparative anatomy. The museum's director was chosen by a board made up of all the professors at the Jardin des plantes, and funding came in the form of public money provided by the Ministry of the Interior.[17]

By contrast, government support for the Hunterian Museum was limited to incidental grants, whereas the regular expenses were borne by the Royal College of Surgeons, from the fees of its members. The original Hunterian Museum, which opened in 1813, consisted of a single hall with characteristically drum-shaped skylights and was built with the aid of two parliamentary grants totaling £27,500. Additions to the collections rendered this space inadequate, and twenty years later, in 1833, the decision was made to demolish virtually the entire building of the Royal College of Surgeons and enlarge the exhibition space by adding a second gallery to the main hall plus an adjoining smaller exhibition room. For this major phase of renovation no further parliamentary grant was obtained. The entire period of rebuilding lasted from April 9, 1834, to February 14, 1837. At this time, the Hunterian had for display 10,000 preparations in spirit, 8,000 dry preparations, among which were some fossils, and 2,000 skeletal specimens.[18] Another decade or so later the issue of space once more became urgent, and after several years of lobbying, a parliamentary grant of £12,000 was obtained, which made possible the construction of a third exhibition room, completed in 1852.[19]

Throughout this period of museum expansion, Owen assumed an increasingly high profile. By the end of the 1830s, having finished the physiological catalogues, he turned his attention to comparative osteology and also to vertebrate paleontology (see table 1), both strong divisions of the Parisian museum at the time he had visited Cuvier. Owen's plan was to turn the Hunterian Museum into a national museum, not just of pathology and physiology, but especially of osteology and paleontology—in other words, a national museum of the life sciences. It was at this stage, during the 1840s, that a major rift began to develop within the Hunterian Museum. The trustees were, on the whole, sympathetic to Owen's plan, whereas members of the Museum Committee and of the Council of the Royal College of Surgeons felt that Owen was moving the Hunterian legacy away from the purpose of surgical training.

In a December 15, 1845, report to the Museum Committee, in which Owen presented an outline of his plan for a national museum of comparative anatomy, he complained that the absence of an adequate collection of comparative osteology, equivalent to that of the Paris and other Continental museums, "begins to excite comment, and cannot be regarded without concern by those who are the natural promoters and fosterers of comparative anatomy in this country." The collection of recent skeletons in the Hunterian Museum should be significantly expanded and combined with the vertebrate fossils from the British Museum. The separate existence of these two collections represented "an obstacle to the scientific application and due fruition of such collections, which cannot, probably, much longer be permitted to exist." The lack of space at the Hunterian was becoming acute, and as a result donations of recent osteology were now being made to Bloomsbury, with only superfluous duplicates passed on to Lincoln's Inn Fields. A major addition to the present museum of three times the space of the major hall was needed to accommodate "the growing movement for a National Collection of Comparative Anatomy in the direction of the College." Money from the public purse would be a necessary condition for this construction project.[20]

The trustees reacted favorably to Owen's expansionist ideas and, on February 28, 1846, resolved to ask the government for financial support "with a view to render the Hunterian Museum worthy of being the great national depository of the sciences with which it is connected."[21] That Owen indeed wanted to move away from a purely medical purpose for the collections is in no doubt. He stated: "The Museum of the Royal College of Surgeons holds the place and fulfills the purpose of the National Depository of the Collections illustrative and forming the groundwork of the Sciences of Physiology, Pathology, and the Comparative Anatomy of Animals both existing and extinct." The extra space for which he campaigned was primarily intended to accommodate an expanded collection of recent and fossil vertebrates to serve "the requirements of the Geologist, for the comparison and determination of the fossil remains of extinct animals on which the deductions of his science mainly rest."[22] As it was—Owen added—the museum already fulfilled an essential function in the research of geologists and paleontologists.

Initially, the college had gone along with this cultivation of comparative osteology and paleontology. After all, the membership derived much national and international prestige from it. T. H. Shepherd, who produced a number of the illustrations for *London Interiors*, painted a striking watercolor of the inside of the main exhibition hall of the Hunterian Museum, showing Owen surrounded by a number of visitors, amid a dramatic display of skeletons, particularly those of two mammalian fossils (fig. 3).[23] In 1844 Carl Gustav Carus, personal physician to the king of Saxony, visited Britain. In London, he made a beeline for the Hunterian Museum and for Owen, "its director, augmentor and interpreter."[24] Carus was especially impressed by Owen's paleontological discoveries and by the fossil bones of the *Glyptodon clavipes* and the *Mylodon robustus*, strategically located at the entrance of the main hall. The Yale geologist Benjamin Silliman, who visited Europe in 1851, also marveled at the osteological treasures of the museum, such as the Irish elk and the moa.[25]

Increasingly, in the course of the 1840s, unease was expressed inside the college about the new emphasis on recent and fossil osteology. At times, the Museum Committee declined Owen's proposals for the acquisition of certain fossils and at regular intervals criticized the conservator for spending too much time on fossils and too little on human anatomy. In 1854, for example, the Museum Committee expressed the wish that the arrangement of the specimens be altered to give more room and prominence to pathology and to normal human anatomy.[26] The surgeons, worried by the fact that Owen managed to reach across the heads of the councillors to two major bodies of trustees, of the Hunterian and of the British Museum, began to

FIGURE 3. The main exhibition hall of the Hunterian Museum, as it looked in the early 1840s. The conspicuous display of the skeletons of the mylodon on the left and the glyptodon on the right is an indication of the prominence given to vertebrate paleontology under the conservatorship of Owen. Steel engraving from a drawing by T. H. Shepherd (Anon., *London Interiors*, vol. 1, opposite 129).

reassert their control. On January 6, 1848, the Museum Committee reported that Owen's plans would involve great expense, "not only in buildings, but in management," and that if government support were forthcoming, it might "in some respects be advantageous to the College," but—the Museum Committee added—an expansion of the museum should not alter "the present mode of carrying out the Hunterian Principles, and make it therefore diverge too much from physiology as connected with pathology."[27] When at last, on May 30, 1851, the resolution of the trustees of February 28, 1846, was brought before the Treasury in a memorial of the Council of the Royal College of Surgeons, the modest sum of £15,000 was requested, allowing for no more than one extra exhibition room. The sum was granted and the hall completed the following year.[28]

Move to the British Museum

In the end, Owen's drive of the mid-1840s did not result in the desired expansion of the Hunterian Museum from a purely medical collection into a national collection of comparative anatomy. Neither the Council of the Royal College of Surgeons nor the trustees of the British Museum gave their consent. The former stuck to the traditional, medical interpretation of Hunter's legacy, and the latter remained faithful to the heterogeneity of Hans Sloane's bequest, the original British Museum collection. A Royal Commission of Inquiry into the Bloomsbury establishment (1847–49, reporting in 1850) under Francis Egerton, the Earl of Ellesmere, recommended that natural history should remain at the British Museum, to stay part of the total contents and connected to the library.[29]

This recommendation expressed the view of several of the leading trustees; but it also reflected the wish of the naturalists themselves, Owen by now on their side. His plans to unify the Hunterian collection of skeletons with the vertebrate fossils in the British Museum could be more suitably realized at the latter place. If his ideal was to organize a museum of comparative anatomy similar to Cuvier's department at the Muséum d'histoire naturelle, then the British Museum was the place to do this, not a medical museum with its specific mandate to instruct surgeons. The *Literary Gazette*, too, argued that the home of paleontology was not Lincoln's Inn Fields but Bloomsbury. The College of Surgeons was an instance "in which we have the profoundest natural history acquirements in the wrong place."[30] Owen's labors at the College of Surgeons—the correspondent believed—had been brought to a successful termination. In other words, now that the British Museum was not coming to Owen, Owen had to go to the British Museum. As G. H. Lewes exasperatedly asked in his essay on Owen for *Fraser's Magazine*: "We have a magnificent collection in the British Museum, and an unrivalled expositor in Professor Owen—why are the two separated?"[31] Others who held the same Whig/Liberal convictions concurred. The Ellesmere Commission had highlighted the inadequacy of scientific representation at the British Museum. In the early part of 1856 this inadequacy became acute when Antonio Panizzi was promoted to principal librarian and thus became administrative head of the entire establishment. He was a strong-willed man who thought little of natural history and "who would at any time give three mammoths for an Aldus [i.e., a book printed by the Renaissance scholar-printer Aldus Manutius]."[32] Having Owen join the staff of the British Museum to take charge of the entire department of natural history would add a scientific counterweight to Panizzi and his bias toward the library and the sculpture galleries, and at the same time would give Owen the museological scope he needed. As Thomas Babington Macaulay wrote to his fellow trustee the Marquis of Lansdowne:

> I cannot but think that this arrangement would be beneficial in the highest degree to the Museum. I am sure it would be popular. I must add that I am extremely desirous that something should be done for Owen. I hardly know him to speak to. His pursuits are not mine; but his fame is spread over Europe. He is an honour to our country, and it is painful to me to think that a man of his merit should be approaching old age amidst anxieties and distresses.[33]

On May 26, 1856, Owen was offered the newly created post of superintendent of the natural history department, the several "branches" of which were upgraded to "departments." The asked-for salary of £800 was granted and Owen accepted. Adam Sedgwick expressed a widely held view when he wrote to Owen in a congratulatory letter that his move to Bloomsbury would be beneficial to British science: "An *Imperator* was sadly wanted in that vast establishment," he added.[34]

Thus, when Owen left the Hunterian for the British Museum, he carried with him the ideal of a national establishment of natural history. Moreover, he was by this time an old hand at dealing with problems of space and an expert lobbyist of bodies of trustees and government officials. No sooner had Owen taken up his new post than he joined John Edward Gray, keeper of zoology since 1840, in a campaign for more space to accommodate the scientific collections, which were suffering from neglect in the jam-packed British Museum building. Soon Owen took complete charge of this campaign, turning it into a drive for a museum of natural history that would reflect the imperial stature of the British nation. It took more than two decades, but at long last Owen did succeed in providing adequate space for Britain's popular

FIGURE 4. The Natural History Museum in South Kensington, in 1880, a year before its official opening to the public. The founding of this monumental "cathedral of science" was principally due to Owen and his Liberal parliamentary allies, particularly W. E. Gladstone, and was accomplished in the face of concerted opposition by an alliance of convenience between Darwinians and Tories. (*The Graphic* 21 [Mar. 27, 1880]: 332.)

collections of natural history when, on Easter Monday, 1881, the new Natural History Museum in South Kensington was officially opened to the public (fig. 4) and when as many as 16,000 visitors crowded the spacious exhibition galleries.[35]

The natural history collections had been on public display in the temple-like Bloomsbury building since 1831, having been moved there from the less commodious Montagu House. All too soon, however, a stream of new objects—antiquities, ethnographic items, dried plants, stuffed animals, and fossils—from private collectors, Near Eastern excavations, African expeditions, and colonial surveys had congested the Bloomsbury temple, even its basement, stairs, and portico. Such a heterogeneous hoard might have been regarded as a perfectly congruous collection in the middle of the eighteenth century, when the British Museum had originated from Hans Sloane's estate; but in the Victorian era, with its growing regard for specialized expertise, these natural objects and human artifacts made uneasy bedfellows.

Even before the completion of the British Museum building in 1845, overcrowding and neglect were the order of the day. Yet the founding of the Natural History Museum in South Kensington was not the inevitable one-and-only solution to the problem of overcrowding in Bloomsbury. Three other distinct plans for dealing with the natural history collections were each championed in

Parliament, defended in testimony to parliamentary committees, and urged on the government in memorials signed by groups of scientists.

The first and most obvious of these was to keep everything together in Bloomsbury and simply expand the existing museum. This option appealed to those who regarded the Bloomsbury temple as a seat of honor and who wanted natural history to occupy that seat. The same option proved attractive to the people who thought little of natural history, because they felt that its objects deserved no more than the always-limited space available in Bloomsbury. The second plan was radical separation, severing "nature" from "art" and additionally breaking up the natural history collection into its component parts of botany, zoology, and mineralogy, each of these to be placed in a specialized metropolitan institution. This was the option that Owen had pursued in the mid-1840s, in the hope that the fossils from the British Museum would be relocated to the Hunterian, to form a national collection of the life sciences. As we saw, nothing came of this, and Owen switched to the policy of keeping natural history intact.

The third proposal for the natural history collections was to leave natural history in Bloomsbury and remove the antiquities department to a different site. This idea was put forward by the prince consort following consultation with Owen, shortly after the latter's appointment to the British Museum. The "trade-and-wage-classes" lived eastward, close to Bloomsbury. As they were first and foremost interested in the natural history collections, these should be left where they were. The "classically educated," to whom the antiquities were of primary interest, lived westward. There, in Kensington, a new National Gallery of Art should be erected, to accommodate both ancient and modern sculptures and modern paintings.[36]

Even though this option seemed to denigrate natural history by associating the subject with "working men, their wives, and children," it nevertheless suited Owen and his colleagues, as it left them in a prestigious location, in direct contact with the library. But the interest of the public or, for that matter, the opinions of the scientists carried less weight than the preferences of the nonscientific trustees and of Panizzi, and the National Gallery Site Commission of 1857, moved for by Lord Elcho, reported against the removal of the antiquities from Bloomsbury.[37] After all, as the *Times* commented at a later stage, the "doomed part" of the collections, to be removed from the "venerable institution in Bloomsbury," would naturally be that part with which the archeological and literary tastes of the British Museum trustees and of the principal librarian least agreed.[38]

The credit for founding a separate but unified national museum of natural history must go to Owen. In the late 1850s and early 1860s he put forward a grand and detailed scheme for a national museum to represent an epitome of the three kingdoms of nature: plants, animals, and minerals. He presented his views (1) in a "Report" of February 10, 1859, sixty copies of which were privately printed for the use of the British Museum trustees; (2) in a lecture entitled "On the Scope and Appliances of a National Museum of Natural History," delivered at the Royal Institution on April 26, 1861, and printed in the *Athenaeum* (July 27, August 3 and 10, 1861); and (3) in a booklet, *On the Extent and Aims of a National Museum of Natural History* (1862; the second edition, also of 1862, has changes on pages 83–86), that was an enlarged version of his Royal Institution talk. Owen envisaged that his museum would contain a central "index museum" in which indigenous British collections and a limited array of type specimens would be displayed, primarily for the benefit of the general public. In the various galleries he wanted as complete a display as possible of genera and even of species, mainly to satisfy the needs of scientific visitors.

To dramatize the need for space, Owen argued in some detail for the desirability of a whale gallery, in which the largest-known specimens of all whale genera and species would be displayed; small objects could be accommodated in any museum, "but the hugest, strangest, rarest specimens of the highest class of animals can only be studied in the galleries of a national one."[39] Owen also wanted to exhibit all known species of the largest land mammals, the elephants, arguing that in a national museum a naturalist should be able to study both generic and species differences. "To pare down the cost of a National Establishment of Zoology, by excluding the bulky specimens from the series, is to take away its peculiar and exclusive function as an instrument in the advancement of natural science."[40] Because England was the main colonial power in the tropics, it had a special obligation to include in its national museum the large tropical mammals. Equally, the museum should contain the huge, paleontological monsters of the past: mastodon, megatherium, dinornis, and many others. Owen asked also for laboratory space, a library, and a lecture theater, and he estimated that, *in toto*, a single-storey building covering ten acres was needed or a two-storey building of five acres.

For a while, Owen hedged as to whether or not this additional space should be acquired by expanding the existing Bloomsbury building; in his "Report" to the trustees of February 10, 1859, he specifically added that the building plan was "not intended to advocate any particular form or site of building, or style of architecture."[41] To remain near the existing museum would enhance the prestige of natural history, but its interests were served best by the provision of adequate space: "I love Bloomsbury much; but I love five acres more."[42] The very scale of Owen's plan tended to favor the option of a separate museum at a new locality. Ever since the early 1850s, the Treasury had designated for museum development the suburb of Kensington, and the realistic alternatives were Bloomsbury or Kensington.

In making the move to South Kensington come true, a marriage of convenience temporarily brought together the two autocratic empire builders at the British Museum, Owen and Panizzi. The latter, who was responsible for the building of the magnificently domed Reading Room, encouraged the establishment of a separate museum of natural history in order to get rid of Owen and his scientific staff. When Owen offered a noncommittal assessment of the two possibilities—Bloomsbury or Kensington—in his booklet of 1862, Panizzi systematically demolished the first but encouraged the second option.[43]

Owen's wavering was intended to avoid alienating any faction among his supporters, some wanting to keep natural history in the Bloomsbury place of honor, others desiring to liberate natural history by reestablishing it in South Kensington. That Owen himself preferred to have a museum of his own, beyond the control of Panizzi, is apparent from a letter to one of his parliamentary supporters, the Liberal MP for Kilmarnock, Edward Pleydell-Bouverie, asking him, in the event that the parts of Owen's booklet in favor of the Bloomsbury site were quoted in the Commons, to draw attention to the parts in the booklet that expressed reservations about the site.[44]

By Word of Mouth

Given the political dimension to his objectives, Owen needed to get his message across to a much larger audience than his immediate circle of colleagues. It was not enough to acquire specimens and to display and catalogue them. Nor did it suffice to describe the specimens in papers delivered to the restricted coteries of scientific societies and publish these in their proceedings and transactions, whether of the Royal, the Geological, or the Zoological Society. As Lewes realistically observed: "The public no more reads Professor Owen than it reads Newton."[45] A few of the Hunterian and British Museum trustees, such as William Buckland, Roderick Murchison, and Philip de Malpas Grey-Egerton, did attend meetings of the scientific societies. Yet for the advancement of Owen's museum plans it was necessary that he reach a far wider audience—an audience that had to include the politically powerful *ex officio* trustees, as well as the scientific ones, and a public whose opinion could be influenced to sway that of the politicians. Reaching such an audience could be achieved only by public lectures, and Owen combined his curatorial duties and scientific studies with an energetic program of lecturing (table 2).

TABLE 2. Richard Owen's Metropolitan Lecture Courses

Hunterian Lectures (1837–55)	
1837	History of Comparative Anatomy and Physiology; Classification; Blood; Muscles; Bones and Teeth
1838	Digestive and Circulatory Systems: Digestion; Absorption; Circulation; Respiration
1839	Excretory and Tegumentary Systems: Excretion and Renals; Skin and Teeth
1840	The Comparative Anatomy of the Generative Organs and the Development of the Ovum and Foetus in the Different Classes of Animals
1841	Comparative and Fossil Osteology
1842	Comparative Anatomy and Physiology of the Nervous System
1843	Comparative Anatomy and Physiology of the Invertebrate Animals
1844	Comparative Anatomy and Physiology of the Vertebrate Animals
1845	Organisation of the Invertebrate Animals, Compared in the Ascending Scale according to Their Classes
1846	Osteology and Neurology of the Vertebrate Animals
1847	Anatomy of Fishes
1848	Comparative Anatomy and Physiology of the Vertebrated Animals with Warm Blood
1849	Generation and Development of the Invertebrated Animals
1850	Generation and Development of the Vertebrate Animals
1851	Comparative Osteology
1852	Anatomy and Physiology of the Invertebrate Animals
1853	Anatomy and Physiology of Fishes and Reptiles
1854	Comparative Anatomy and Physiology of the Vertebrated Animals with Warm Blood
1855	Structure and Habits of Extinct Vertebrate Animals, Illustrated by the Hunterian Series of Fossil Remains
1856	[Lectures suspended]
Jermyn Street Lectures (1857–61)	
1857	Osteology and Palaeontology, or the Frame-work and Fossils, of the Class Mammalia
1858	Fossil Birds and Reptiles
1859	Fossil Fishes
1860	[No lectures]
1861	Fossil Reptilia
Fullerian Lectures (1859–61)	
1859	Physiology and Affinities of Fossil Mammals
1860	Fossil Birds and Reptiles
1861	Comparative Anatomy, Physiology and Fossil Remains of the Class of Fishes
London Institution Lectures (1861–64)	
1861	Mammals
1862 (Spring)	Birds
1862 (Winter)	The Class Reptilia
1864 (Jan.)	Classification, Organisation and Fossil-Remains of Fishes

Owen was a committed lecturer who strongly believed in the importance of public presentations. It was only by means of oral discourse, he argued, that a museum's collections could be fully taken advantage of. The scientist might want to use the collections for his specialized interests, to identify and interpret particular objects of natural history; but if the national collections were to serve the diffusion of the scientific learning contained in them, oral exposition was a *sine qua non.* One way of accomplishing this was to organize guided tours of the museum during which a conservator would hold forth on the meaning of the objects on display. More effectively, lecture courses could be offered in which the collections would serve to illustrate general principles.[46]

In his evidence before the Ellesmere Commission, Owen criticized the lack of provisions for public lectures at the British Museum. Politicians and the public alike could justly require that the collections of natural history be used as "an instrument in the diffusion of the principles of the science, and that not only by published works but by oral discourses."[47] Subsequently, in his treatise on the purpose and scope of a national museum of natural history, Owen reiterated that "public lectures at the Museum illustrative of the collections" were indispensable.[48] His sketch of the layout of a national museum showed a centrally located lecture theater. In Alfred Waterhouse's final product, however, no lecture hall was included; it was the only major feature of Owen's ground plan of 1859 that was not realized in the architecture of 1880–81.[49]

The lack of institutionalized lecture courses at the British Museum was probably the single biggest drawback to Owen in moving from the Hunterian to the British Museum. During his years at the Royal College of Surgeons, he had built up something of a lecturing empire. One of the conditions under which the government had entrusted Hunter's museum to the Company of Surgeons was "[t]hat a course of lectures not less than twelve in number upon comparative anatomy, illustrated by the preparations, shall be given twice a year by some member of the Surgeon's Company."[50] Until 1837 it remained the custom to devote one course of twelve lectures to comparative anatomy and another to surgery. These Hunterian Lectures were separate from the once-a-year Hunterian Orations, endowed in 1813 in memory of Hunter and of others who had devoted themselves to surgery. For the Hunterian professors, lecturing duties came on top of busy surgical practices in various London hospitals, and they never held the position for very long, a few years at most. Moreover, they tended to lecture on the subjects of their own specialized interests, and as a result the college failed to provide a systematic series of lectures covering the entire range of the museum's contents.

In 1836 Owen was appointed Hunterian professor on the rotating scheme, but the following year he was made the first sole and permanent Hunterian chairholder. The two different courses, anatomy and surgery, were combined into a single course of twenty-four lectures, exclusively concerned with comparative anatomy and physiology.[51]

Owen's Hunterian audience comprised medical students, members of the college, members of the Council of the Royal College of Surgeons, and visitors. Those who came to sit "at the feet of the new Gamaliel"—as J. Willis Clark put it—formed a wider spectrum than just his scientific friends, as "by-and-by a Hallam, a Carlyle, and a Wilberforce might be seen there side by side with the lights of medicine and surgery."[52] "These lectures, more than anything that he wrote, made Owen famous, and procured for him a passport into society."[53] On one special occasion, Robert Peel came to listen; on another, Prince Albert attended. Owen much valued the presence of such nonmedical listeners which he showed by for example the fact that he might pitch his expositions at a level sufficiently elementary for comprehension by "noblemen, bishops and such non-professional gentlemen."[54]

The importance Owen attached to having a platform from which to lecture is indicated by the fact that early in 1856, before the date of his formal resignation from the Royal College of Surgeons, the possibility that he continue giving his courses entered into the negotiations for his appointment to the British Museum. The previous year, Murchison, one of the museum trustees, had succeeded Henry De la Beche as director general of the Jermyn Street establishment that combined the Geological Survey, the Museum of Practical Geology, and the Metropolitan School of Science Applied to Mining and the Arts (previously called the Royal School of Mines). The school, as a teaching institution, was born from the museum's collections, and as such perfectly fitted Owen's philosophy. Moreover, in order to increase the popularity of the fledgling Jermyn Street enterprise, Murchison was keen to have Owen's name and fame connected with it. In consultation with the Board of Trade and the Treasury, the decision was made "to place the Lecture-Room at the disposal of Prof. Owen, at such times and on such subjects as are compatible with the objects of the Museum and School in Jermyn Street, and as may be found useful in the illustration of the Collections of the British Museum."[55] As a result, Owen's lectures at the Hunterian Museum were continued at the Museum of Practical Geology from 1857 until 1861 (table 2).

The first course of twelve lectures, in which Owen discussed the exciting discoveries and issues relating to fossil mammals, was a resounding success. Many old friends attended, such as Charles Lyell, William Buckland's son Frank, and the African explorer and missionary David Livingstone.[56] The

FIGURE 5. Owen lecturing at the Museum of Practical Geology in 1857 to a large and mixed audience. Owen's popularity as a lecturer enhanced his clout in soliciting government support for his plans to establish a national museum of natural history. (*Illustrated Times* 4 [Apr. 18, 1857]: 252.)

lectures were extensively reported, and the *Illustrated Times* depicted "Owen lecturing" in front of a packed audience of men and women (fig. 5). "Many of the most distinguished men in London set aside their work at the busiest time of the day [2 PM] in order to be present there."[57] Given the fact that the lectures were delivered on a weekday afternoon, the intended audience was professional and middle class. Owen himself confessed that the success of his lectures exceeded his "utmost expectations."[58] Public interest was kept up until the last lecture of the 1857 series. Arrangements were made for publication, and Owen's entry "Palaeontology" in the *Encyclopaedia Britannica* (1858) and his textbook *Palaeontology* (1860) were among the spin-offs. Several powerbrokers in museum matters came to listen, for example, the Duke of Argyll, Lord Lansdowne, Dean Milman, and, of course, Murchison himself, who reacted in a positively glowing manner to Owen's introductory lecture, which, he believed, demonstrated that Owen's promotion to the British Museum was wholly appropriate: "I never heard so thoroughly eloquent a lecture as that of yesterday; and I can assure you that I have not in the course of my life been more gratified than by the proofs which Owen gave of his admirable qualifications for carrying out those higher behests which, as a Trustee of the British Museum, it has been my pride to have warmly assisted in promoting."[59]

Just as important as the size and composition of his audiences were the press reports of Owen's lectures. With the *Lancet* Owen had an uneven relationship, hostile to the extent that he was part of the much-criticized establishment of the Royal College of Surgeons, friendly to the extent that he was at odds with its council. Yet several of his Hunterian courses were reported in the *Lancet*. The *Medical Times and Gazette*, by contrast, was unwaveringly favorable and regularly brought Owen's lectures to the attention of its medical readership. Even his lectures at the London Institution received generous coverage.[60] The *Athenaeum*, the *Literary Gazette*, and increasingly also the daily press, both national and provincial papers, turned Owen's public engagements into sensational events, greatly adding to his public profile and to the efficacy of his efforts to advance the museum movement.

The Road to Albertopolis

The state of the British Museum and Owen's scheme for a national museum of natural history aroused strong feelings in political circles, and throughout the 1860s the issues were repeatedly debated in Parliament. The most vehement exchanges occurred during 1860–63.[61] A variety of topics came up for debate, for example, gas lighting, evening opening, Sunday opening, public lectures, staff salaries, and damage to specimens from dust and soot. On the main question, namely of overcrowding and whether the natural history collections should be removed from the Bloomsbury building and relocated in South Kensington, the Commons was divided approximately along party lines, the Liberals backing Owen, the Conservatives opposing him.

Owen's strongest supporter was the Liberal chancellor of the exchequer William Ewart Gladstone, and the most outspoken critic of Owen's plans was the MP William Henry Gregory. Gregory had entered Parliament as a Conservative MP for Dublin (1842–47). After a period largely devoted to horse racing, he reentered Parliament as a Liberal-Conservative member for the Irish county Galway (1857–71). His politics became gradually more Liberal, and upon the death of Viscount Palmerston in 1865 he formally joined the Liberal Party. His early personal friendship with Benjamin Disraeli was never superseded, however, by any close relationship with Gladstone, whom he disliked. Gregory's strong opposition to Owen was restricted to the early part of the 1860s, and his involvement in museum affairs was primarily based, not on an interest in natural history, but on his love of art and archeology.

In 1866, the Tories returned to power under Lord Derby, and Disraeli succeeded Derby as prime minister in 1868. Once in office, the Conservatives no longer opposed Owen's museum concept. In a variety of ways the acquisition, study, and display of natural history specimens from around the world, on the grand scale of the South Kensington plan, suited Disraeli's imperial policies. Owen, rather sardonically, ascribed the changed situation to "the acceptance by Mr Gregory of the government of a tropical island."[62] In fact, Gregory was not appointed governor of Ceylon until 1871. Although Disraeli dropped his opposition to the South Kensington scheme, he took no direct action to get the plan off the ground. It was not until Gladstone's first term as prime minister (1868–74) that Parliament voted the necessary means for the realization of the new museum (without much further opposition or debate). In 1871 the design by Alfred Waterhouse received final approval of the trustees; in 1873 the building began to "rise out of a hole" beside Cromwell Road; and in 1880 a finished product was handed over at a total cost of £400,000, the price of the land not included.[63]

To some extent, the parliamentary debates of the early 1860s reflected the opportunism of party politics. There was more to the British Museum debates, however, than a Tory attempt to frustrate the Liberal/Whig government of the day. Owen's proposal was part of the larger issue of scientific emancipation and reform, which, in the form of the planned natural history museum, forced the parliamentarians to put a price on the scientific enterprise and allocate a social niche to the emerging class of professional scientists. The latter found a natural ally in the Liberal Party, especially in its newer membership of middle-class merchants and industrialists.[64] The Conservatives, by contrast, were more inclined to safeguard the ascendancy of the landed classes and the aristocracy.

The Kensington site in particular, to which Owen's plan had been linked more or less from the start, had a strong ideological connotation of Whiggishness. It represented part of the cultural heritage of Peel's progressive liberal Conservatism with its middle-class values of free trade, moderate reform, industrial enterprise, scientific research, and education, much of which had been incorporated in Gladstone's Liberalism. Peel had been something of a father figure to Prince Albert, who had played a leading part in the inception and organization of the Great Exhibition of 1851. With the profits of this hugely successful event, land had been bought in Kensington for a projected cultural complex focused on scientific and technological advancement, sometimes referred to as "Albertopolis." The prince consort cared little for aristocratic company and preferred the company of literati, scientists, educationalists, and so on. Culture took the place of class, and the Kensington estate became the hub of a "brave new world."[65]

Owen belonged to the Peelite-Liberal camp. He had been actively involved in the 1851 Great Exhibition in a variety of capacities and was a friend of the prince consort, invited to the palace to instruct the royal children (in the heat of the parliamentary debates Owen's Kensington plan was at times referred to as a "court job"). Owen's friends included such Whigs as T. B. Macaulay and Lord John Russell; the prince consort and Macaulay had each called for Owen's promotion to the post of superintendent at the British Museum; Viscount Palmerston had accompanied Owen to examine the proposed Kensington site for the new museum; and Gladstone asked Owen for guidance, in person and through correspondence, in his efforts to provide a Kensington seat for the nation's natural history collections.[66]

Thus, opposition to the removal of the natural history collections to "Albertopolis" was more than a concern for the inconvenience of the location. It expressed a Tory dislike of the middle-class entrepreneurial parvenus to whom scientific culture was a means of social advancement, and who had found a political niche in the Liberal Party. Gregory had at times expressed his loathing of them: "He did not wish to see all the institutions of the country fall into the grasp of that craving, meddling, flattering, toadying, self-seeking clique that had established itself at Kensington; that had been doing a good business there, and now wanted to extend its operations." And he read to the House a clipping which referred to the representatives of "science and art" as "parasites who have fastened themselves upon so many of the institutions which have already gathered about the great Kensington estate, or who are expectants of the new 'kingdom come' to be established there."[67]

Reform of Museum Management

Conservative opposition to Owen's scheme was increased by the fact that the proposed move from Bloomsbury to Kensington went hand in hand with a demand for the reform of the British Museum's system of management by trustees and for its replacement in the new, separate museum of natural history by a form of scientific self-rule in the person of a director. This demand demonstrated that the road to "Albertopolis" led away from traditional authority and patronage, away, too, from the prestige of title and fortune, and toward the recognition of professional merit.

In the course of the half century 1831–81, when the natural history collections were lodged in the Bloomsbury museum, no fewer than four parliamentary inquiries (by two select committees and two royal commissions) were held that dealt either exclusively or in part with the condition and management of the British Museum. Each of these inquiries was initiated under a reformist,

Whig/Liberal government: the first, in 1835–36, under Melbourne; the second, in 1847–49, under Russell; the third, in 1860, under Palmerston; and the fourth, in 1870–74, under Gladstone.[68] In tandem with several Whig/Liberal MPs, the scientists themselves emphatically and recurrently demanded reform of the management of the British Museum. They did this in testimony before the parliamentary committees and also in a series of five memorials addressed to the government: in 1847; two in 1858; in 1866; and the last in 1879.[69]

There were at this time two basic models of administrative control of public institutions of "science and/or arts." One was management by trustees, as in the case of the Hunterian Museum and the British Museum; access to public funds was gained via *ex officio* government trustees. In the other model, administrative control was placed in the hands of a director who was directly responsible to a government minister. The famous example of the latter form of institutional administration was the Parisian Jardin des plantes, but a growing number of London institutions, too, were governed in a similar way, such as the Royal Observatory, Kew Gardens, the Museum of Economic Geology, and the new establishments in South Kensington. Variations on the two models existed. In the Hunterian Museum, for example, the Museum Committee stood between the trustees and the curatorial staff. At the Royal Observatory a board of visitors diminished the exclusiveness of the director's relationship to the responsible government functionary.

The Board of Trustees of the British Museum was composed of forty-eight members. One trustee was directly appointed by the Crown. Twenty-three were *ex officio* trustees who were prominent members of the government, the church, and the judiciary, and this category also included the presidents of the Royal Society, the College of Physicians, the Society of Antiquaries, and the Royal Academy. Nine trustees were representatives of families that had made major donations to the British Museum, such as, for example, the Sloane, Cotton, Harley, and Elgin families. A further fifteen members were elected trustees, chosen by the others. Nearly all family and elected trustees were aristocrats, among whom were several dukes, marquises, and earls. Museum appointments came under the patronage of three *ex officio* trustees, the so-called principal trustees: the archbishop of Canterbury, the Lord High Chancellor, and the Speaker of the House of Commons. In its *Report*, the Ellesmere Commission of 1847–49, "appointed to inquire into the constitution and government of the British Museum," concluded: "such a Board of Trustees, to any one who considers the individuals who compose it, with reference to their rank, intelligence, and ability, would give assurance rather than promise of the most unexceptionable, and, indeed, wisest administration in every department."[70]

Yet by 1847 not even a single "token naturalist" had been included among the elected trustees. This rankled the emerging class of specialists in botany, zoology, paleontology, and other subjects of natural history. As the *Quarterly Review* commented, in an article on the British Museum by Richard Ford, best known as an expert on Spanish culture, a breeze was gathering: "The Naturalists, who were to have been conciliated by the election of Dr Buckland as trustee on the next vacancy, and who then would have been quiet and dumb as dormice, waxed exceedingly wroth when the honour was conferred on Mr. Macaulay, who canvassed for it."[71] In 1847, when Murchison was president of the British Association for the Advancement of Science (BAAS), a memorial "respecting the Management of the British Museum" was presented to the prime minister, Lord John Russell—the first of the five memorials. The signatories complained that the interest of natural history was not adequately represented by the British Museum's trustees and maintained "that the qualifications of these gifted individuals do not necessarily include an interest in, or the ability to judge of, many of those measures which may best promote Natural History. . . . these distinguished men are unable adequately to direct the vast and rapidly increasing Natural History Departments of the Museum."[72]

The BAAS memorialists at this time did not reject the model of administration by trustees but suggested its reform by proposing that a knowledge of natural history be recognized as one of the requisites for election to a trusteeship of the British Museum. Among the fifty-seven signatories were, apart from Murchison, the presidents of the Linnean, Geological, and Entomological societies, and such well-known names as Greenough, Buckland, Daubeny, Sedgwick, Lyell, Darwin, W. J. Hooker, Phillips, and many more. Significantly, the president of the Royal Society, the marquis of Northampton, an *ex officio* and an elected British Museum trustee, was not among them. The following year, in 1848, Buckland was elected to the Board of Trustees. In the early 1850s Philip de Malpas Grey-Egerton and Murchison himself were added, an aristocrat-paleontologist and a gentleman-geologist respectively. Owen was among the scientists who signed the 1847 memorial, and he was part of the deputation that presented it to Russell.[73]

In his testimony before the Ellesmere Commission of 1847–49, Owen expanded upon the memorial, complaining that the department of natural history was less adequately represented on the Board of Trustees than were the departments of literature and of antiquities, and that this had led to serious gaps in the museum's riches, such as the lack of a proper conchological collection. He recommended that any scientific trustees be organized in a separate committee, similar to the Hunterian Museum Committee, which was composed of people "who have taught anatomy and surgery, and have themselves been curators of anatomical museums." Furthermore, Owen described in considerable detail the administrative setup of the Jardin des plantes, in Paris, where each department was headed by a "professor-administrator," directly responsible to the cabinet minister in charge of education. The *Quarterly Review* mocked Owen's testimony: "Conchologists could not help becoming crustaceous when Britannia, who rules, or did rule, the waves, was short in shells and seaweeds."[74]

A decade hence, and having moved from the Hunterian to the British Museum, Owen more explicitly suggested that a new and separate museum of natural history be managed by a director, without an overseeing body of trustees. In his 1858 presidential address to the BAAS he drew attention to the successful administration of Kew Gardens, the Museum of Economic Geology, and the South Kensington Museum.[75] That same year, a handful of scientists, headed by Huxley and including Darwin, again memorialized the government in the person of the chancellor of the exchequer—the second of the five memorials—requesting that any new scientific museum "be placed under one head, directly responsible to one of Her Majesty's Ministers."[76] This request was reiterated in yet another memorial—the fourth—in 1866, which was signed by twenty-five "Darwinians."[77] One of the signatories, the ornithologist P. L. Sclater, brought this memorial to the attention of the BAAS in 1870. He cited Kew Gardens and the Royal Observatory as model institutions and criticized the fact "that the actual government of our natural-history collections is at present vested in persons who have no special qualifications for the task."[78]

The most comprehensive recommendations for administrative reform were made in the early 1870s by the Devonshire Commission on Scientific Instruction and Advancement of Science (1870–74). It concluded that the objections to the system of administration by trustees were, with respect to the natural history collections, well founded and not "attended by any compensating advantages."[79] It recommended (1) that the move to Kensington be accompanied by a change in the administrative system; (2) that a director should be appointed by the Crown, immediately responsible to a minister of state and in charge of the entire administration of the establishment; (3) that Owen be made the first director; and (4) that a board of visitors be constituted, nominated in part by the Crown and in part by the various metropolitan scientific societies. When, by the end of the decade, the commission's recommendations still had not been implemented, the BAAS, in 1879, again addressed the government—the fifth memorial—pressing its case for reform.[80] Yet the Treasury continued to drag its feet on the issue.[81]

At this time the chancellor of the exchequer was Stafford Henry Northcote, serving in Disraeli's second cabinet. The reluctance of the Conservatives to remove the natural history collections from under the aegis of the British Museum trustees was not surprising. The clamor for administrative reform was an integral part of the wider reform movement which had also manifested itself in other institutions. As Roy MacLeod has argued, it represented the gradual advance, not of any radical politics, but of the Peelite tradition of change combined with accommodation, effectively incorporated in Gladstone's Liberalism.[82] During the reform decades of 1830–50 the shrill criticism voiced by Charles Babbage, David Brewster, William Hamilton, and the *Edinburgh Review* was assimilated piecemeal by Oxbridge's latitudinarian Anglican scientists, who were consolidated in the BAAS and various metropolitan scientific societies—just as the criticism by Thomas Wakley and the *Lancet* of the Royal College of Surgeons was in part accommodated by the college's reformed charter of 1843, extending the franchise to a new body of fellows.

Conflict with the Darwinians

Surprising as this may seem, opposition to Owen's museum expansion plans came not only from Tory traditionalists but also from a clique of fellow naturalists associated with Darwin. By the middle of the 1850s, Owen had become the most publicly visible scientist of the British empire, and the tide of his popularity had risen higher than ever. By the end of the decade, however, this tide began to turn, as vexation, jealousy, and fear among several of his metropolitan colleagues took on an organized form of opposition. All along, Owen had been involved in scientific squabbles—about priority, about nomenclature, and about issues of interpretation. But this time clouds of a different and darker hue began gathering over Owen's head, as contemporaries such as Darwin and Lyell, and young Turks such as Hooker Junior and Huxley, banded together to stem the progress of Owen's institutional self-advancement. What they tried to thwart was Owen's plan for a separate museum of natural history in South Kensington. If they had won the day, London's grand cathedral of natural history would never have been built.

Owen's conflict with the Darwinians is most commonly read as a cognitive clash—a disagreement over "evolution." It was less the intellectual weight of Owen's objections to natural selection, however, than the strength of his position and the fear and loathing this engendered that lay behind the struggle that was to burst upon the public in the wake of Darwin's *Origin of Species.* It is a matter of factual documentation that the tear in the fabric of scientific England that split Owen off from his Darwinian detractors had opened up well before any disagreement over natural selection surfaced. It is possible to pinpoint the beginning of this split by looking at the sequence of memorials presented by the scientists to the government on the question of the British Museum.

In the 1820s and 1830s the suggestion had at times been made by, for example, young Lyell or by the ornithologist and liberal MP N. A. Vigors that the Continental example of placing natural objects and human artifacts in separate museums should be followed.[83] During the following two decades strong opposition had arisen to this idea. As everyone acknowledged, the library formed the core of the Sloanian heritage.[84] But why should natural history be severed from this stem of cultural legitimation? During the first half century or so of its existence (i.e., the second half of the eighteenth century) the British Museum collections had consisted of manuscripts, printed books, and natural history objects. In other words, natural history had right of primogeniture over the younger department of antiquities. Why should the elder collection be evicted from the Bloomsbury estate?

Accordingly, the scientists tended to show an exaggerated and at times obsessive preoccupation with the national library and its importance to their research activities. In the 1847 memorial they emphasized that "we do not contemplate a separation of the Natural History collections from the other departments of the British Museum, as we well know that the cultivation of natural science cannot be efficiently carried on without reference to an extensive library."[85] Owen was one of the people who signed this memorial, as was Darwin, along with Hooker Senior, Lyell, and others. All of them were also among the 114 signatories to the first of the two 1858 memorials, in which they opposed the idea of severance. They objected not only to the establishment of a new museum but also to the suggestion that the collections be broken up and dispersed among the specialized museums of the Royal School of Mines, the Linnean Society, the Zoological Society, and other potential metropolitan storehouses.[86] Dispersal had been recommended by, for example, Richard Ford in the *Quarterly Review* in 1850, as a denigrating way of getting rid of natural history.[87] Owen, as we saw above, supported the idea of breaking up the natural history collections among various institutions in 1845 but then, in his testimony before the Ellesmere Commission of 1847–49, abandoned it.

The issue of piecemeal dispersal came before the trustees in June 1858, when they considered the suggestion that the Banksian Herbarium be transferred from the British Museum to Kew Gardens, in the charge of the Hookers, Senior and Junior. Before the Sub-committee of Natural History, Owen argued that perhaps Kew should receive the botanical collection from the British Museum but that, otherwise, natural history, including mineralogy, should be kept intact.[88] Darwin effectively expressed the same opinion, stressing the need to keep the natural history collections at the British Museum: some weighty arguments might be advanced in favor of Kew, but "I think it would be the greatest evil which could possibly happen to natural science in this country, if the other collections were ever to be removed from the British Museum and library."[89] Lyell registered unqualified opposition to the idea of severance: "I heard with the greatest concern of the proposal of removing any part of the Botanical Collection from the British Museum to Kew," he wrote to Murhison.[90] The latter himself was keen to keep natural history in the seat of honor in Bloomsbury and give the entire museum over to science.[91]

The first of the two 1858 memorials ended in a grand appeal to the government not to yield to the argument that, because on the Continent objects of art and of nature were exhibited in separate establishments, the same separation should be copied in London. "Let us, on the contrary, rejoice in the fact, that we have realised what no other kingdom can boast of, and that such vast and harmoniously related accumulation of knowledge is gathered together around a library, illustrating each department of this noble Museum."[92] The *Quarterly Review*, however, in a discussion of the scientists' memorial, stated perceptively: "we believe that all their arguments may be summed up into one which they have not expressed—the dread that the separation would diminish their importance." The anonymous reviewer was John W. Jones, keeper of printed books at the British Museum and later to succeed Panizzi as principal librarian; he pointed out that a removal from the British Museum would liberate natural history, and in particular its superintendent, from the humiliating control of Panizzi, who was the head of the museum: "Let the natural history be elevated to its rightful dignity. Let it form an independent institution, with Professor Owen at its head, and let him have a temple of his own instead of being a lodger."[93]

This issue of the *Quarterly Review* appeared in July 1858, and Owen soon took up its suggestion. In September 1858, in his presidential address to the BAAS, he cautiously began to promote the idea of removal to a separate museum, and a few months later, in February 1859, he produced his first, detailed plan for an independent national collection of natural history.[94] No sooner had Owen indicated that he was coming round to the idea of a separate museum than Huxley started a campaign to thwart it. As Adrian Desmond has observed, Huxley belonged to a fast-rising middle class of new-style biologists who nurtured a strong anticlerical bias and who became the crusaders in the cause of Darwinian evolution. They maneuvered to capture control of the key scientific posts in the metropolis, for example, in the Royal Society and the University of London.[95] The greatest obstacle to the attainment of this goal was Owen, regarded by many of his contemporaries as the greatest living naturalist, whose Peelite allegiance was entirely at odds with Huxley's aggressive and radical stratagem. To block Owen was, temporarily, a higher priority than to promote the cause of natural history, even if this required duplicitous tactics.

Huxley had signed the memorial of July 1858 objecting to the removal or dispersal of the natural history collections; but in November of the same year, shortly after Owen had come out in support of separation from the British Museum, Huxley composed a further memorial to the chancellor of the exchequer, supported by a small number of colleagues, Darwin among them. In this memorial Huxley accepted what Darwin only four months earlier had described as "the greatest evil," namely, removal from the British Museum, and in addition argued that the natural history collections should be broken up and distributed among various metropolitan institutions. The main botanical collection, the Banksian Herbarium and the fossil plants, should be moved to Kew Gardens, where Huxley's close friend J. D. Hooker was assistant. For zoology there should be established a popular museum and a scientific one, possibly in one and the same building. Furthermore, Huxley *cum suis* recommended that the new Museum of Economic Zoology be further developed and that the disposal of the mineral collection also be considered.[96] Although this was not spelled out, mineralogy would, if removed, probably be deposited in the Museum of Practical Geology, where Huxley was professor and curator. The most obvious potential beneficiaries of this plan were Hooker and Huxley.

In spite of this double somersault away from the July 1858 memorial, Huxley rather disingenuously insisted in his testimony before the Gregory Committee of 1860 that "I am at a loss to understand in what respect, or where, the inconsistency lies between the two documents."[97] In reality, however, Huxley had changed his stance both with respect to where the natural history collections should be located and in the matter of administrative reform (from trustees to a single director), lest Owen be elevated to the post of museum director: "I am much inclined to believe, as matters now stand," Huxley stated before the Gregory Committee, in support of the system of governance by trustees, "that it is a very important circumstance to have a body of educated and liberal gentlemen, men of the world and of standing in society, interposed between men of science and men of letters on the one hand, and mere officials on the other."[98]

Huxley also opposed Owen in the latter's demands for spacious museum accommodation. Space was directly related to the issue of prestige. To Owen, a museum was first and foremost a place for scientific research and only secondarily an institution for the entertainment and education of the visiting multitudes. The prestigious Jardin des plantes and also the Hunterian Museum exemplified his views. The *Lancet* supported Owen's ambitious South Kensington scheme, which it regarded as "not one iota too much."[99] Such a research institution needed comprehensive collections, even of the largest mammals, and therefore space.[100] A national museum of natural history, Owen enthused, had to be an establishment "constituting a material symbol of civilisation and appreciation of science, for the realisation of which the resources of the greatest commercial and colonizing empire of the world give it peculiar advantages and facilities."[101]

In arguing for his whale gallery Owen cited the generous provisions for whale skeletons in Dutch and French national museums. The largest possible specimens should be exhibited to gratify the curiosity of the common visitor, but more importantly, a representative species of each genus ought to be present for the benefit of the naturalist. Owen proposed to display a *Balaenoptera* (e.g., blue whale), a *Balaena* (e.g., bowhead whale), a sperm whale, a bottle-nosed whale, a grampus, and a dozen skeletons of narwhals, dolphins, and porpoises. By happy coincidence, a 70-foot-long whale became beached in 1863 near Caithness, in Scotland. Owen traveled north to inspect the stranded giant, determined that it was a full-grown female sperm whale (cachalot), and bought the carcass on behalf of the British Museum for £30. Owen then used the purchase for political ends, chastising his parliamentary adversaries, in a lecture on "the capture of cachalot," for their stinginess in opposing a whale gallery.[102] This sperm whale was the first such whale possessed by a museum in Britain, and its skeleton was later exhibited in the central hall of the Natural History Museum in South Kensington, in precisely the same spot where today the dinosaur *Diplodocus carnegiei* stands, as its successor icon of a national collection (fig. 6).

Owen's parliamentary supporters, such as Richard Monckton Milnes, reiterated the demand for space in the Commons by calling for a serious scientific museum, not "a mere raree-show." "What we want to do is to rival that magnificent establishment, the Jardin des Plantes; an establishment which has satisfied the aspirations of all scientific men, but which it never occurred to any Frenchmen to unite with the Louvre."[103] In an age of science, the nation needed a first-rate scientific museum, whether the public would visit it or not. Owen's adversaries in Parliament, on the other hand, eagerly used the ammunition provided by Huxley (initially referred to as "Professor Huxtable"), who criticized Owen's elaborate scheme. The whale gallery, for example, would not interest the public; the whales would be difficult to obtain; the skins, even if procured, would produce an intolerable stench.[104] To alleviate the

FIGURE 6. The cathedral-like central hall of London's Natural History Museum during the 1880s when the main exhibit was the skeleton of a whale, later replaced by the dinosaur *Diplodocus carnegiei.* (F[lower], *General Guide to the British Museum (Natural History)*, 16.)

overcrowding of the British Museum, Huxley estimated that two (or two and a half) and not Owen's ten acres would suffice, even taking into account "every specimen that is likely to be obtained in the next 50 years."

The issue of the purpose of a museum (whether for public instruction or for scientific research) and of museum organization (what to exhibit and what to put in store) was a serious one, discussed by J. E. Gray and A. C. L. Günther, keepers of zoology at the British Museum, and later by W. H. Flower, who in 1884 succeeded Owen as director of the new Natural History Museum.[105] During the crucial early 1860s, however, the issue of museum organization and space was directly related to that of the purpose and prestige of natural history as a scientific subject and a profession. To argue that a limited display of selected specimens would best serve the interest of the working classes, as did a number of MPs who opposed Owen's scheme, was equivalent to downgrading natural history by implying that its primary purpose was little more than popular edification. Owen's demand for space, on the other hand, was part of a strategy to elevate the social status of scientific work. Huxley's abetting of the parliamentary detractors of natural history can be explained only by his desire to block Owen. The *British Medical Journal* regretted Huxley's lack of appreciation of "the expanded ideas of Mr. Owen."[106]

In his *Autobiography*, Darwin reminisced that Owen had become his enemy out of jealousy at the success of *The Origin of Species.*[107] As we shall see, Owen's reaction to Darwin's magnum opus was more multifaceted than that, but what concerns us here is that Owen's attack on Darwin's *Origin of Species* appeared in the *Edinburgh Review* for April 1860, nearly a year and a half after Huxley, Darwin, and others had attempted to sabotage Owen's plans for what was to become his most monumental accomplishment.[108] The founding of the Natural History Museum was the result of Owen's vision combined with Gladstone's political pertinacity. It was not merely an architectural expression of the popularity of natural history. The new museum, its dimensions, and even its location were the fruits of reformist ideals in the Peelite tradition. They represented not only a triumph over right-wing opposition from Conservative politicians who resented the growing authority of science within the nation's cultural institutions but also a victory over left-wing obstruction, primarily from Huxley, Darwin, and their confederates.[109]

Doing the Work of Empire

Thus, institutional power politics was at the root of the infamous conflict between Owen and Darwin. Let me reiterate that it was less the threat of Owen's objections to natural selection than the dread of his institutional power that lay behind the struggle that was to burst upon the public in the wake of *The Origin of Species.* By contrast, Desmond suspects that the "politics of evolution" was hidden behind Owen's every move, virtually from the beginning of his employment at the Hunterian Museum. Owen's choice of animal subjects for study, he believes, was determined by an ideological program and a grand strategy designed to counter evolutionary theories. For example, Owen's work on the duck-billed platypus was supposedly "an attempt in part to counteract [Robert E.] Grant and [Étienne] Geoffroy [Saint-Hilaire]—to show that the strange duck-billed mammal (which Grant believed laid eggs) was not transitional at all, or bird-like in the way demanded by Geoffroy and his supporters."[110]

The fact that some of Owen's studies could be, and were, used to counter Geoffroy is not denied here. What I do deny is that the politics of evolution had a programmatic significance for Owen's work and that a hidden agenda to counter the French transmutationists and their metropolitan followers explains Owen's choice of subjects. A grand strategy did underlie Owen's preference for specific topics, but it was not of an antitransmutation kind; it was pro-museum, in that his papers reflected the policy of specimen acquisition and were part of a planned, systematic enrichment of the collections to establish national hegemony, international parity, and imperial grandeur. The grand strategy was not to counter Geoffroy or, much later, Darwin but to build the most precious collection of museum objects to force the hands that held the public purse strings. Furthermore, to the extent that scientific theory came into his choice of topics, the positive demonstrating of Cuvierian functionalism was a far more significant factor than the negative disproving of transmutation theories.[111] No theory of evolution ever ranked so high as to be a programmatic determinant in Owen's scientific work, either positively or negatively, either when arguing for his own theory of divinely preordained evolution or when refuting materialist transmutation.

A major reason for acquiring, displaying, and describing particular specimens was to enhance the ranking of a museum. Adding a rare specimen, a very large specimen, or a specimen from a newly explored region expanded the museum's sphere of public influence. The scientific interpretation could be controversial or doubtful and could change over the years, but a rare specimen was always an adornment to the collection. As Owen repeatedly proclaimed, every European nation possessed its national museum of natural history, and the quality of its contents bore a direct relationship to the level of civilization of each national community. To collect, to display, to describe, and to lecture about objects of natural history increased the institution's and, by association, the nation's "intellectual wealth." In 1842, Owen told his Hunterian audience:

"Collections of natural objects, selected for their significance, rarity, or beauty, have ever been regarded as the signs and ornaments of civilized nations; and, though at first viewed with feelings of curiosity and wonder, they soon became recognized as important aids to the acquisition of intellectual wealth."[112]

The acquisitions policy, determining the size and the composition of the collections, could be an effective strategy of campaigning for government aid in support of architectural expansion. By adding valuable specimens to the contents of his museum, Owen increased pressure on the authorities to provide money for adequate space. Acquisitions constituted a basic determinant of his scientific oeuvre in that these influenced, programmatically, the very availability of objects on which he could work. The primary reason for studying one rather than another topic of natural history was often museological and not always—as habitually assumed—a matter of cognitive, scientific interest. An example was *Archaeopteryx*, of which the British Museum secured a specimen for its collections on Owen's recommendation. Owen then described the "feathered fossil" in a paper to the Royal Society, read on November 20, 1862.[113] The physical possession of this rare specimen, as a symbol of institutional sway, was the primary significance of the acquisition. The *Archaeopteryx* represented, first and foremost, an object of competitive museum building. Its chief importance did not derive from its bearing on evolution theory as the apparent transition from reptile to bird but was the same as, for example, that of the dodo, of which Owen was very keen, too, to acquire a specimen for the museum and about which he wrote a classic memoir at just about this time.[114]

Although the official rhetoric was that a national museum of natural history aimed to be "a more or less complete epitome of the three kingdoms of Nature—Animals, Plants, and Minerals,"[115] the acquisitions policy was far more politically charged than that. From the very start to the very end of his museum career, Owen was constantly and deeply involved in the international competition for national museum enrichment, especially the acquisition of objects of colonial natural history.

The beginning of Owen's professional life coincided with a significant new influx of rare museum specimens. These were the trophies of voyages of discovery that were embarked upon not long after 1815, following the end of the Napoleonic Wars. Initially, the main geographic destination was the Arctic, but soon other regions were set sail for, in particular the Antarctic and then Africa. Following the eighteenth-century precedent of Joseph Banks's participation in the *Endeavour* exploration of the southern seas, naturalists were recruited to the scientific voyages for the specific purpose of observing and sampling the natural world. Darwin sailed around South America on the

Beagle (1831–36); Joseph Hooker traveled to Antarctic latitudes with the discovery ships *Erebus* and *Terror* (1838–43); and Huxley explored the continental rim of Australia aboard the *Rattlesnake* (1846–50).

In order to gain as many objects of natural history as possible from these and other sources, various individuals and societies drew up guidelines for collecting.[116] In 1821 Buckland helped enlarge his museum collections with his "Instructions for Conducting Geological Investigations, and Collecting Specimens," the last instruction being to wrap each specimen separately, pack the whole with moss or hay in a strong box, and ship this to "Rev. Professor Buckland, Museum, Oxford."[117] A decade later, the Geological Society, in an effort to divert part of the stream of both indigenous and foreign rocks and fossils to its headquarters, issued its own "Instructions for the Collection of Geological Specimens" (1831).[118]

Not long afterward, Owen staked his claim by preparing a small booklet entitled *Directions for Collecting and Preserving Animals* (1835), which incorporated some manuscript instructions for collectors left by John Hunter.[119] He circulated it "to enlist the services of our adventurous countrymen in the cause of Natural History."[120] Owen believed that not only the scientific voyages but also the commercial ones, in particular the whaling trips, could be used on behalf of natural history. In 1839 he wrote to J. W. Lubbock, who was treasurer of the Royal Society and involved with the society's participation in the voyages of exploration, to suggest that the whaling industry be enlisted in the quest for colonial specimens and that prizes be awarded to those who returned with the rarest objects or the best records of observations.[121]

The relevant government departments were willing participants. In 1832 the Admiralty printed *Hints for Collecting Animals and Their Products*, written by Buckland's friend and Owen's patron William Broderip. The culmination of this tradition of composing lists of instructions for travelers to foreign parts was the Admiralty's *Manual of Scientific Inquiry*, edited by John Herschel on behalf of the Royal Society. It was a voluminous handbook with contributions by some of the best-known names of contemporary British science and scientific exploration.[122] Owen contributed the chapter on zoology, which was an expanded version of his earlier Hunterian guidelines. Separately printed as the *Manual of Zoology*, this went through several editions. For specific expeditions, Owen added additional instructions, for example, in 1858, when he was asked to advise the Zambesi expedition and the expedition to Vancouver Island.[123]

Owen's systematic involvement in colonial natural history can be illustrated by several parallel series of papers he wrote on marsupials (both recent and extinct), on monotremes, on the dinornis and other flightless birds, and on fossil reptiles from South Africa, each series a reflection of the programmatic enrichment of the museum by specimens from Australia, New Zealand, and, later, from the Cape and other African regions. The Australian monographs, some ninety in total, were the most numerous and earned Owen the epithet of "father of Australian paleontology."

Unlike the British Museum, the Dutch and the French national museums of natural history employed professional naturalists as collectors. In 1835 there were no fewer than eight professional *naturalistes voyageurs* attached to the Jardin des plantes, trained collectors who were dispatched to various parts of the world to collect natural history specimens, at the government's expense.[124] In 1838 the stratigrapher W. H. Fitton, just returned from a trip to the Netherlands, wrote to Owen that he was impressed by the collections there, "and I am sorry to say generally that the use the Dutch have made of their *colonies*, for the benefit of Natural History, puts England to shame."[125] The Dutch, at the Rijks museum van natuurlijke historie, engaged two types of naturalist-collectors; some were in paid employment, whereas others were given a small subsidy or nothing at all but were under contractual obligation to give first choice of their specimens to the museum, which would buy at market value and assist in the sale of the rest.[126]

The British navy and its officers were by no means the only international collecting arm used by Owen to enrich his collections. The services of civil servants, missionaries, colonial governors, settlers, and travelers were constantly engaged.[127] In a retrospective on his museum work, Owen commented:

> The annual additions of specimens continued to increase in number and in value year by year. I embraced every opportunity to excite the interest of lovers of natural history travelling abroad, and of intelligent settlers in our several colonies, to this end, among the results of which I may cite the reception of the aye-aye, the gorilla, the dodo, the notornis, the maximised and elephant-footed species of dinornis, the representatives of the various orders and genera of extinct Reptilia from the Cape of Good Hope, and the equally rich and numerous evidences of the extinct Marsupialia from Australia, besides such smaller rarities as the animals of the nautilus and spirula.[128]

Australian Possessions

The person from whom Owen received many of the Australian marsupials and monotremes was his longtime friend George Bennett, who had obtained his membership diploma in the Royal College of Surgeons in 1828, then visited Australia in 1829 and again in 1832. Each trip produced a stream of new specimens destined for the Hunterian Museum. The first trip provided

Owen with the rare pearly nautilus, and the second netted him no fewer than 510 specimens, supported "by the copious and accurate observations on the locality, temperature, time of day, and other circumstances connected with the capture of each specimen."[129] In 1836 Bennett returned to Australia, this time to stay, settling in Sydney as a medical practitioner and participating as secretary and curator (1835–41) in the establishment of the Australian Museum.[130]

The Bennett specimens, combined with the ones from the Zoological Gardens, allowed Owen to participate in a European debate on the reproductive processes of both marsupials and monotremes. One of the questions he, with Bennett's help, tried to answer was whether the echidna and the platypus lay eggs (are oviparous) or bear live offspring (are ovoviviparous) and whether they possess mammary glands. The oddest monotreme, the *Ornithorhynchus paradoxus*, had been named in 1800 by Johann Friedrich Blumenbach, based on a specimen obtained from Joseph Banks.[131] In 1800, and again in 1802, the College of Surgeons gave notice of British claims to the antipodean regions with papers on the peculiarities of the platypus presented to the Royal Society by Everard Home.[132] Soon, however, the scene switched to the Jardin des plantes, where Henri de Blainville, Cuvier, and Geoffroy Saint-Hilaire all concerned themselves with the features and taxonomic position of the odd Australian fauna. Cuvier, most conventionally, placed the antipodean mammals in existing groups, the monotremes with the edentates and the marsupials with the carnivores. Both Blainville and Geoffroy set the monotremes apart, as a link connecting birds and mammals—Blainville in support of his creationist belief in a linear chain of being, and Geoffroy in support of his Lamarckian theory of transmutation; but whereas Blainville kept the monotremes within the class of the mammals, although in a subclass of their own, Geoffroy, most drastically, argued that the monotremes were not mammals at all and should be ranked as a new and fifth vertebrate class.[133]

The spotlight then turned back to Germany, where Johann Friedrich Meckel published a splendid monograph, *Ornithorhynchi Paradoxi Descriptio Anatomica* (1826), in which he reaffirmed his earlier announcement of 1824 that mammary glands are present in the duck-billed platypus. A protracted controversy followed the appearance of this memoir, in which Meckel was supported by the leaders of German physiology, Karl Ernst von Baer and Johannes Müller.[134] At London's University College, Robert E. Grant entered the fray in 1829 with a letter to Geoffroy endorsing the latter's belief that the platypus is oviparous.[135] This led Geoffroy to restate that monotremes are not mammalian and that the vertebrates must be divided into five classes, mammals, monotremes, birds, amphibians, and fishes. The Germans, however, believed that the mammalian character of the monotremes did not depend on whether or not the young were hatched from eggs or were born live, as in each case mammary suckling was possible. In 1831, Home, in his dying breath on the subject, denied the presence of mammary glands in the *Ornithorhynchus.*[136]

Owen's friends later credited him with the discovery, in 1832, of mammary glands in the monotremes,[137] but his work added up to no more than an extended confirmation of Meckel's view. As to the issue of how they are born, Owen argued that the monotremes are ovoviviparous or implacental, like the marsupials.[138] This issue remained undecided until much later when, in 1884, the naturalist W. H. Caldwell, traveling in Queensland, came upon incontrovertible evidence and sent a telegram, via the Sydney chemist Archibald Liversidge, to the BAAS, at the time in session in Montreal: "Monotremata oviparous, ovum meroblastic," which, when passed on across the Atlantic, became corrupted and published in the *Times* as "viviparous" and "mesoblastic ovum."[139]

Although Owen contributed to keeping the monotremes firmly within the class of mammals, one cannot say that he tried to steer away from their "transitional" character. In fact, he emphasized the "primitive" nature of both marsupials and monotremes, showing that the absence of a proper corpus callosum in their brains was reminiscent of cerebral features in the lower, oviparous classes. What his studies did show, beyond any doubt, however, was that the Hunterian collections ranked with those of Berlin and Paris, symbolizing London's place at the hub of an empire. Over the decades, while the taxonomic question lost much of its controversial implications, the value of the platypus as a trophy of imperial prowess remained; witness the fact that in 1851 in the nave of the Crystal Palace, as part of Britain's self-presentation as a colonial power, two specimens of the platypus were exhibited.[140]

Owen's first major supplier of Australian fossils was Thomas Livingstone Mitchell, who was the surveyor general of New South Wales (1828–55). It was Mitchell who initiated the systematic exploration of the caves and fissures in Wellington Valley, with their riches of Pleistocene vertebrate fossils preserved in characteristically red-colored deposits. In 1830 Mitchell communicated his early results in a paper to the Geological Society, having submitted the bones to the Hunterian Museum for examination by William Clift, who identified them as belonging mainly to kangaroos and other indigenous marsupials.[141] On a return visit to England, in 1835, Mitchell enlisted the assistance of Owen, who contributed a letter to Mitchell's *Three Expeditions into the Interior of*

Eastern Australia (1838) in which Owen identified and named a number of extinct marsupials. He christened the remains of two kangaroo species, "at least one-third larger" than today's *Macropus major*, as *Macropus Atlas* and *Macropus Titan*, not anticipating that later he would receive the remains of even larger extinct kangaroos. He also identified a new marsupial genus, which he named *Diprotodon* because of its two chisel-shaped incisors in both the upper and the lower jaw.[142]

Other fossil fragments in Mitchell's red earth breccia seemed to prove the Pleistocene existence in Australia of such pachyderms as elephant and hippopotamus. In 1842 a further donation of fossils from Mitchell reached Owen, who excitedly concluded that the new specimens "incontestably establish the former existence of a huge proboscidian Pachyderm in the Australian continent, referable to either the genus *Mastodon* or *Dinotherium*."[143] Unlike marsupials, which are adapted to periodic droughts, the pachyderms frequent marshes, swamps, or lakes, and Owen concluded that, when these animals roamed the Australian continent, the climate was more humid and that the extinction of the large quadrupeds might have been due to climatic change. Almost two years later Owen was alerted by the German explorer of Australia Ludwig Leichhardt to the possibility of a misidentification[144] and was "enabled to correct my error, and to show that the supposed Dinotherian remains were really those of an adult individual of the same Marsupial genus and species as the immature fragment of lower jaw on which the *Diprotodon australis* was founded."[145] Mitchell named a peak in Queensland Mount Owen, and Owen gave an extinct *Diprotodon*-like giant the name *Nototherium Mitchelli*.[146]

Thus, the Australian Pleistocene mammals, similar in size to African pachyderms or South American megatherioids, had been marsupials, just like today's fauna, confirming the law of faunal continuity or succession of types. Owen changed his climatological explanation of their extinction and surmised that it might have been "the hostile agency of man" (i.e., hunting by the Aborigines) that had led to the demise of the giant marsupials.[147] It took several decades of patient collecting before the *Diprotodon* and other giant marsupial quadrupeds could be fully restored, and during Owen's museum career their skeletons never attained the iconic celebrity possessed by the megatherium from South America or the dinornis from New Zealand.

However keen Owen was to lay his hands on exotic specimens, he himself never participated in any explorations, either at sea or on land. His furthest trips, to Egypt, were in the company not of naturalists and explorers but of his museum patrons and were for the purpose of seeing the sights of human civilization. Yet Owen considered himself a full participant in the work of empire, arguing that there were two types of naturalist, one who goes out into the field and one who works in the museum, each complementing the other.[148]

After all, the discovery and exploration of new colonies did not just mean travel through virgin territories or simply the mapping of unfamiliar tracts of ground. Nor did dominion over such territories mean merely the establishment of missionary and trading posts or the imposition of colonial administration. It also meant that the natural world, in its manifold aspects, had to be described, named, sampled, and taken possession of. The naturalist at home was therefore an indispensable ally of the explorer abroad, and the natural history museum, where the foreign fishes, reptiles, birds, mammals, insects, and fossils, in their thousands, were stored, classified, interpreted, and exhibited, became a prominent symbol of British dominion, not only over the natural world in general but more particularly over specific colonial possessions. The India Museum, for example, which was founded in 1801 by the East India Company, functioned as a concrete representation of the company's commercial activities.[149] More generally, museums were the architectural embodiment of claims to both intellectual and political mastery.[150]

In this way museums could become a showcase of empire where visitors were able to see the trophies of conquest. Owen used the metaphor of colonial expansion to describe his scientific work, calling it an "annexation of fresh territory to the Empire of the Known."[151] The natural history collections became a vivid means of "re-enacting and extending the work of empire," as Harriet Ritvo puts it.[152] The range of specimens symbolized the extent of imperial sway and reflected the scope of British mercantile power around the world. This symbolism was in place at the very start of Owen's career, when he was given permission to dissect any animal that had died in the London zoo. Humphry Davy, who was the main driving force behind the establishment of the Zoological Society, designed the zoo in Regent's Park as a microcosm of Britain's overseas possessions and, in Desmond's words, as "a visible affirmation of London's global preeminence."[153]

In their metonymical representation of empire, the collections of natural history could be used to attract political support. Owen ended his booklet *On the Extent and Aims of a National Museum of Natural History* with the following exhortation:

> The greatest commercial and colonizing empire of the world can take her own befitting course for ennobling herself with that material symbol of advance in the march of civilization which a Public Museum of Natural History embodies, and for effecting which her resources and command of the world give her peculiar advantages and facilities.[154]

Owen counted among his good friends missionaries and explorers, in particular David Livingstone, Richard Burton, and P. B. Du Chaillu. He was actively involved in the affairs of the Royal Geographical Society and the Royal Colonial Institute. He reviewed books concerned with exploration; and he willingly offered his work on the fossils of Australia, New Zealand, and South Africa for use by the advocates of an empire when colonialism had become a controversial issue. Seated at the hub of a colonial network of specimen supply, Owen was the keeper and interpreter of the imperial collections.

3

Gothic Designs

As we have seen, Owen's museum setting was a first-order determinant of the topics on which he worked. It can throw light also, I believe, on his predilection for particular scientific theories and the fact that a rift ran right through his collected works, in the sense that individual studies were either functionalist or transcendentalist. This split indicated not just that these two epistemologies variously impacted on Owen's thinking but above all that each was linked to one of two constituencies, both of which were needed for the fulfillment of Owen's institutional ambitions. More specifically, the split reflected an opportunistic running with the Oxbridge hares and hunting with the London hounds.[1]

As stated at the end of chapter 1, the networks of patronage in particular, which were crucial to the realization of Owen's museum plans, provide an overarching framework of explanation for his choices of theory. For his Oxford- and Cambridge-educated patrons, Owen reinforced the Paleyan design argument by demonstrating the efficacy of Cuvierian functionalism. Some of the most sensational functionalist feats were Owen's reconstructions of prehistoric "monsters," that is, of the very objects that gave content and definition to natural history museums. The functionalist dictates of Oxbridge, those cities of Gothic architecture, inspired the reconstruction of Owen's most renowned and Gothically bizarre museum objects. Let us examine the functionalist strand in Owen's work before following, in the next chapter, the transcendentalist one.

Oxbridge Patronage

The two groups of men most directly relevant to the realization of Owen's career ideals were the trustees of the Hunterian Museum and the trustees of the British Museum. In a sense, the two bodies formed a single one: not only were they similarly constituted, but some trustees served on both boards. As the Earl of Ellesmere, chairman of the Royal Commission of Inquiry on the British Museum, rhetorically asked when taking Owen's evidence on May 23, 1848: "The constitution of the Board of Trustees [of the Hunterian Museum] very nearly resembles that of the British Museum?"[2] The cooperation, the goodwill, and the friendship of these men were a *sine qua non* for the success of Owen's institutional politics, and he assiduously cultivated their company. He has been accused of snobbery, of hobnobbing with the high and mighty in church, state, and polite society.[3] Without denying the truth of the allegation, it should be added that the grandees with whom he mixed were, by and large, the very people whose voice and vote counted in museum matters.

Yet although Owen was *in* their society, he was not *of* them; neither by family background nor by education did he *belong*. He never joined the ranks of the trustees, but remained their paid employee. Moreover, he participated in the drive for the reform of museum management by trustees, with the ultimate goal of replacing them by professorial directors, that is, by people such as Owen himself. To keep the support of the trustees and, simultaneously, to work toward abolishing the very system of management by trustees required all the ambiguity for which Owen is so notorious. His museum politics was conducted along a double track; one kept him on the right side of the trustees; along the other he strove for autonomy for himself and for a metropolitan group of upwardly mobile scientists, all lacking the advantages of special education or family background.

The fundamental point to be made in this context—fundamental because it will prove essential for understanding Owen's scientific oeuvre—is that each of the two tracks or, to abandon the metaphor, each of the two parties to which Owen was connected had its own political agenda and additionally promoted a different scientific epistemology. To be able to see this, we need to define the parties in question in some detail. Let us first consider the trustees, and in particular the ones who were most active and powerful in the furtherance of Owen's institutional schemes. These were not a random collection of individuals, connected merely by a shared interest in Owen and his work. On the contrary, they belonged to a coterie of like-minded friends well before Owen appeared on their horizon. This select circle was composed of gentlemen and noblemen who were educated at Eton, Harrow, and Winchester and who went

FIGURE 7. William Buckland in 1832, the year in which his attention was drawn to Owen because of the latter's *Memoir on the Pearly Nautilus.* Buckland's patronage gave Owen access to a circle of Oxford-educated museum trustees on whose support his advancement was crucially dependent. Engraving from an oil painting by Thomas Phillips, 1832 (courtesy Department of Earth Sciences, University of Oxford; used by permission of University of Oxford).

on to Christ Church or Oriel College at Oxford. The initial center of this circle was William Buckland, the idiosyncratic geologist and canon of Christ Church who in 1845 was appointed by Peel to the deanery of Westminster (fig. 7). Once he had moved to London, Buckland became a trustee of the Hunterian Museum (1847–56) and, moreover, in 1848 was the first naturalist to join the ranks of the elected trustees of the British Museum (among the *ex officio* trustees was the president of the Royal Society, and consequently Joseph Banks and Humphry Davy had preceded Buckland as scientifically qualified trustees).

Crucial to Owen's advancement were, in addition to Buckland, several of Buckland's friends and pupils. These included the aristocrat-paleontologists Philip de Malpas Grey-Egerton and his lifelong friend since undergraduate days William Willoughby, Viscount Cole, who in 1840 succeeded to the title of Earl of Enniskillen; both were educated at Christ Church during the third quarter of the 1820s and attended the lectures by Buckland, just then riding the crest of a tide of popularity produced by his cave studies. The two friends were keen collectors of fossils, especially of fossil fishes, and their collections were later acquired for the British Museum (Natural History). Both entered Parliament, and Enniskillen, after 1840, sat in the House of Lords. Both men joined the Board of Trustees of the Hunterian Museum several years before Buckland did and served on it for many years, Egerton from 1840 until 1881 and Enniskillen from 1840 until 1886. In 1851 Egerton was also elected a trustee of the British Museum.

Probably the most deeply devoted Owen supporter of this circle was a near contemporary of Buckland, to whom Buckland himself owed much—William John Broderip, magistrate and naturalist, educated at Oriel College, and one of its reform-minded "Noetics."[4] Also educated at Oriel was a further Buckland pupil, Samuel Wilberforce, who preceded his geological mentor as dean of Westminster before being advanced in 1845 to the bishopric of Oxford, in which see he became one of the most influential figures of the established church. Like other members of the circle, Wilberforce attended not only the lectures by Buckland but also those by Owen. Wilberforce was elected a trustee of the Hunterian Museum, to fill the vacancy created by the death of Peel, a position that he held for twenty-two years (1851–73).

Another product of Christ Church and Buckland's lectures was H. W. Acland, who as an undergraduate befriended Buckland's protégé John Ruskin. In 1845 Acland was appointed Lee's reader in anatomy at Christ Church, from which position he resigned upon his nomination in 1857 to the regius chair of medicine in the university. Since the incumbents of these positions were *ex officio* trustees of the Hunterian Museum, this was a typical instance of Owen being placed under the trusteeship of someone who was his inferior both in years and in intellectual accomplishments. As Acland's biographer notes: "It was from the lips of Richard Owen and under the inspiration of the great Hunterian collection in the College of Surgeons that Acland had received his first lessons in Comparative Anatomy."[5] Especially Acland's involvement in the museum movement was especially deep: he was the main figure to convert Buckland's Oxford heritage into the architectural result of the University Museum.

Adopted into this coterie of liberal Anglican Oxonians was Roderick Murchison. Born into a family of Scottish Highland landowners, he became attracted to geology mainly through his contact with Buckland's circle. Murchison, who with Buckland and Philip de Malpas Grey-Egerton was one of the early naturalist-trustees of the British Museum, served also as Hunterian trustee (1856–71).

Less directly connected to the Bucklandians, but more powerful as a politician, was Francis Egerton, whom we have already encountered as the Earl of Ellesmere, in his capacity as chairman of the Royal Commission of Inquiry on the British Museum of 1847–49. He, too, was educated at Christ Church before entering politics in the Liberal/Conservative tradition. Most powerful of the political patrons of natural history was yet another product of Christ Church, Robert Peel, MP for Oxford until 1829 and twice prime minister (1834–35, 1841–46). Of the same stable was his political heir, William Gladstone, four times prime minister (1868–74, 1880–85, 1886, 1892–94). On the Conservative side of this network of liberal Anglican Oxonians was Robert Inglis, who in 1829 defeated Peel as MP for Oxford over the issue of Catholic emancipation and, although an old-fashioned Tory, was a committed patron of various learned societies.

Francis Egerton's father, George Granville Leveson-Gower, who was Marquis of Stafford and, during the last year of his life, Duke of Sutherland, served as a Hunterian trustee (1809–33), whereas Francis's elder brother was a trustee of the British Museum; Francis Egerton himself was also active as a British Museum trustee. Peel, Gladstone, and Inglis all served on both boards, Inglis being one of the most active in terms of attendance. Gladstone's commitment has been chronicled above; about Peel, Owen wrote the following from personal experience: "As a Trustee of the Hunterian Museum Sir R. P. made himself personally acquainted with the contents, their general arrangement, and the mode in which they were catalogued, by more than one long and detailed inspection and on all occasions manifested an interest in that Museum and a desire to render every aid in its extension and improvement."[6]

Christ Church antecedents by themselves were not a guarantee of friendly support for Owen, witness Owen's parliamentary opponent of the early 1860s, Gregory, who, too, was a product of Eton and Christ Church. It was the additional connection with Buckland or Peel or with both that tended to indicate friendship for Owen, and it was the Christ Church background in general that in many instances led to political office.

The control these people exerted over museum matters amounted to a virtual stranglehold. To sum up: apart from Broderip, all these men were at some time trustees of either the Hunterian or the British Museum, and the majority of them served on both boards. To this fact should be added that they also represented the so-called working trustees, that is, the trustees who took a genuine interest in museum affairs and actually attended meetings and served on subcommittees or royal commissions and parliamentary committees of inquiry. Philip Egerton in particular was pervasively and energetically active.

Many illustrations of the personal closeness that developed between members of the Christ Church network and Owen can be cited. Two dramatic ones relate to Buckland and to Broderip respectively. When early in 1850 the symptoms of the mental illness which debilitated Buckland during the last half decade of his life grew excessive, it was to Owen that Murchison confidentially turned, telling him that "the Dean" had committed "much greater excesses than ever in respect to his own person—beating his head and scratching himself so as to produce alarm." Buckland's youngest children believed that "Papa must be acting," but the older ones were horror-stricken. Such scenes should not be permitted, Murchison insisted, and he urged Owen to team up with Benjamin Brodie and take action to remove Buckland from his family and put him under medical care.[7]

Broderip died three years after the demise of his old fossil-hunting companion Buckland, in 1859. Broderip and Owen had been so close that when the news of Broderip's death reached Philip Egerton, he sent his condolences to Owen, stating that "no one will suffer more than you who were almost a part of himself."[8] Apparently, when Broderip lay *in extremis* and a doctor entered to attend, the dying naturalist hallucinated, thinking it was Owen, "and faintly uttered 'dear Owen!'—the last words he spoke."[9] More joyful, if less poignant, instances of Owen's social contact with such Oxonians as Enniskillen, Peel, and other network members are sprinkled through the pages of the "Lives and Letters" of distinguished Victorians.

We have already seen how Owen's museum plans were championed by various members of the Christ Church network, and throughout the following chapters we shall have occasion to connect Owen's scientific work to the patronage of some of the same people. It will be enough here to trace a few steps in Owen's personal advancement that were brought about by the Buckland connection. A major step up the social ladder for Owen was the award, in a letter dated November 1, 1842, by Prime Minister Peel, of a civil list pension amounting to £200 a year. The fact that the total sum allocated for scientists was no more than £300 underlines the special nature of this royal honor. On a visit to Cambridge toward the end of 1841, at William Whewell's, Owen had met the chancellor of the exchequer of Peel's newly formed cabinet, Henry Goulburn, and also Lord Brougham and asked them about the possibility of a civil list pension.[10] Yet it was less the Trinity College (Cambridge) connection than the Christ Church (Oxford) one that led to the desired outcome. Buckland, having asked Owen to provide him with a concise but detailed summary of his past work and future plans, submitted this document to his old acquaintance Peel, with a cover letter dated January 12, 1842, that emphasized Owen's national reputation, equivalent—he asserted—to that of the astronomer G. B. Airy or the chemists John Dalton and Michael Faraday, and stressing, too, Owen's international Cuvierian rank.[11]

On October 4 that same year, Buckland sang Owen's praises in a further communication to Peel, turning up the already fairly shrill pitch: "Owen has for some years been without an equal in this country, and I know not his superior in the world."[12] Peel obliged and, almost apologetically, wrote to Owen on November 1, 1842: "The amount within my control for the present year (so far as science is concerned) is very limited. It does not exceed £300 in the whole, but as I know no public claim preferable to yours I shall have great satisfaction in proposing to H.M., with your consent, that an annual pension from H.M. Civil List of £200 shall be granted to you."[13]

Owen responded instantly, and in his letter to Peel, also dated November 1, he appeared overwhelmed by the honor: "Your goodness will pardon me if my feelings render me unequal to thanking you as I ought. The manner in which you have deigned to make the offer far outweighs in my estimation the handsome provision which will enable me to pursue my studies with renewed ardour, and to show by increased exertion my gratitude for the Royal favour."[14] Whewell congratulated Owen on the successful outcome of Buckland's petition and added: "I am afraid I cannot please myself with the thought of having had much to do with this satisfactory event, . . . but I am quite content to rejoice in what is done, without wishing to have any other concern in it than the sympathy of a friend."[15]

The relationship with Peel opened up to Owen a new world of social contacts which took him beyond the society of fellow scientists and into that of the highest and mightiest in the land, Prince Albert and other members of the royal family included. Peel was well known for the parties he organized and for the fact that he included among his guests the best and brightest from "science and the arts." Owen became a favored house guest, which allowed him to mix, on an equal footing, with other leading figures from the scientific world, such as Buckland, Robert Brown (the redoubtable botanist of the British Museum), the Marquis of Northampton (president of the Royal Society), and the bishop of Norwich (president of the Linnean Society), and also with distinguished "arts" people, such as Charles L. Eastlake (director of the National Gallery), the historian Henry Hallam, the painter Edwin Landseer, the civil engineer John Rennie, the poet Samuel Rogers, and the society wit and canon of St. Paul's Sidney Smith, and in addition with a medley of aristocratic "stars and Garters and Ladies fair and gay."[16] This particular list of people was cited by Owen from among the dinner and after-dinner guests at one particular social event in 1844, organized by Peel for the king of Saxony. "It was a proud and gratifying event to me I must confess to be included in the dinner-list,"[17] Owen wrote to his sister Maria, sketching a detailed plan of the seating arrangement around the dining table which showed Owen flanked by Hallam on his left and on his right by Carl Gustav Carus, the king's personal physician and a leading anatomist of the Romantic school of *Naturphilosophie.*

An admired feature of Peel's country seat, Drayton Manor, near Tamworth in Staffordshire, was the portrait gallery, which was divided into three compartments by marble pillars, with pictures of the greats from politics, the fine arts, and the humanities and science displayed in double rows. The portraits in the scientific category included Buckland, Cuvier, Alexander von Humboldt, and Justus von Liebig. In 1844, while organizing his gallery, Peel wrote to Buckland: "I should very much like to have, as a pendant to that of Cuvier, the portrait of Professor Owen."[18] The ground having been prepared by Buckland, Peel proceeded to ask Owen directly for his portrait, and Owen suggested a particular portrait, painted earlier that same year by Henry W. Pickersgill, whom William Broderip had taken along to the historic lecture by Owen in which the dinornis was "unveiled." Owen wrote to Peel: "No artist ever worked, I believe, with more zeal and determination to succeed: and my esteem for the painter of Cuvier and Humboldt left no place for regret as to the time which he thought it requisite I should devote to him."[19] Pickersgill's portrait of Owen showed him standing, in the act of lecturing and holding a dinornis bone. Peel was anxious, however, to have Owen in a sitting pose, and a second version was produced which was hung right at the entry of the first compartment of Peel's gallery, flanking one side of the entry, with Cuvier's on the other side. "I am much satisfied with the light and the place in which my own is hung," Owen confided to his sister Eliza.[20]

After the death of Peel, Gladstone took his place over the years as Owen's most powerful political patron. Just as Peel had supplemented Owen's Hunterian salary with a civil list pension, so Gladstone supplemented Owen's British Museum pension, early in 1884, following his retirement, by £100 annually. At about the same time, Owen received a knighthood, also on Gladstone's recommendation. Long before, there had been complaints about the fact that Owen had not been awarded a title. In 1873 *Vanity Fair,* for example, ended its description of Owen by commenting that "if men who have done service to mankind and honour to the English name (otherwise than by acquiring property) had any claim to distinction he [Owen] would now, after close upon seventy years' labour, illustrate some one of those titles which illustrate many men of a very different calibre."[21] Two and a half months later Gladstone offered Owen a choice of distinctions, and Owen asked for the Companion of the Bath, the smallest ribbon of that order. At the same time, the Earl of Leicester, the unremarkable Thomas William Coke, was honored with the Garter.

The press cried foul in a piece entitled "The Earl and the Professor," accusing Gladstone of misappropriation by having awarded the earl ("who has done nothing") this highest distinction and the professor ("the illustrious man of science") a lowly one. The criticism was unjust; the reason for the disparity was not Gladstone's lack of appreciation of Owen's merit but the fact that Owen himself was not prepared to pay the substantial honor fees; his CB came gratis, whereas the Garter went with fees amounting to £1000, substantially more than Owen's annual salary.[22] In fact, as long ago as 1845, Peel had offered Owen a knighthood, but Owen had declined, thinking it proper to wait until after he had succeeded in founding a national museum of natural history. Upon his retirement, however, Owen was gazetted a Knight Commander of the Bath (KCB); and while Buckland had been Owen's advocate with Peel, this time Lyon Playfair, the parliamentary spokesman for science, argued Owen's case with Gladstone and provided the prime minister with a summary of Owen's qualifications, written by Owen himself.[23]

London Clubs

Owen's knighthood came at the end of his museum career and thus too late to have strengthened his hand in the political maneuvering for institutional power. What must have helped—although the evidence is circumstantial—was his much earlier admission to membership in certain clubs, the social composition of which was similar to that of Peel's house parties. A multitude of clubs existed in Victorian London. John Timbs's *Club Life of London* of 1866 presented "sketches of One Hundred Clubs," as distinct from coffeehouses and taverns.[24] The exclusiveness of the more prestigious clubs endowed them with an aura of power, and membership in these could be a way of affirming or conferring social status. By being able to meet and mix with the membership of the select clubs, one became part of the social networks that dominated Victorian sociopolitical and intellectual life. Owen's election to the Athenaeum, in 1840, was therefore a milestone of social recognition. It made him part of a select "association of individuals known for their scientific or literary attainments, artists of eminence in any class of the fine arts, and noblemen and gentlemen distinguished as liberal patrons of Science, Literature, or the Arts."[25] It should be remembered that the person who proposed Owen for election under Rule II was Buckland's pupil Viscount Cole.

In contrast to the large clubs with their own magnificent clubhouses, such as the Athenaeum, the Reform, or the Travellers, the small dining clubs provided for more intimate gatherings and for greater ideological coherence. This was precisely why T. H. Huxley, in 1864, organized his own, very small X Club, which from 1886 existed within the Athenaeum.[26] Given Huxley's penchant for imitating Owen—as a competitor rather than a follower—the idea for his X Club may well have sprung from Owen's enviable place on the inside of metropolitan club life. Not long after his election to the Athenaeum, Owen was invited to join two literary dining clubs, both exclusive, one being the most prestigious of its kind. Early in 1844 Owen received the announcement of his election to the dining club of the Literary Society. The person who proposed him was none other than powerful museum trustee Robert Inglis, and it brought Owen into regular socio-intellectual contact with his sponsor and, in addition to such Peel visitors as Eastlake and Hallam, with a Who's Who of Victorian London. In addition to Inglis, several of these people were trustees of the Hunterian Museum, the British Museum, or both.[27]

The following year, Owen was elected to the most prestigious dining club of all, variously called "The Literary Club," "Dr. Johnson's Club," or simply "The Club," which had been founded in 1764 by Joshua Reynolds and Samuel Johnson. When by 1800 the Club had reached maturity, the number of members was restricted to forty, the dinner meetings took place once every fortnight during the sitting of Parliament, and the meeting place was the Thatched House in St. James's Street, where many other such clubs met, until in 1863 the building was demolished, and its patrons had to look for other accommodation. It was said "that Bishops, even Lord Chancellors, were known to have knocked for admission unsuccessfully,"[28] although Tennyson professed never to have heard of the Club when he was invited to join.[29]

Originally, it was a club for professional authors, but through the nineteenth century its membership changed, with fewer authors by profession and more titled members. Owen was elected at the age of forty, which was unusually young. Tennyson was not elected until 1865, John Tyndall not until 1871, and J. D. Hooker not until after his stint as president of the Royal Society, in 1878, the year before Owen rose to the top of the membership list and became "father of the club," that is, senior member by date of election. At this time he reminisced, in a letter to the poet Henry Taylor, that the Club had been a representative body and that he—Owen—belatedly had replaced Oliver Goldsmith, as the latter was considered a naturalist on account of his *History of the Earth and Animated Nature*, which went through several new editions during the first half of the nineteenth century.[30] In 1886 Owen was elected the first honorary member of the Club, a new category comprising those members "who are precluded from habitual attendance."[31]

The membership of virtually every club paled by comparison with that of "The Club," which was a veritable galaxy of stars in the Victorian firmament. It was crowded with great and, in many instances, aristocratic names from the worlds of politics, religion, law, scholarship, and, gradually, science. Because Owen was an active member for just over four decades, he saw several generations of members come and go: "I have seen and heard generations, like bright clouds—adorning and fading, and passing away!" Owen wrote in 1879.[32] Most importantly, just as the Board of Trustees of the Hunterian Museum resembled that of the British Museum, so the membership of both boards enjoyed a numerous representation in the Club. Dinner conversation did not always consist of light chit-chat; one meeting was described by Owen as "a very brilliant intellectual evening."[33]

Several of his regular dining companions were powerbrokers in museum matters, as trustees of either the Hunterian or the British Museum or also as members of parliamentary inquiries, and a majority were sympathetic to or represented the Peelite cause. The Duke of Argyll served on both boards; so did Charles J. Blomfield (bishop of London), Gladstone, George Grote (the historian of Greece), Hallam, the Marquis of Northampton, and Lord John Russell. Lord Romilly (master of the rolls) and Spencer Horatio Walpole (home secretary in several Derby ministries) were *ex officio* trustees of the British Museum; Lord Stanley (fifteenth Earl of Derby) was a family trustee; and the politician David Dundas and the statesman the Marquis of Lansdowne were elected trustees. There were other trustee names on successive membership lists, but the men mentioned here are those whom Owen himself recorded as the ones he met over dinner. Lord Stanley and Walpole both served on Gregory's 1860 Select Committee on the British Museum. Owen's membership in these clubs therefore not only matched his social aspirations but was of material significance in the advancement of his museum plans. Here he mingled in an atmosphere of relaxed intimacy with the people who walked the corridors of power. It should be noted that the dates of his elections to these clubs coincide with the period when he began formulating his plans for a national museum of natural history.

Although the Oxford circle was crucial to the realization of Owen's career plans, their success also depended to a significant degree on the patronage of various Cambridge figures. Several trustee-members of both the Literary Society Club and the Club were alumni of Trinity College, Cambridge, most importantly Charles Blomfield and the Marquis of Northampton, "double trustees" of the Hunterian and the British Museum. Two other

Owen stalwarts during the parliamentary museum debates of the early 1860s were Trinity men: Richard Monckton Milnes and Edward Pleydell-Bouverie. Milnes had been one of the Cambridge "Apostles" and was eager to join the Board of Trustees of the British Museum. This did not happen until 1881, after the death of his enemy, Panizzi. Milnes, as Baron Houghton, was elected to "The Club" in 1875.[34] None of the Cantabrigians, however, was quite as effective in promoting Owen's museum cause as the Oxonians, because they lacked what characterized Buckland's Christ Church circle, namely the combination of paleontological expertise and high political office.

Prominent among Owen's scientific friends at Cambridge were the Woodwardian professor of geology Adam Sedgwick and the master of Trinity College William Whewell. Owen and Whewell, both of whom were Lancastrians and had attended the same grammar school, enjoyed a specially close bond. Sedgwick and Whewell were central figures in what Susan Cannon has called "the Cambridge network," which loosely combined various forces of latitudinarian Anglicanism. It was at the level of scientific societies that Owen's respective Oxford and Cambridge support merged into a unified Oxbridge patronage, in particular in London's Geological Society and in the peripatetic British Association for the Advancement of Science. Here the Christ Church and Trinity College circles interlocked in the promoting of a program of research in paleontology, carried out primarily by Owen. Jack Morrell and Arnold Thackray, in their study of the first decade of the BAAS (1831–41), have demonstrated the ascendancy of the Cambridge network in this society.[35] Yet members of the Christ Church network made their presence felt, too. Although the control they exerted was especially strong in the museum world, it went well beyond this to include the Geological Society and the BAAS. Before 1858, for example, when Owen himself was president of the BAAS, Buckland, Philip Egerton, Francis Egerton, Murchison, Wilberforce, Inglis, Peel, and Enniskillen all served as the association's annual presidents or vice presidents.[36]

The hands that pinned on Owen's lapel the highest scientific honors that the metropolitan societies had to offer were manicured by an Oxbridge education. On February 16, 1838, Whewell, as president of the Geological Society, awarded Owen its highest honor, the Wollaston Medal. Oxford and Cambridge dominance was less pronounced in the Royal Society, but there Buckland's influence was decisive, too, when in 1846 the Marquis of Northampton, as president, conferred on Owen the Royal Medal. Again, in 1851, when Owen received the Royal Society's ultimate accolade, the Copley Medal, Philip Egerton was on the council, and the citation by the president, the Oxford alumnus the Earl of Rosse, closely resembled passages in Broderip's review of Owen for the *Quarterly*.[37]

Functionalist Dictates

This support for Owen did not come without strings. By pushing Owen and his museum cause, the Oxford and Cambridge patrons expected him to carry out his vertebrate morphology and paleontology in line with a particular dictate, namely that his work be illustrative of the scientific epistemology of natural theology. Buckland's letter to Peel recommending Owen for a civil list pension ended with the telling qualification "that every new discovery he [Owen] makes excites in him such feelings as a mind constructed like that of Paley is alone competent to enjoy."[38] An equally telling fact was that Owen's portrait complemented Cuvier's in the Drayton Manor gallery.

To the Oxbridge scientists and patrons of science, the promotion of scientific inquiry represented part of liberal Anglican reform and took place under the flag of the traditional Paleyan design argument. They actively championed a reinvigorated version of natural theology to serve the scientific dons in their campaign to broaden science education at England's ancient universities. At the time, Oxford and Cambridge were still Anglican institutions. Many college fellows, including the geology professors, were clergymen. Buckland, as a canon of Christ Church, enjoyed material security in the embrace of the ecclesiastical establishment, and Sedgwick acquired financial independence as a prebendary of Norwich Cathedral. Much of the teaching was geared to the education of the future clergy of the established church. During the first half of the nineteenth century, over 60 percent of all Oxford students and some 50 percent of Cambridge graduates were destined for Holy Orders.[39]

To justify the pursuit of scientific interests and to make science an integral part of the curriculum, its Oxbridge practitioners presented the study of nature as a source of design arguments and thus as a form of natural theology. Proofs of design served as justification for the inclusion of new scientific subjects by demonstrating the usefulness of these subjects to the moral and religious education of the future clergy.

Thus, natural history was presented as a source of proofs of "the power, wisdom, and goodness of God" as manifested in the natural world. All the inaugural and introductory lectures in scientific subjects that were established at Oxford and Cambridge during the early part of the nineteenth century were, by and large, public affirmations of the conviction that the natural sciences afford an abundance of instances of Paleyan—or, rather, of divine—design. Virtually the entire Oxbridge network of Owen's patrons was committed to the functionalist approach in studying nature.

Paleontology in particular was prepared for its role of ecclesiastical servitude by being based on the epistemology of Cuvier's functionalism. The

Cuvierian approach provided the perfect means of making natural history part of the Paleyan tradition of natural theology. Both Buckland in his *Vindiciae Geologicae* (1820) and his Bridgewater Treatise (1836) and Sedgwick in his *Discourse on the Studies of the University* (1833 and several later editions) contributed significantly to the functionalist interpretation of the geological past; others, such as William Daniel Conybeare and Whewell, backed them up. In fact, historical geology helped perpetuate the popularity of the design argument in England, and the functionalist synthesis of the period can be seen as representing natural history in its state of Anglican, Oxbridge subjugation.

Owen's scientific research was inseparable from his museum plans and must be looked at in the light of the constraints explicitly or implicitly imposed by the patronage needed for the fulfillment of his institutional ambitions. A majority of his Oxbridge acquaintances—trustees of the Hunterian and the British Museum and fellow members of prestigious clubs—expected him to do their scientific bidding by validating the design argument in his studies of vertebrate morphology or in anything else his scalpel or pen touched. To oblige the patrons of his museum plans meant, purely and simply, scientifically to develop the Paleyan design argument. The precise nature of the functionalist approach and how it shaped Owen's scientific work in detail will now be explored.

The British Cuvier

The epithet by which Owen became widely known was that of "the English Cuvier" or, more comprehensively, "the British Cuvier." What exactly did this mean? To Owen, it meant more than just one thing. First, it simply signified that he was the most prominent naturalist in Britain, just as Cuvier had been in France. It was in this sense that Meckel had been called "the German Cuvier," that Agassiz would be called "the American Cuvier," and that Robert E. Grant's supporters, during the 1830s, hailed him as "the English Cuvier," some time before Owen won that title. By donning the mantle of a famous predecessor or of a much-lauded contemporary, Owen did not so much express a doctrinal allegiance as an aspiration to occupy a position of cultural eminence. However, the epithet also had a second meaning, at least in Owen's case. It indicated that he was anointed the custodian of the functionalist design argument, and in particular that he was the person around whose shoulders the mantle of Buckland was being wrapped.

Cuvierian functionalism in Britain was, for the most part, not directly taken from Cuvier but was known in the form of what Buckland and his colleagues had made of it. Buckland himself, of course, knew Cuvier's work *in extenso*. The two men visited each other, exchanged publications and specimens, and kept in touch via an intermediary, the naturalist Joseph Pentland, who spent the period 1820–32 in Cuvier's workshop.[40] In his famous *Reliquiae Diluvianae* Buckland grafted Cuvierian catastrophism—the theory that the history of the earth is punctuated by sudden inversions of land and sea—outlined in the preliminary discourse to the *Ossemens fossiles*, upon British diluvialism by equating the last of Cuvier's cataclysms with the biblical deluge.[41] As the popularity and validity of diluvialism waned, by around 1830, the functionalism of the *Ossemens fossiles*, rather than its catastrophism, was highlighted and used to shore up the traditional design argument by extending it to the new science of geology; vertebrate fossils in particular became the latest marvels of teleology. In the process, "function" as the final cause or purpose of "form" became overemphasized as the central tenet of the Cuvierian doctrine.

The BAAS, established in 1831, functioned as an effective platform for promoting Oxbridge natural theology.[42] Buckland was president of the first full meeting, held at Oxford in 1832, and his highly acclaimed lecture on the megatherium was nothing less than a sermon in which he preached the Oxonian version of Cuvier's functionalism. He prefaced his presentation with a eulogy of the recently deceased grand old man, "that exalted and most illustrious naturalist," whose name made up a trinity with those of Aristotle and Pliny. Thus, no sooner had Cuvier given up the ghost than Buckland usurped the Cuvierian heritage on behalf of Paleyan natural theology and the cause of liberal Anglicanism. In front of an audience of men and women (the gender mixture symbolizing the latitudinarian character of both the BAAS and natural theology), Cuvier's work was turned into explicit evidence of God's attributes; it had demonstrated, Buckland asserted, "the unity and universal goodness of the great Creator."[43]

Buckland further proclaimed that Cuvier's tomes were filled "with proofs of wise design, in the constant relation of the parts of animals to one another, and to the general functions of the whole body."[44] This incorporation of Cuvier's paleontology into the English design argument was completed by Buckland in his Bridgewater Treatise *Geology and Mineralogy Considered with Reference to Natural Theology*, which appeared in 1836 after half a decade of arduous labor by its author.

Several of Buckland's colleagues, at both Oxford and Cambridge, joined in the effort of harnessing Cuvier's horse to Paley's cart. Sedgwick, for one, in his *Discourse on the Studies of the University* (1832), left no doubt as to where he stood on the issue. Geology shows that, in every instant of change in the fossil record, "the new organs, as far as we can comprehend their use, were exactly suited to the functions of the beings they were given to."[45] Whewell, no less, championed the functionalist cause. In his *History of the Inductive Sciences* (1837) he pronounced: "The study of comparative anatomy is the study of the adaptation of animal structures to their purposes."[46] Without the functionalist approach—he ventured to suggest—Cuvier would not have succeeded in his epoch-making paleontological reconstructions.

By contrast, Edinburgh's David Brewster, in an essay entitled "Life and Works of Baron Cuvier" and published in the *Edinburgh Review*, put far less emphasis on the functionalist aspect of Cuvier's work.[47] It was the Oxbridge people, rather than the Edinburgh diaspora, who invented the "English Cuvier." In 1841 Owen completed the delivery of his major "Report on British Fossil Reptiles" to the BAAS. The *Literary Gazette*, which devoted over twenty columns to the "Report," added that "when Dr Buckland happily characterised its author as a worthy successor of Cuvier, a general burst of applause broke from every part of the audience."[48] Whewell occupied the presidential chair that year. A decade later, in 1851, the Broderip-Owen duo, in their summary of Owen's oeuvre for the *Quarterly Review*, reminded the readers, right from the start, that Owen "is recognized throughout Europe as the Cuvier of England."[49] That same year, Owen received the highest honor that the Royal Society can bestow upon its fellows: the Copley Medal. The president, William Parsons, third Earl of Rosse, a former Oxford man and BAAS president in 1843, pointed out that Owen's publications had begun to appear by the end of Cuvier's life; "the mantle of that great man seems to have descended, at his death, upon the shoulders of our distinguished countryman."[50] In 1866 Frank Buckland, having seen a French man-of-war in Boulogne that was called *Cuvier*, entered in his diary: "When will the day arrive that our Admiralty will pay a similar compliment to English science, and launch the 'Professor Owen' gunboat from the slips at Portsmouth?"[51]

Owen was allowed to enter the privileged world of Anglican Oxbridge through the portals of Cuvierian functionalism, with Buckland assuming the role of gatekeeper. Owen's dependency on such patronage did not necessarily subvert his scientific integrity, but his work must be read in the light of it. Three examples of his Cuvierian work—on the nautilus, the dinornis, and the Crystal Palace "monsters"—will now be looked at in some detail.

Owen first came to Buckland's attention in 1832 when the former sent his prospective patron a prepublication copy of *Memoir on the Pearly Nautilus*, adding in a cover letter: "Since the decease of the lamented Cuvier, there is no one whose opinion on this work I look for with more anxiety than your own."[52] Buckland reacted with alacrity to "Mr. Owen's admirable work" and the "masterly" manner of Owen's investigation.[53] The nautilus is a cephalopod, a relative of squids and octopuses, which lives in the front cavity of a shell composed of a spiral succession of air-filled chambers. Although specimens of the shell are not uncommon, it is rare to find the soft parts still attached. In 1831 the Royal College of Surgeons received just such a rare specimen, which had been caught by George Bennett off the New Hebrides in the Pacific Ocean. Owen immediately set to work and produced the first of his remarkable series of lastingly valuable monographs. This and related studies culminated in his entry "Cephalopoda" for Todd's *Cyclopaedia of Anatomy and Physiology* (1836). These studies were followed by a series of papers to both the Zoological Society and the Royal Society, and the latter organization awarded Owen its Royal Medal in 1846 for his research on belemnites.[54] This work also led to one of several unedifying quarrels with Gideon Mantell.[55]

Anthony Carlisle called the memoir "an excellent specimen of Hunterian-Cuvierian Natural History,"[56] and Buckland enthused about "the value of Professor Owen's highly philosophical and most admirable Memoir upon this subject; a work not less creditable to the author, than honourable to the Royal College of Surgeons, under whose auspices this publication has been so handsomely conducted."[57] A description of the nautilus was an obvious coup, not just for the Royal College of Surgeons, but also for England. The interest and quality of Owen's first memoir were such that it was translated into both French and German. It stung various Continental colleagues into action, in particular the French and the Dutch, who both had colonial possessions in the Indo-Pacific where the nautilus lives, in relatively deep water.

The greatest contemporary excitement about the nautilus memoir was not generated in the arena of Europe-wide colonial competition, however, but at home in the restricted circle around Buckland. The nautilus, it must be remembered, was a "living fossil," a present-day relative of such extinct cephalopods as the ammonites. Buckland's interest in these was legendary. On one field trip he found a specimen of *Ammonites bucklandi*, without the inner whorls, and thrusting his head through the hole he rode home, dubbed by his friends the "Ammon Knight."[58] As early as 1820 Richard Whately, the later archbishop of Dublin, composed "Elegy Intended for Professor Buckland," which opened with the words

> Mourn, Ammonites, mourn o'er his funeral urn,
> Whose neck ye must grace no more;[59]

At the time of Owen's memoir, Buckland was working on his Bridgewater Treatise, and probably its most lively and original part became the section on "proofs of design in the mechanism of fossil chambered shells." Here he applied Owen's anatomical work to the functional interpretation of the fossil nautiloids and the related ammonites. In the "Concluding Observations" of his memoir, Owen had already discussed the problem—though without being able to provide a conclusive answer—of how the pearly cephalopod moves through the water. In 1833 Buckland and Owen joined forces to figure out how the nautilus and, by analogy, its fossil relatives may use the air-filled chambers as floats to rise to the surface or sink to the bottom.

Buckland turned the nautilus into an instance of Nieuwentyt-like functional mechanics by comparing the shell to the hydraulic apparatus of the "water-balloon." He speculated that the buoyancy of the air chambers counterbalanced the weight of the shell, giving the combined body and shell approximately the weight of water. To rise or sink, only a slight modification in buoyancy would be required, for which the organism possesses a "hydraulic apparatus for varying the specific gravity of the shell." This is the siphon that runs through the air chambers and connects them to the frontal dwelling chamber, where it communicates with the pericardium. Buckland hypothesized that the animal, by ejecting a pericardial fluid, into the siphon can compress the air in the chambers and increase the specific gravity of the shell, and vice versa. By analogy, the fossil ammonites of long ago must have possessed the same mechanism to rise and sink through the waters of the primordial seas: "From the similarity of these mechanisms to those still employed in animals of the existing creation, we see that all such contrivances and adaptations, however remotely separated by time or space, indicate a common origin in the will and design of one and the same Intelligence."[60]

Monster Models

However glittering the success of the *Memoir on the Pearly Nautilus* was in its application to Buckland's functionalist paleontology, it paled in comparison to the luster of the Cuvierian triumphalism generated by Owen's reconstruction of the moa, a giant wingless bird, allied to the ostrich, brought to extinction in relatively recent times as a consequence of the Maori settlement of New Zealand. Owen's work on the moa became the *locus classicus* of the Cuvierian tradition in Britain. It had something in common with Buckland's hyena den theory in that in both cases the initial evidence consisted of nothing more than mere fragments of bones. Buckland used gnawed, broken pieces of bone, found in caves, to argue that these caverns had been the lair and larder of hyenas. In other words, from fossil fragments he managed to infer the living habits of extinct, prehistoric species.[61] Yet Buckland did not reconstruct the complete animals of which the fossil bones were remnants.[62] Owen performed such a feat in the case of the moa and, in doing so, appeared to prove the

Cuvierian dictum that from a single fragment the complete bone, and from this the entire skeleton of an animal, can be derived. In his *Ossemens fossiles*, Cuvier waxed lyrical about the success of his method by means of which fossil fragments from the gypsum quarries of Montmartre had been turned into various whole skeletons, identifiable as those of extinct pachyderms: "At the voice of comparative anatomy, every bone, and fragment of a bone, resumed its place. . . . each species was, as it were, reconstructed from a single one of its component elements."[63]

The story of the reconstruction of the moa began in 1839, when Owen was shown what looked like an ox's marrowbone that had been brought to London from New Zealand. It was six inches long and five and a half inches at its smaller circumference, with both ends broken off. Primarily on the basis of its surface texture, Owen argued that the fragment, which was part of the shaft of a femur, belonged to the class of birds, in particular to the order of the wingless ostriches. He conjectured "that there has existed, if there does not now exist, in New Zealand, a Struthious bird, nearly, if not quite, equal in size to the Ostrich."[64] Owen presented his conclusion to the Zoological Society, but the publications committee, which included the ornithologists N. A. Vigors and W. Yarrell, only reluctantly accepted Owen's paper. Was it possible—they worried—that such a huge bird had found subsistence on as small a tract of land as New Zealand, and could a mere fragment of bone be sufficient evidence? Yet the paper was published in 1840, the responsibility for it "resting exclusively with the author."[65] Already in late 1839, Owen had a hundred preprints made, and he distributed these among friends in New Zealand to stimulate a search for corroborating evidence.[66]

Confirmation did come early in 1843, apparently unrelated to Owen's circular, when the first of two boxes of bones, sent to Buckland by a missionary friend from New Zealand, arrived in London. At this stage in the unraveling of the plot, Owen's Oxonian supporters became part of the story and turned "the greatest zoological discovery of our time"[67] into a myth of Cuvierian functionalism. It became an instance of the Broderip-Buckland duo putting Owen once again before the wagon of Oxbridge functionalism. Buckland had the box forwarded to Owen; on January 19, 1843, it was ritually opened in the presence of Broderip, and revealed were the bones of a large cursorial bird. From that date on, in letters and published writings, Broderip enlarged upon the accuracy of Owen's original paper. "Every word comes true to the letter," he gushed to Buckland;[68] and in the *Quarterly Review* he recalled seeing in 1839 the original fragment and Owen's predictive sketch of what the undamaged, complete bone must have looked like; Broderip kept the sheet of paper on which the shape of the unknown bone was drawn. "When a perfect bone arrived, and was laid on the paper, it fitted the outline *exactly*."[69] In Gruber's words, to the lay public as well as to many experts, Owen's feat of reconstruction seemed an arcane exhibition of his craft, similar to an astronomer's accurate prediction of hitherto-unknown planets and satellites; the organic world appeared to be ruled by laws no less than the realm of planetary physics.[70]

In reality, Owen's estimate of the size of the bird had been rather conservative, and the moa of early 1843, which Owen called *Dinornis novae-zealandiae*, turned out to be substantially larger than an ostrich.[71] From Broderip's legend it was but a small step, however, to the popular myth that Owen had reconstructed an entire bird from a mere fragment of bone. *The Life and Correspondence of William Buckland* recounted this belief: "By the process of severe philosophical induction, and not by mere guesswork, he [Owen] was enabled to describe the bird with the utmost accuracy from the inspection of the solitary small fragment of the thigh."[72] Admittedly, there were contemporary grumblings that Owen's discovery was neither as accurate nor as original as it was made out to be. Such grumblings, known from an 1848 diary entry by Gideon Mantell,[73] have found amplification in more recent "demythologizing" studies of Owen's moa work,[74] in which the role of the retired naval surgeon John Rule has been highlighted. Rule was the person from whom Owen obtained the original fragment and who told him that it was part of the femur of a very large, possibly extinct, bird: "Herewith I desire to offer for sale a portion or fragment of a bone, I believe the largest and most rare that has been found, part of the femur of a bird now considered to be wholly extinct."[75]

Irrespective of Rule's avian suggestion, Owen's was a brilliant piece of work because of the predictive accuracy of his reasons for attributing the bone, not just to a bird, but to a genus of ostrich-like wingless birds, even though the bone lacked the characteristic avian air cavity and the relative lightness that struthious bones also possess. This is certainly the way in which Yale's Benjamin Silliman, who was well acquainted with both parties, judged the matter in the diary of his London visit in 1851.[76] Mythologized or not, Owen's reconstruction of the moa triggered much fruitful exploration. Various colonial administrators and dignitaries helped to transport crates filled with bird bones to England. New moas, both smaller and larger than ostriches, were discovered, some as tall as twelve feet. The *Dinornis elephantopus* was followed by the *Dinornis robustus*, and the *Dinornis giganteus* by the *Dinornis maximus*.[77] Gruber rightly locates the significance of Owen's moa work in its function as an intellectual engine of colonial natural history. The start of moa research was also "the beginning of the sustained search for additional remains that served as an important focus for New Zealand natural science for the next forty years, culminating finally in the exhibition of the reconstructed Moa skeletons which were the main attractions at New Zealand's exhibit during the Colonial Exhibition in London in 1886."[78]

The second of the 1843 boxes of bones from New Zealand had already allowed Owen to establish, late in 1843, that five distinct species of dinornis had existed, ranging in stature from ten feet to the seven and a half feet of an average ostrich, to the size of a cassowary, a dodo, and a bustard.[79] "I suspect, in short," he summed up with a rare display of jocularity in a letter to H. T. De la Beche, "a fine family of fern-eaters, enjoying New Zealand in fee simple, until the Malay Colony arrived, who lived as long as they could on *Dinornis* soup with fernroot sauce; and, when the stupid big birds were all slain, (for they could neither fly nor swim), the Maoris then took to eating one another."[80] The moa remained one of Owen's favorite topics, on which he continued to work until the end of his publishing career. He combined it with studies of such other flightless birds as the dodo, which had also been a topic of Oxonian interest, much encouraged by Broderip and Buckland.[81] The largest of the femurs in the boxes of 1843 was held by Owen in the Pickersgill portrait for Drayton Manor, part of Peel's "gallery of modern worthies" (fig. 8).[82] The moa bone also proved to be the key to the gates of Buckingham Palace, as the prince consort let it be known, via Buckland and Peel, that he wanted to meet Owen and see the dinornis.

Essential in all this was that the giant birds "stepped out of the past into the present"[83] across the threshold of Cuvierian functionalism. It is this fact that explains why the Oxonian camp created the myth of "an entire bird out of a single fragment of bone." To the functionalists, the timing of Owen's prediction could hardly have been better. In 1839, the year that Owen described the fragment, transcendentalist criticism of Cuvier, and by implication of Oxbridge natural theology, was beginning to sound alarmingly loud. An attack on Cuvier's method by his successor Henri de Blainville was splashed across the pages of the *Lancet* and supported in no uncertain terms by Edinburgh's Robert Knox in his English translation of Blainville's lectures on comparative osteology. Blainville specifically attacked the Cuvierian notion that from a single bone an entire skeleton could be reconstructed. This, he asserted, was a mere pretense, and he challengingly stated: "In my opinion, no one has ever made good this pretension."[84] Knox added a belligerent footnote, stating that all along he had publicly inveighed against this boast of the Cuvierian school and "endeavoured, to the utmost of my power, to stem the torrent of assertion, amounting to absolute nonsense, which had set in, in this country, on the subject."[85]

FIGURE 8. Owen (ca. 1846) wearing the robes of Hunterian Professor at the Royal College of Surgeons and holding in his left hand a *Dinornis* femur. From merely a fragment of a similar thighbone he deduced, in apparent confirmation of Cuvierian functionalism, the former existence in New Zealand of the now-extinct group of flightless birds, the moas. Engraving after a daguerreotype (Timbs, *Yearbook of Facts in Science and Art*, 1852, frontispiece).

Owen's moa work seemed the right answer at just the right time, and if his reconstruction was not exactly Cuvierian, in that he merely inferred from the original fragment the form of a complete bone, and not of a complete bird, a little mythologizing of his accomplishment would readily make up the difference. That the moa story was part of the controversy over the validity of functionalism is certain. After the ritual opening of the first box early in 1843, Broderip wrote to Buckland how correct and precise Owen's prediction of 1839 now proved to be, adding: "All this not from any guess but from secure philosophical induction. This is not only another proof of the powers of our great physiological friend; but it comes well in aid against the sneers that de Blainville, Geoffroy, and other French physiologists have lately directed against the followers of Cuvier, and the principle of building up the whole skeleton from a bone or even the fragment of a bone."[86] Owen himself, in a letter to Silliman, presented his moa bone as a stick with which to beat Cuvier's detractors.[87]

Throughout the 1840s the widening ripples produced by the impact of Owen's "magnum bonum"–to use one of Sydney Smith's dinner table quips[88] –spread beyond the inner circle of the metropolitan scientific societies into the wider world of London's literati. It took another decade, however, and the reconstruction of even more spectacular extinct monsters, including the dinosaurs, before the public at large caught notice of Owen's Bucklandian work. Yet at this time, before the late-nineteenth-century discovery of the sensational North American dinosaur skeletons by the "bone barons" E. D. Cope and O. C. Marsh, it was not the dinosaurs by which prehistoric life was most popularly known, but several prehistoric mammals, among them the now almost forgotten megatherioid mammals from South America, represented by such genera as megalonix, megatherium, and the somewhat less bulky mylodon.[89]

Cuvier was the first, in 1795, to identify the megatherium as a gigantic ground sloth, the size of a hippopotamus or rhinoceros. Fossil remnants of this monstrosity were much sought after by museum curators across Europe. The first partial skeletons were used to mount a megatherium for display as early as 1789, in the Royal Museum of Natural History in Madrid; but London received more than its fair share of the fossil loot from the Argentinean pampas, primarily thanks to the British consul at Buenos Aires, Woodbine Parish, who donated a partial skeleton to the Royal College of Surgeons in 1832 and a more complete one in 1841, the latter identified by Owen as a mylodon. To this rapidly growing store of paleontological treasures in the Hunterian Museum was added a megatherium cranium with teeth, brought back by Darwin from South America. In 1845 the British Museum, too, obtained its own partial skeleton of a giant ground sloth.

The skeleton of 1832 was described by Clift,[90] but when Darwin added his fossils to the collection upon his return from the *Beagle* voyage in 1836, Owen entered the fray that had developed over both the precise taxonomic affinity of the animal and its foraging habits. Several naturalists, including Cuvier and Buckland, had speculated that the bony armor that was found in conjunction with the megatherioid remains indicated that the giant sloth had been covered by a tessellated cuirass.[91] Parish noticed, however, that in the vicinity of in situ megatherium bones there were giant drumlike shells analogous to those of today's much smaller armadillos, suggesting that another monster beast might have existed alongside the megatherium. He asked Owen to contribute a note to his *Buenos Ayres* (1838), and Owen identified the remains, in part on dental evidence, as that of glyptodon, a giant armadillo.[92]

Owen made the megatherioid specimens of 1841 and 1845 the subject matter of two brilliant memoirs, *Description of the Skeleton of an Extinct Gigantic Sloth* (1842) and *Memoir on the Megatherium, or Giant Ground-Sloth of America* (1861); and when he was invited to deliver the Croonian Lecture for 1851, he chose the British Museum megatherium as his topic.[93] These memoirs were type specimens of Bucklandian comparative anatomy in which a detailed description of form was used to infer function, eating habits, and habitat. Cuvier had reasoned that the giant sloth was a vegetarian and used its claws to dig up roots. The German naturalist C. H. Pander and his artist coauthor Edouard d'Alton, instructor to Albert von Sachsen-Coburg, the later prince consort, went further and speculated that the animal was a fossorial, burrowing beast and lived underground like some giant mole. The Danish naturalist P. W. Lund, too, went out on a limb and reasoned from the megatherium's relationship to the present-day sloth that it was actually arboreal and climbed tree branches to feed off the leaves.[94]

Buckland echoed Cuvier's opinion and, in addition, he managed to turn the awkward-looking animal from an ill-adapted colossus into a perfect Paleyan contrivance. He chose the giant ground sloth as the subject of his famous BAAS lecture of 1832 and expanded upon this in the Bridgewater Treatise. Some naturalists had described the beast as an ill-designed monstrosity; "I select the Megatherium," Buckland proclaimed, "because it affords an example of most extraordinary deviations, and of egregious apparent monstrosity."[95] His chapter in the Bridgewater Treatise on "this Behemoth of the Pampas" became a celebrated example of relating form to function—of relating teeth, claws, and other peculiarities to digging up and feeding on roots. What looked like a distorted giant proved to be, after all, God's own megatherium, "every bone of which presents peculiarities that at first sight appear imperfectly contrived, but which become intelligible when viewed in their relations to one another, and to the functions of the animal in which they occur."[96]

In his megatherioid studies, Owen closely followed the Bucklandian line and, as in the case of the dinornis, made his memoir a defense of Cuvier to counter the antifunctionalist propaganda that the megatherium was not a sloth but was related to the armadillos, that it was not a vegetarian but fed on meat, at least for part of its nourishment, and that it was a burrower, if not for concealment, then for digging up ants.[97] Owen gave short shrift to such bizarre deviations from the Buckland-Cuvier position. In a study of the megatherioid teeth from Darwin's collection, he noticed their similarity with the teeth of today's sloths and concluded that leaves would have been its staple food. Based on an array of further anatomical observations he hypothesized that the mylodon could have used its tail as a third hind leg to form a tripod; raised up on this the beast would have been able to push over or wrench out trees with its forelegs, having first scratched away the soil from around the roots with its awesome claws.[98]

Buckland lightheartedly countered that the beasts would have run the frequent risk of getting their heads smashed in by the falling trees. No sooner had this objection been raised than the mylodon skeleton of 1841 arrived, of which the skull exhibited two major fractures, one healed and the other partially so. This remarkable confirmation of the "anticipation of the possible contingencies of the Professor's hypothesis, and the hypothesis itself," provided Broderip with a further anecdotal illustration of Owen's skill as a functionalist restorer of fossil monsters.[99] Ever since, in the Crystal Palace gardens and in museums, the giant ground sloth has been mounted raised up on its hind tripod with its forelegs up against a tree trunk (fig. 9).

Museum Icons

The megatherioid memoirs were mere offshoots of a more systematic program which was in part financed by the BAAS, under the patronage of the magic Oxbridge circle. A substantial sum of the limited BAAS budget was repeatedly voted to Owen through the period 1838–45 to make possible a series of reports on British fossil reptiles, mammals, and birds; fossil fish had been assigned to Agassiz.[100] These reports were an exceedingly laborious compilation of the relevant material that had been collected and described in Britain since the early part of the century. They incorporated much of Buckland's paleontological work but also that of Conybeare, Mantell, Parkinson, and others. In producing this impressive summary and synthesis, Owen read the literature and inspected specimens in museums around Britain and across the Continent. The first part of the two-part report on reptiles contained the descriptions of no fewer than sixteen plesiosaurus species and ten ichthyosaurus ones, the majority of these newly named by Owen.[101] In the second part he summarized the available data on crocodilians, lizards, chelonians, ophidians, and batrachians; and, most fascinatingly, he added the new order of the dinosaurs, which included the then-known three genera, megalosaurus, iguanodon, and hylaeosaurus, described during the decade 1822–32, the first by Buckland and the latter two by Mantell.[102]

The reports constituted a tour de force and, as we saw, culminated in Buckland's conferring the title of "the British Cuvier" on Owen. They were followed by occasional memoirs on fossil reptiles, published under the auspices

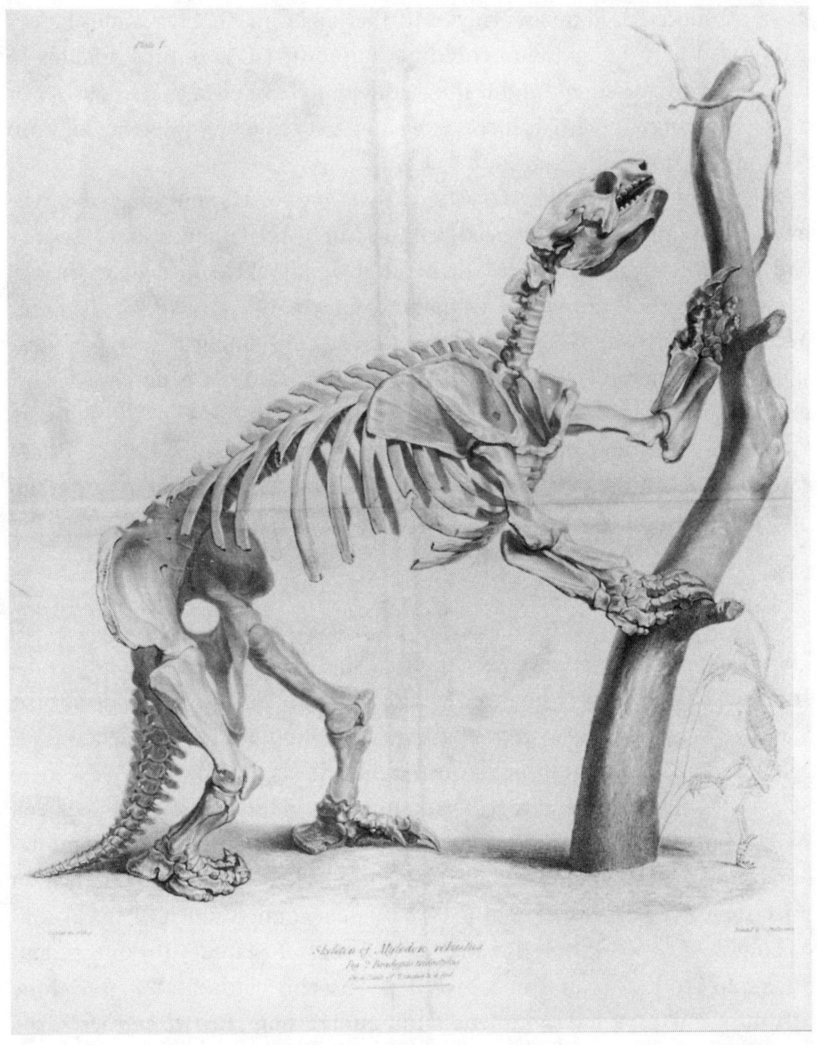

FIGURE 9. Skeleton of *Mylodon robustus* in a lifelike position as interpreted by Owen. For the purpose of scale, its present-day relative, the comparatively diminutive *Bradypus tridactylus*, is shown in the lower-right corner. (Owen, *Description of the Skeleton of an Extinct Gigantic Sloth*, pl. 1.)

of Owen's own Palaeontographical Society, a learned printing club instituted in 1847, "[f]or the illustration and description of British fossil organic remains."[103] In 1852, when the Crystal Palace was moved from Hyde Park to its permanent site in Sydenham, the prince consort suggested that models of prehistoric monsters should be placed in the grounds. Owen was the obvious expert for the commissioned sculptor, Benjamin Waterhouse Hawkins, to turn to for consultation in building the life-size models. Fossil saurians had been "restored to life" on paper, in sketches by H. T. De la Beche and, more artistically complete, by the Romantic artist John Martin.[104] "The animals, like the geologists, seemed generally to have been engaged in combat."[105]

Before the appearance of Owen's reptilian *catalogue raisonné*, the size of the dinosaurs had been overestimated at fifty to seventy feet long. Now the estimates were reduced to some thirty feet for the megalosaurus, and twenty-eight feet for the iguanodon, although Hawkins made the models slightly larger. Their shape to a large extent was inferred from that of the closest living relatives: the varan for the carnivorous megalosaurus, and the iguana for the herbivorous iguanodon. In accordance with mistaken popular belief, Hawkins gave the iguanodon a horn on its snout, as Martin had done in "The Country of the Iguanodon," the frontispiece to Mantell's *The Wonders of Geology* (1838), making the dinosaur look somewhat like a rhinoceros. This was not on Owen's advice; to his credit, Owen had correctly observed that the spurious horn was in fact a claw-bone, or, more precisely, a spiky thumb.[106]

In a characteristically Victorian instance of drama, a dinner was organized in the nearly completed iguanodon. On New Year's Eve 1853, which fell on a Saturday, at 4 PM, at the invitation of Hawkins and of the directors of the Crystal Palace Company, twenty-two people gathered for dinner inside one of the largest model dinosaurs (fig. 10). In spite of the inconvenience of the date—the *Illustrated London News* reported—"Mr. Hawkins had one-and-twenty guests around him in the body of the Iguanodon on Saturday last; at the head of whom, most appropriately, and in the head of the gigantic animal, sat Professor Owen."[107] In fact, no more than a dozen people were actually seated within the dinosaur's hulk, and the others were arranged around an adjoining table. The event was rife with functionalist symbolism. Nameplates were hung high of Cuvier, Buckland, Mantell, and Owen—the paleontologists who were most closely associated with the study of the dinosaurs and at the same time were prominent functionalists. In case any doubt about the precise nature of the event had been left in the public's mind, Owen rose to remove it, commenting "upon the course of reasoning by which Cuvier, and other comparative anatomists, were enabled to build up the various animals of which but small remains were at first presented to their anxious study; but which, when afterwards increased, served to develop and confirm their confident conceptions—instancing the Megalosaurus, the Iguanodon, and Dinornis as striking examples." On the page which carried this story, the feat of Owen's reconstruction of the dinornis was retold.[108]

The Crystal Palace caper demonstrated that the functionalist approach had increasingly become Owen's popular party dress, not his professional working clothes. *Vanity Fair* added to a cartoon describing Owen as "Old

FIGURE 10. Owen at the head of a dinner table inside the iguanodon model at the Crystal Palace, Sydenham, on New Year's Eve 1853. This theatrical event represented the Victorian zenith of Cuvierian functionalism and its paleontological tenet that from mere fragments of extinct forms of life their entire structure can be deduced. (*Illustrated London News* 24 [Jan. 7, 1854]: 22.)

Bones": "His monsters became the delight and the wonder of an idle people, and having found a permanent place at the Crystal Palace, they to this day remain the best-known works of their designer."[109]

The iguanodon dinner was in some ways the culmination of the early-nineteenth-century Cuvierian tradition. Yet ironically, the very confidence with which flesh and additional bones had been added to the few iguanodon fragments actually known served to disprove the myth that the whole animal could be accurately reconstructed from any of its constituent parts. The iguanodon model was seriously flawed in a number of respects, most obviously in its quadrupedal position. Only the discovery of much more complete skeletons, in 1878, at Bernissart in Belgium, revealed the bipedal nature of this particular race of dinosaur.[110]

A controversy has developed over why Owen coined the order of the dinosaurs. Adrian Desmond has put forward the argument that Owen in doing so was motivated by a desire to undercut the transformist ideas of his colleague Grant. By making the dinosaurs as mammalian as possible, Owen implicitly made modern reptiles appear less advanced than their fossil predecessors, thus refuting any simple chain of progressive transmutation.[111] Pauline Dear has countered that Owen's order of the dinosaurs was the product of contemporary advances in taxonomic practice. Buckland and Mantell had relied heavily on dental characteristics, and to these Owen added skeletal features, in particular "vertebrae as a fundamental criterion in reptilian taxonomy."[112] Owen's own definition of what constitutes the "dinosauria" illustrates this point: "An order of extinct reptiles, characterised by cervical and anterior dorsal vertebrae with par- and diapophyses [certain protuberances], articulating with bifurcate ribs; dorsal vertebrae with a neural platform, sacral vertebrae exceeding two in number; body supported on four strong unguiculate [with nails or claws] limbs."[113]

I agree with Dear that the leitmotiv of Owen's dinosaurian work was not fear of Grantian evolution. Simplistic lineages of transmutation already seemed effortlessly refuted by the sequence of actual stratigraphic occurrence.[114] Although the followers of Cuvier and those of Geoffroy were divided over the issue of transmutation, it did not form the focus of their intellectual differences. At this time functionalism was the contentious issue. The dinosaurs were put by Buckland before the cart of natural theology, and Owen was the coachman to steer them from the BAAS meeting of 1841 in Plymouth to the Crystal Palace gardens in Sydenham, where Hawkins stabled them in 1853. Any antitransformist applications of Owen's reptilian reports were, admittedly, a welcome side effect, part of Owen's overall Cuvierian mission; but this was not a formative concern.

Moreover, by defining the dinosaurs, he created a new icon of museum display—the most famous and enduring, as it turned out—adding to the ones that, during this same period, he was in the process of sculpting: the megatherium, the glyptodon, and the dinornis. These came on top of the ichthyosaurus and plesiosaurus and of course the longer known-mammoth and extinct pachyderms, already on display in many a self-respecting museum. The symbolic worth of these extinct monsters was vividly illustrated in the main exhibition hall of the renovated Hunterian Museum. At the left of the entrance was placed the mylodon, raised up against a tree trunk, and on the right stood the bulging drum of the glyptodon, two pillars, as it were, that helped support the weight of Owen's museum program (fig. 3).

Such fossils, by being made the object of public admiration and wonder, had two important consequences for the museum movement. First, they encouraged further overseas explorations. Second, the exhibited monster models became the characteristic images of a proper museum collection and increasingly defined what constituted an up-to-date Victorian museum of natural history. They were the equivalent for natural history of what the treasures from Mesopotamia were to the antiquities department of the British Museum, described by Owen's friend Layard in his famous *Nineveh and Its Remains* (1849). The resurrection from the geological past of Owen's model monsters was, in its significance for his museum work, not unlike the excavation at Nimrud and the arrival at the British Museum in early 1852 of the two colossal human-headed winged lions.[115] All around the Western world, metropolitan, provincial, and university museums found legitimacy in the possession of the actual bones of the fossil monsters or their casts.

That Owen was perfectly aware of the iconic value of his Cuvierian reconstructions is apparent from a bizarre article for the *Edinburgh Review* that he wrote—anonymously and in the third person—about his own megatherioid work. A surprising feature of this review article is that in 1882, when it appeared, none of the publications under discussion was more recent than 1864, and the oldest carried a date of 1838. Why review such long-extant literature? The occasion of the review was not the appearance of the articles and memoirs but the opening of the Natural History Museum in 1881. The fact that Owen decided to select "ancient animals in South America" (i.e., the giant sloth and armadillo) as the subject matter for this review article indicates how crucial they were as museum objects (fig. 11). He boasted that his new museum's megatherium skeleton was more perfect than the original one in Madrid, and he praised himself as "the officer to whom mainly the public are indebted for this storehouse of the national treasures of natural history."[116]

The review was one of several in which Owen not only played the role of an anonymous critic of his own work but referred to himself in the third person, even to the point of stating that he had been "favoured to listen" to a lecture by himself.[117] This bad habit began in 1851 as part of his cooperation with Broderip when the two men wrote a summary of Owen's publications for Lockhart's *Quarterly Review*. Initially, it was proposed that Broderip write the essay, "but I found," Owen confided, "it would be easier and perhaps clearer if I did it myself. The list astonishes me! I wonder how Lockhart will manage, for it is already condensed to the utmost, and it looks enough for two long articles in the 'Quarterly.'"[118] The review was indeed broken up into two parts, the first entitled "Progress of Comparative Anatomy" (1852), the second "Generalizations of Comparative Anatomy" (1853), each title implying that the subject was, by and large, identical with Owen's contributions to it. Owen further shaped this twin review of his own work by both minor and major alterations to the proofs.[119]

RIDING HIS HOBBY.

FIGURE 11. Owen caricatured riding a megatherium skeleton. The megatherium was a popular museum icon that helped strengthen the drive for a separate natural history museum. ("Riding His Hobby" by Frederick Waddy, *Cartoon Portraits and Biographical Sketches of Men of the Day*, opposite 36.)

Parting Company with Lyell

The moa reconstruction was a high point in Owen's remarkable record of resurrecting extinct species, yet it also led to his biggest blunder in vertebrate paleontology. From the late 1820s on, discoveries began to be reported of fossil footprints in formations of the Permo-Triassic New Red Sandstone, in Scotland, Germany, and Connecticut. Buckland excitedly included the subject of fossil footprints, or "ichnology," in his Bridgewater Treatise.[120] After all, what a marvelous extension of the validity of the Cuvierian dictum—that an organism can be reconstructed from one of its parts—it would be if this could be done merely from its footprints.

The most spectacular of the European prints were attributed to an animal which the German naturalist J.-J. Kaup named *Chirotherium*, because the impressions looked like those of human hands. Probably emboldened by the discovery of the Stonesfield opossum, Kaup speculated that the prints were made by marsupials, pushing the origin of the mammalian class even further back into geological history.[121] As was to be expected, Grant objected and suggested that reptiles had made the imprints. Owen objected as well and suggested, in two papers to the Geological Society in 1841, that the footprints were the work of batrachians, salamander-like creatures of a low rank among the reptiles for which he coined the name *Labyrinthodonts*.[122] This instance alone should be a warning against any speculation that Owen's work represented a systematic campaign to undercut Grant.

The Connecticut footprints, described by, among others, the Amherst College geologist Edward Hitchcock,[123] and given ample press coverage by his Yale colleague Benjamin Silliman in the latter's *American Journal of Science*, were three-toed ichnolites, up to ten to fifteen inches in length, and looked very different from the *Chirotherium* impressions. The Americans interpreted their discovery as the footprints of large birds, *Ornithichnites*—incorrectly, as it turned out later in the century, when it became apparent that these ichnolites were made by bipedal dinosaurs. Presuming the footprints to be avian required accepting a very early occurrence of birds. Owen's initial reaction was wisely to caution that the tridactyl imprints might well have been produced by reptiles. But then Mantell was approached by the Americans and asked for his expert advice; he was given "a very full and beautiful series of these tracks,"[124] on which he reported to the Geological Society early in 1843, adding to his already-existing reputation as the discoverer of presumed bird remnants in the Cretaceous Wealden formation. At this stage Lyell intervened, giving it as his opinion that the ichnolites, in spite of their enormous size, were indeed avian, because the recently reconstructed New Zealand moa demonstrated that gigantic birds, big enough to have produced the Connecticut Valley impressions, had in fact roamed the surface of the prehistoric earth.[125] Broderip hinted at the same possibility in a letter to Buckland, written shortly after the historic opening of the moa box: "There is no end to the interest of this animal and mark well what a chapter it will open in the book of *Ornithichnites*."[126]

Owen now threw caution to the wind and acquiesced to the prevailing opinion. It seemed most reasonable, he wrote to Silliman, "to conclude that the *Ornithichnites* are the impressions of the feet of birds, which had the same low grade of organization as the *Apteryx* and the dinornis of New Zealand, and these latter may be regarded as the last remnants of an apterous race of birds, which seems to have flourished at the epoch of the New Red Sandstones of Connecticut and Massachusetts."[127] And when Hitchcock presented to the Royal College of Surgeons a series of twenty-seven *Ornithichnites* slabs and casts, Owen placed them with the moa bones.[128]

The fascination with the presumed bird impressions lasted for a long time; by the late 1850s it reached a peak when Hitchcock published his definitive *Ichnology*, when several magazines extensively reviewed "fossil footprints,"[129] and when Owen's *Palaeontology* textbook pictured the moa and the Connecticut Valley footprints in one and the same illustration to demonstrate the presumed connection.[130] The quality of the impressions made it possible to speculate about the nature of *Ornithichnites* skin, which, in Owen's view, resembled that of an ostrich, not of a reptile.

Such anomalies were grist to the uniformitarian mill: "One fact of this kind is of more value," one journal writer commented, "than a host of negative evidence, and it is triumphantly appealed to by those geologists who contend for the uniformity of the course of nature from the earliest epochs to the present time."[131] Lyell made eager use of both the Stonesfield opossum and the Connecticut Valley ichnolites. The footprints in particular were a promising feature because they seemed to prove that highly developed vertebrates had existed as early as the Paleozoic era.

By about 1850, W. E. Logan described a hitherto-unknown type of ichnolite from the very ancient Potsdam Sandstone in Canada.[132] Owen was asked to pronounce on the nature of the tracks, just as Lyell was preparing for the second of his two consecutive anniversary addresses to the Geological Society; and Owen's flirtation with Lyellian uniformitarianism continued. In the first address, of 1850, Lyell had outlined his objections to progressivism derived from inorganic phenomena.[133] In the second, of 1851, he turned his attention to organic fossils and emphatically denied that their stratigraphic distribution showed a progressive trend.[134] This second address ended with the quotation of a letter by Owen identifying the Potsdam tracks as reptilian, thus pushing back the earliest evidence of this class of vertebrates to the lower part of the Paleozoic. The sandstone impressions, Owen pronounced, "accord best with those of the land or freshwater tortoises."[135] This gave cause for worry to the advocates of progressive development, even to those who defined it in terms of adaptations to environmental change. When the following year the Cambridge mathematician and geologist William Hopkins succeeded Lyell as presidential speaker, he gave vent to this concern.[136]

Hopkins himself was an ardent progressionist who spent a large part of his address arguing in support of the twin theories of a cooling earth and of fossil progress, siding with his fellow Cantabrigian Sedgwick. To be featured in the illustrious Lyell address pleased Owen, but shortly after Hopkins's address, he changed his mind and reidentified the tracks as those of primitive crustaceans, most likely of *Limulus*. A note to this effect was added to the printed version of the address by Hopkins, who ended with the following sigh of relief: "The argument drawn from the Professor's former conclusion against the doctrine of the *progression of animal life* is, of course, much weakened by the result of his more complete investigation."[137] This reidentification was prominently referred to in the review of Owen's oeuvre for the *Quarterly*.[138]

At this time, too, Owen wrote a review for the *Quarterly* of the latest editions of Lyell's *Principles* and *Manual* and of the two antiprogressionist addresses to the Geological Society. He did so with the express intention "to take up the gauntlet which Sir Charles has thrown down."[139] Owen bestowed fulsome praise on Lyell's textbooks but then turned his attention to the issue of progressive development and point by point refuted Lyell's objections. This was Owen's finest, most systematic review article, only slightly disfigured by the customary references to himself. He tactlessly, but accurately, characterized Lyell's rejection of the overwhelming evidence for progressive development as the special pleading of the lawyer that Lyell was rather than the balanced assessment of the "philosopher" that he should be.[140] One of Lyell's many points was that the reason we do not find mammalian fossils in the older fossiliferous rocks is that their "facies"—as we would call it today—is marine, diminishing the likelihood that mammals would have become trapped within them. Owen retorted by citing the case of the whales, as he had done in his *History of British Fossil Mammals* and was again to do in his *Palaeontology*: no order of the mammalian class is at the present day represented by so many and so widely dispersed individuals as that of the whales, which are all marine; some attain the largest size known among past and present vertebrates, and the fact that the skeleton is ossified and has many vertebrae must have increased its chance of fossilization; if whales had existed in Paleozoic or Mesozoic oceans, it would be inconceivable that they should not have left an abundance of fossil evidence; given their scanty remains in Mesozoic rocks, followed by an abundance in the Tertiary Suffolk Crag, the only rational conclusion is that the nondiscovery of whales in the older fossiliferous strata indicates that they did not then exist.[141]

In a letter to Owen about this review, Lyell, inferring "from internal evidence that it is yours," thanked Owen for praising the textbooks and promised to "try and weigh your arguments impartially and dispassionately."[142] It had become clear to Lyell, however, that he could no longer count on Owen, and when by the end of the year a report of a reptilian discovery in what were said to be Devonian rocks near Elgin, in Scotland, reached London, Lyell turned to Mantell for confirmation. In fact, the rocks proved to be of a later, Triassic, age, but a controversy developed all the same between Mantell and Owen over the priority of description of the Elgin reptile.[143] In the literature, Owen's claim has drawn the short straw, but as Michael Benton has shown, in a fine piece of historical detective work, Owen's description was in fact the first to appear, in spite of conspiratorial machinations by Lyell and Mantell.[144]

Thus, in his *Quarterly* review "Lyell—on Life and Its Successive Development," Owen neatly combined a defense of fossil progress with a parading of his favorite museum icons, the whales, and this at a time when, donning the gown of the British Cuvier, he placed before the British public such other icons of a natural history museum as the megatherium and the dinosaur. In the course of the 1850s Owen turned the whales into a standard argument against the uniformitarians, in his paleontology textbook as well as in such public lectures as the Fullerian course of 1859, and he simultaneously used the gigantic size of the cetaceans to lay claim, in his 1859 "Report" with plan for a natural history museum, to a spacious palace for his collections.[145]

This strongly antiuniformitarian, anti-Lyellian stance of the 1850s served to turn the attention of Owen's supporters away from the fact that in the course of the 1840s he had significantly shifted his ground. Very shortly after he had been declared the new standard-bearer of Oxbridge's Cuvierian tradition, in 1841, he had begun to ally himself with London's Germanizing circle and adopted several of the tenets of *Naturphilosophie.* By defending the faith of a progressive development of life on earth against its uniformitarian objectors, Owen disguised his own diminished commitment to the Cuvierian, environmental interpretation of geological progress. The appearance of Chambers's *Vestiges* in 1844 and the English translation of Oken's *Lehrbuch der Naturphilosophie*, published at Owen's instigation in 1847, increasingly made the taxonomic model of progress seem synonymous with the notion of organic evolution. Sedgwick fumed against this evolutionary pantheism; so did Hugh Miller; and F. C. Bakewell felt compelled, unlike his father, who had confidently advocated progressionism, defensively to emphasize in his popular *Geology for Schools and Students* (1854) that progress is an environmental process and that it provides no grounds for the belief in transmutation.

No longer could such language be heard coming from Owen's lips. On the contrary; while keeping up the appearance of Cuvierian loyalism by waving the progressionist flag at the Lyellians, Owen subtly adopted the German, transcendentalist approach to paleontology, adding it to the end of his review

"Lyell—on Life and Its Successive Development." He did this by applying von Baer's embryological law to the fossil succession, demonstrating that progress can be defined as a radiating pattern from the more general to the more specific. Already in his Hunterian course of 1841, Owen had subscribed to von Baer's law, which superseded the simplistic notion of recapitulation and stated that successive embryonal stages of higher animals resemble the permanent, adult stages of lower ones, not because they recapitulate these, but because more heterogeneous or special structures arise out of more homogeneous or general ones. To this Owen added that the embryonic phases through which a higher animal passes resemble the adult states of the early, extinct members of its class. Thus, the extinct forms represent the more general structures, whereas the present-day forms show the more special ones. In the embryos of modern bony fish, for example, the caudal vertebrae form a series of cartilaginous centers which later become blended to form the base of a vertically extended symmetrical tail fin. In all Paleozoic fishes the early, embryonal state persists, and the tail fin, by the retention of the series of terminal vertebrae in the upper lobe, is unsymmetrical. In birds, too, the embryos of living species show separate terminal vertebrae which gradually grow together; but, as the discovery of the *Archaeopteryx* was to show, Mesozoic birds retained in adulthood the embryo condition of a series of distinct vertebrae.[146]

These and similar instances provided Owen with his own, anatomical criterion of geological progress. A higher or more advanced organization is that which departs from the general type. In this context, Owen's fascination with intermediary forms begins to make sense. Ever since his study of Darwin's South American fossils, and his systematic description of teeth in the *Odontography*, intermediary extinct types had fascinated him. The Darwin material had included the *Toxodon*, with its mixed features reminiscent of a rodent, a pachyderm, and a cetacean; its description had earned Owen his FGS (fellow of the Geological Society). Among the saurians, the ichthyosaurus combined fishlike vertebrae with a dolphin-like exterior and various crocodilian features. Thus, the more general form combines features which exist separately in later species, these having diverged from a common type by a process of adaptive radiation. Owen cited many examples, from the invertebrates, the vertebrates, and especially the reptiles and mammals. The palaeotherium was an Eocene herbivore which combined features now separately present in the horse and the tapir. The palaeotherium adhered more closely to the mammalian type by having a second and fourth toe, which in the horse are reduced to splint bones; a similar argument could be made for its dentition. The Miocene mastodon had two incisors in the lower jaw, whereas in today's African and Asian elephants these are absent; several more departures of dentition from the normal type occur. "The above cited and analogous facts indicate that in the successive development of the mammalia, as we trace them from the earliest tertiary period to the present time, there has been a gradual exchange of a more general for a more special type."[147]

Although in places Owen retained the language of functionalism by pointing out that the new, more specialized forms were adapted to particular purposes—the coalesced cannon bone of the horse, for example, allows the animal to run fast—his criterion of progress was a radical departure from the environmental one and situated Owen much closer to German transcendentalism than to Oxbridge natural theology. He never went quite as far as Agassiz, however, who delighted in speculations about a triple parallelism of fossil succession, taxonomic scale, and embryonic development. With the flair of an Okenian naturalist, Agassiz coined the term "prophetic type" for the intermediary forms which combine features later separately present in new forms. The pterodactyl, for example, was a prophetic type which, although truly reptilian, foreshadowed in its anterior extremities features of birds and bats.[148] This issue, however, was connected no longer primarily to the functionalist concerns of Owen's Oxbridge patrons but to a developing transcendentalist discourse among an emerging group of professionals based at various metropolitan scientific institutions.

4
The Vertebrate Blueprint

During a dinner at the Literary Club in 1849, Owen met the French statesman and historian François Guizot, whose government had been brought down in the revolution of the previous year and who had himself gone into exile. In the presence of Peel and several other dignitaries of state, church, and the arts, Owen was presented to the deposed royalist premier as "the Cuvier of England." This drew from Owen the aside: "I wish they would be content to let me be the Owen of England."[1] It would be a mistake to interpret this lament as a sign of modesty on Owen's part; nor did it mean that Owen tried to back into the limelight by feigning to be burdened by yet another compliment. Owen's desire to be regarded as "the Owen of England" indicated that by the end of the 1840s he felt uncomfortable with the epithet of the "English Cuvier" because much of his work was no longer Cuvierian in approach but had become Okenian. Through the 1840s Owen had increasingly begun to side with the so-called philosophical anatomists, the followers of German *Naturphilosophie* who believed that "form" rather than "function" offered the right solution to the problem of organic diversity. Their epistemological approach had its own institutional location, and whereas Owen had one foot in the functionalist camp, the other was firmly planted among the transcendentalists.

A Metropolitan Scientific Culture

Let me stress by repeating the important point that, although Owen's Oxford and Cambridge friends promoted him, both directly with money and honors, and indirectly by looking after the interests of natural history, and although Owen did oblige his paymasters with a series of spectacular functionalist studies, he never became truly one of them. Acceptance into Peelite society did not change the fact that Owen's own social background and his own ultimate purposes were different from those of his patrons; and this difference predisposed him to adopting a scientific epistemology that went beyond the functionalism of natural theology.

Many authors have attributed the demise of the design argument to Darwin and his theory of evolution by natural selection.[2] However, the edifice of natural theology had already begun to crumble during the two or three decades which preceded *The Origin of Species*, as a generation of London naturalists abandoned the functionalist epistemology of the Bridgewater Treatises and switched to a different kind of scientific explanation, imported from Edinburgh and the Continent. In doing so, they gave metropolitan science its own philosophical mooring. It may seem contradictory, but Owen was the leading proponent of the nonfunctionalist, non-Oxbridge, metropolitan form of scientific explanation.

The impetus for this epistemological divergence from natural theology came from the new scientific institutions in the metropolis. By these I do not mean such Oxford- and Cambridge-dominated bodies as the Geological Society or any of the other scientific and learned societies that were founded during the nineteenth century but the following research and teaching establishments: the Royal Institution, founded in 1799, which provided a base for Humphry Davy, W. T. Brande, Michael Faraday, and other chemists; the Hunterian Museum, acquired also in 1799, where Owen spent the first half of his career; the Museum of Practical Geology, founded in 1840 and expanded eleven years later to incorporate the School of Mines, where Edward Forbes and T. H. Huxley were among the early lecturers; the British Museum, where the departments of natural history provided a professional niche for various naturalists, in particular Owen for the second half of his career; and the University of London, both University and King's Colleges, founded in 1826 and 1829 respectively, where Owen's most immediate colleagues—Robert E. Grant, William Sharpey, and Robert B. Todd—taught biomedical subjects.

The purpose of these institutions differed fundamentally from that of the Oxford and Cambridge colleges. At the latter, the natural sciences were part of a program of liberal education, not taught for any purpose of professional training. Both William Buckland in his *Vindiciae* and Adam Sedgwick in his *Discourse* were unequivocal on this point. They promoted geology, then the newest of the natural sciences, under an explicitly non- or even antiutilitarian banner and not for any economic purpose.

The word "utility" is used here to refer to material benefit, not to the Benthamite basis of moral philosophy of the greatest happiness of the greatest number. With respect to moral philosophy, members of the Oxbridge network were divided: Buckland seemed sympathetic to the Benthamite doctrine, whereas Sedgwick objected to Bentham's utility and Paley's expediency as a basis for human actions.[3] It was the perception of geology as history—albeit history of an antediluvian world—and thus as a branch of humanistic learning that made it popular as part of the curriculum at the two ancient universities—in addition to geology's use as a branch of natural theology. Buckland bluntly stated in his inaugural lecture, in answer to the question of why geology should be offered a place at Oxford, that "it is founded upon other and nobler views than those of mere pecuniary profit and tangible advantage." He continued: "The human mind has an appetite for truth of every kind, Physical as well as Moral; and the real utility of Science is to afford gratification to this appetite."[4] That Buckland and his Oxbridge colleagues regarded perfect adaptations and design in nature as prominent aspects of those truths, we have already seen.

This antiutilitarian view continued to predominate until well after the recommendations for reform by the "Oxford Commission" of 1850. Both traditionalists and advocates of moderate reform believed that learning for profit was vulgar and had no place in England's ancient universities. Even Acland, in spite of his efforts to promote the biomedical sciences at Oxford, believed that the university's main purpose was to provide a program of liberal studies.[5] For professional training, whether in medicine or in any of the natural sciences, the nation's capital was the more suitable place. As the High Church theologian J. B. Mozley commented, in his *Quarterly Review* condemnation of the Oxford Commission, the road of the chemist, the mineralogist, and the geologist was to the metropolis.[6]

By contrast, the five cited metropolitan institutions were flying the flag of utilitarianism. The explicit aim of the University of London was to provide opportunities for professional training. The same was true of the Royal Institution and the Museum of Practical Geology, both of which exemplified the conviction that scientific expertise was an integral part of a successful industrial economy. In 1851, the year of the Great Exhibition, one of its leading instigators, the prince consort, opened the Museum of Practical Geology, and Lyon Playfair, one of the early lecturers, concluded his introductory address by emphasizing "that it is indispensable for this country to have a scientific education in connexion with manufactures, if we wish to outstrip the intellectual competition which now, happily for the world, prevails in all departments of industry."[7]

The removal of the natural history departments from Bloomsbury to South Kensington made the British Museum (Natural History) part of Prince Albert's new world, too, where scientific inquiry and the interests of industry lived side by side in blissful matrimony. Neither a natural history museum nor a medical museum was of course useful in a directly commercial or industrial sense; yet the purpose of the Hunterian Museum was not a liberal but a professional education, namely of medical students.

Owen's Oxbridge patrons did not object to his participation in the utilitarian razzle-dazzle of the period. On the contrary; they encouraged it, and several themselves served as "Her Majesty's Commissioners, 1851." After all, the metropolis was the right place for applying scientific expertise to industrial and other useful ends. Buckland in particular had shown a penchant for economic geology, especially after he had left Oxford for London, where he enthusiastically took part in the march of material progress. His son Frank followed in Buckland Senior's utilitarian footsteps.[8] What did cause concern—even alarm—was the fact that metropolitan scientists, having found a professional justification for their pursuits, abandoned the functionalist epistemology of natural theology. Given the utilitarian underpinnings of metropolitan science, it did not need the connection with natural theology. To present science as a source of arguments of design was of little use in London and in fact symbolized Oxford and Cambridge domination.

Many metropolitan scientists were not Oxbridge alumni but had received their education at Edinburgh or additionally also on the Continent. From there they brought with them a different epistemology, the idealism of Romantic *Naturphilosophie*, in which the transcendental logic of form rather than functional adaptation was the criterion by which physical reality—especially organic reality—was explained. To a man, those biomedical scientists who advocated the transcendentalist epistemology had an Edinburgh background. They included Robert Knox, "the earliest and most outspoken proponent of idealism in British natural history";[9] his most effective student, Forbes, who introduced transcendental morphology into the BAAS; Grant, who took up the cause of Geoffroy Saint-Hilaire at University College; his colleague Sharpey, an admirer of Owen; Martin Barry, who followed in Karl von Baer's footsteps with two classic papers on morphological unity in the *Edinburgh New Philosophical Journal*; and William B. Carpenter, whose *Principles of Physiology* represented a major inroad of German transcendentalism into London. Another London transcendentalist who had received his education in Scotland and on the Continent was Joseph Henry Green, who introduced a Coleridgean appreciation of Kantian idealism into the Royal College of Surgeons. Owen himself matriculated at Edinburgh University and spent a few months in "the Athens of the North."

Functionalism was the epistemology of Oxbridge science; idealism, that of the metropolitan scientific institutions—at least before the scientific naturalism of the Darwinians gained ascendancy in the course of the 1860s. It was no coincidence that Romantic science entered London through the first of the independent scientific institutions of the nineteenth century, the Royal Institution, where Davy and later Faraday carried out scientific research in an intellectual context that from the outset was in open communication with German Romanticism.

Independent metropolitan culture was embodied in Thomas Carlyle, well known for his "Germanizing." Like Coleridge before him, Carlyle believed that Germany was the contemporary heartland of spiritual renewal. In his "signs of the times" (1829) he characterized the early nineteenth century in England as an "Age of Machinery," not just in industry, but also with respect to thought. The Paleyan design argument, based on the analogy of mechanical contraptions, appeared shallow to Carlyle, and he disassociated himself from the functionalist design argument and the Benthamite basis of moral philosophy, advocating instead a belief in a transcendental source of moral conviction. For Carlyle, Goethe was Germany's spiritual leader, and it was G. H. Lewes, one of Carlyle's admirers, who wrote a classic biography of Goethe in which, among other things, he described Goethe's seminal contributions to idealist morphology.[10]

Carlyle rejected the bourgeois liberalism of his day, which sought to remedy the country's ills by means of external, political measures. A society's progress, in his view, came from the moral strength of individual men. In his lectures "On Heroes, Hero Worship, and the Heroic in History" (1840), he argued that the welfare of mankind, and its advances in civilization, depended on the virtue of a few great individuals, superior in moral character and intellectual ability. These were divinely chosen people, sent at preordained times to enlighten and lead the common masses with their courage and truthfulness. Carlyle's heroes included mythological gods, prophets, poets, priests, literary men, and political rulers; two of his English heroes were Shakespeare and Cromwell.[11] Such a view of history, applied to Carlyle's own times, meant that the problems of the day should be addressed by leaders whose legitimacy stemmed from talent and expertise, not from ties to privileged families or long-established institutions. A new aristocracy of talent had to take the place of the customary one by birth.

This social philosophy had obvious attractions for the emerging generation of metropolitan scientists, who were trying to lay claim to part of the cultural authority monopolized by the members of traditional institutions. Owen, for one, was attracted by it, and while he developed his transcendental morphology, he simultaneously nurtured a friendship with Carlyle, with whom he shared a variety of specific interests. As Caroline Fox recorded in her diary: "His [Owen's] delight in Carlyle is refreshing to witness."[12] Owen joined Carlyle in an admiration of German culture, in particular of Goethe, and also an adoration of such heroes at home as Milton and Cromwell. In 1845, while on a visit to Florence, Owen had a copy made of Cromwell's portrait in the Pitti Palace. No sooner had the shipment arrived in Lincoln's Inn Fields than Thomas and Jane Carlyle went over to inspect the portrait and compare it with their own. A levee of other visitors followed, all keen to see the Florentine portrait.[13]

Such Calvinist idolatry caused Owen some embarrassment when, during his first dinner at Johnson's Club, with its predominantly anti-Cromwellian Oxonian membership, he was asked whether Cromwell deserved a monument in Westminster Hall among the cenotaphs of the kings and queens of England. Owen prevaricated by quoting from Milton's "Ode to Cromwell" and replying that Cromwell already had a monument in the minds of men.[14] Yet that Owen had a genuine and lasting interest in Carlyle and his heroes is not in doubt. In 1860 he wrote to his wife: "I have got '*Sartor resartus,*' '*Chartism*' and '*Wilhelm Meister translated*' by Carlyle."[15] Tired of a day's scientific work, Owen would go home and find relaxation in the reading of Carlyle and other coryphaei of metropolitan literature.[16]

Carlyle, for his part, thought highly of Owen, as we saw in chapter 1; and in a real sense, Owen, with his autocratic, larger-than-life personality and his discoveries of the transcendental meaning of vertebrate morphology, fitted Carlyle's model of the hero. As a Victorian naturalist, Owen combined the features of both poets and prophets of yore, who—Carlyle maintained—were the same in this fundamental respect: "they have penetrated both of them into the sacred mystery of the Universe; . . . 'the Divine Idea of the World.'"[17] Owen's monographs, especially on fossils of vertebrate skeletons, demonstrated that he was one of the anointed few who did see the mysterious truth behind nature's external clothing. Caroline Fox recorded in her diary for June 4, 1864: "The Carlyles had been to see Cromwell's portrait at the Owens, Carlyle grumbling at all Institutions, but confessing himself convinced by Owen's 'Book on Fossils.'"[18] Carlyle's personal description of Owen reads like that of one of his heroes: "He is a man of real talent and worth, an extremely rare kind of man. Hardly twice in London have I met with any articulate-speaking biped who told me a thirtieth-part so many things I knew not and wanted to know."[19]

This view of Owen as a hero-leader of the day had not changed when later Carlyle read Owen's booklet *On the Extent and Aims of a National Museum of Natural History.* Carlyle himself had little interest in museums and believed that Owen should write the ultimate book on natural history, "which would far outshine the biggest *museum* even the British nation could build." Yet he wished Owen well in the pursuit of his museum, if only because the nation should comply with the vision of one of its intellectual leaders.[20]

The metropolitan idealists were, by and large, an unintegrated group of men, not organized in any fraternity that held power. To the extent that the Carlylean naturalists clubbed together, it was in the form of the metropolitan branch of the "Red Lions." The Red Lion Club was formed by Forbes, at the BAAS meeting of 1839, as a protest against "dons and donnishness" in science.[21] Owen soon became one of the most ardent supporters of its London branch, which met monthly in the Cheshire Cheese in Fleet Street.[22]

One of the ablest advocates of German idealism was associated with Owen's own College of Surgeons. When Owen joined the college, the older generation, that was still in power, were medical Paleyites: Abernethy, Bell, Carlisle, Astley Cooper, all advocates of traditional Hunterian functionalism. Gradually, however, throughout the 1830s and 1840s, a younger generation of Edinburgh-educated surgeons infiltrated the ruling body of the college. Prominent among these was Green, who joined the council in 1835, the Court of Examiners in 1846, and served twice as president (1849–50; 1858–59).

This picture of an early Victorian fault line through English science dividing Oxbridge functionalists from metropolitan transcendentalists is of course an oversimplification. Neither "Oxbridge" nor the "metropolis" was characterized by a monolithic composition. Baden Powell, for example, although an Oxford man, was an ardent follower of Owen's transcendental morphology. Several Cambridge people, too, either seriously pursued or flirted with German scholarship. The so-called Cambridge Apostles idolized the German historian of Rome Barthold Georg Niebuhr, shared Coleridge's distrust of natural theology, and indulged in criticism of Paley. Owen's parliamentary supporter Richard Milnes was one of them.[23] William Whewell himself saw a force of reform in a serious study of German scholarship, although in the end he stuck to the supremacy of functionalist thought.[24] As Jack Morrell and Arnold Thackray have pointed out, the scientists were associated with many rival and overlapping groups. There were the Oxford Tractarians, headed by John Henry Newman; the Benthamites, loosely associated with University College, London; and various coteries of naturalists associated with the British Museum and with a range of metropolitan learned societies, the Antiquaries (founded 1707), the Linnean (founded 1788), the Zoological (founded 1826), and the Entomological (founded 1833). In addition, there was the cluster of radical medical practitioners, whose mouthpiece was Thomas Wakley's *Lancet*.[25]

The tensions caused by the epistemological self-assertion of the metropolitan, upwardly mobile Carlyleans over and against the ascendancy of Oxford and Cambridge functionalists runs right through Owen's oeuvre, which was an ambivalent mixture of demonstrations of functionalism and of transcendentalism, reflecting the politics of the museum movement. The former approach was needed to satisfy his trustee-patrons; the latter gave expression to his drive for scientific autonomy. This was the two-track nature of Owen's museum career, referred to above.

I agree with Dov Ospovat that during the pre–*Origin of Species* period, and certainly during the pre-*Vestiges* years, the primary issue that divided British naturalists was not "pro- or contra-transmutation" but "form or function."[26] Here I add to Ospovat's observation that these different approaches had institutional anchorages. The emphasis on "form," by being part of the wider acceptance of German Romanticism, provided Owen and his London colleagues with a suitable philosophical framework for a metropolitan, as opposed to Oxbridge, intellectual culture. There were only two philosophies of nature in Victorian Europe in which religious belief was given a place. One was the increasingly antiquated-sounding natural theology; the other was some form of Romantic *Naturphilosophie.* Owen and many of his Edinburgh- and Paris-educated *confrères* were not irreligious, yet at the same time they did not belong to the establishment of England's ancient universities; they were "MDs," not "DDs."

German transcendentalism symbolized an independence from England's ancient universities; it represented the epistemology with which science was associated in places where it flourished as an independent subject of education and research, both in the German universities and at Edinburgh. Adopting the non-English approach to morphology meant casting off the yoke of Oxbridge domination and staking out a separate, segregated area of scientific culture where the first generation of professional men at the new metropolitan institutions held sway.

The Great Executor

The transcendentalist approach was most prominently associated with the study of organic form. The decades immediately preceding Owen's appointment to the Hunterian Museum in 1827 had been a period of extraordinary changes in comparative anatomy, marking a florescence that extended well beyond the middle of the nineteenth century. This flowering did not start everywhere at the same time; it began in Germany in the 1780s, reached Paris by approximately 1800, and took until the 1830s before producing a London spring. One of the changes was a quantitative one; during the period 1780–1830 far more books and papers on comparative anatomy were published than had appeared during the entire period of its history until then.[27] Another change was the introduction of university courses on comparative anatomy and the concomitant appearance of elementary textbooks on the subject. The Göttingen professor Johann Friedrich Blumenbach began teaching a comprehensive course as early as 1785. Two decades later his popular *Handbuch der vergleichenden Anatomie* (1805) was published. A further change that took place at this time consisted of an increase in emphasis on osteology as the central concern of comparative anatomy. One obvious reason for this was the contemporaneous emergence of paleontology, most spectacularly of vertebrate paleontology. Although the usefulness of comparative anatomy was said to be to human anatomy and physiology and thus to medical training, much of the subject's prestige derived from its successful application to the identification of extinct vertebrates from fragmentary bones and teeth. This was accomplished by Blumenbach, Petrus Camper, P. S. Pallas, J. C. Rosenmüller, and many other, mainly medical men,[28] but most comprehensively by Georges Cuvier. The combination of the latter's five-volume *Leçons d'anatomie comparée* (1799–1805) and four-volume *Recherches sur les ossemens fossiles* (1812) was widely hailed as the start of a new epoch; across Europe he was crowned the founder of comparative anatomy in its modern, scientific form. The Muséum national d'histoire naturelle in Paris became the Mecca of the subject, and every student with aspirations to recognized expertise made a pilgrimage to the museum during his *Wanderjahre.*[29]

Last but not least, these new developments went hand in hand with innovations of theory. When Owen visited Paris in the summer of 1831, he listened to some of the debates that formed the tail end to the infamous clash of the previous year that had pitched Cuvier against Geoffroy at a number of different sessions of the French Academy. Scientifically, at issue was whether or not there exists a "unity of composition in the animal kingdom." It seemed as though this clash was staged to present to all of Europe the two main rival schools of anatomical philosophy, and the Academy debates of 1830 have long provided historians with a focal point for the discussion of early-nineteenth-century biological theory.[30] Toby A. Appel has given us a particularly fine account of the Cuvier-Geoffroy encounter, throwing new light on the event by interpreting it in terms of institutional and political party power play. Geoffroy's biological theories meshed well with antiroyalist sentiments, and it may have been the upsurge in republican unrest of the late 1820s which forced the autocratic Cuvier to confront his rival publicly.[31]

This upsurge of republicanism finally led to the July Revolution of 1830 and to the expulsion of the royal family and the abdication of Charles X. On the day that the news of the events of July 27–29 reached Weimar, the ailing poet

Goethe greeted a visiting friend with the outburst: "Now, what do you think of this great event? The volcano has come to an eruption; everything is in flames." The friend took this to refer to the political commotion in Paris, but Goethe retorted that he was speaking of the historic debates between Cuvier and Geoffroy, in which he regarded Geoffroy as "a powerful and permanent ally." "From the present time," Goethe enthused wishfully, "Mind will rule over Matter in the scientific investigations of the French."[32]

Almost immediately following the protracted Academy debate of 1830, Cuvier went on what proved to be his last visit to England. Here young Owen was entrusted with the task of showing the grand old man around the Hunterian collections; Owen's reward was an invitation for a reciprocal visit to the Paris museum. In spite of such cross-Channel contacts, there was no major British input into the Cuvier-Geoffroy debate, nor, for that matter, was there any from Italy. By the early 1830s the need for an outstanding anatomist to represent Britain in the European arena was painfully felt. Everard Home had produced papers on comparative anatomy which were published in the *Philosophical Transactions* and as *Lectures on Comparative Anatomy*, but they formed an unintegrated collection of case studies; moreover, Home's posthumous reputation was tarnished when in 1834 it became publicly known that much in these studies had been pirated from John Hunter's unpublished papers, the majority of which Home, in 1800, had taken to his house and many years later, in 1823, had destroyed.[33]

Thus, there was no one in Britain who had produced works comparable to Cuvier's massive oeuvre or to the monumental volumes of the "German Cuvier," Johann Friedrich Meckel. To fill this British vacancy became an even bigger challenge when by the early 1830s several representatives of the first generation of comparative anatomists died. Both Goethe and Cuvier passed away in 1832, while the following year proved to be the last in the lives of both Home and Meckel. Neither Cuvier nor Meckel had succeeded in completing and integrating their major projects, in part because of the continuing avalanche of new data. If at this late stage an Englishman were to produce a comprehensive work on comparative anatomy, national honor—Owen suggested—could still be saved.

An unambiguously formulated summary exists of the long-term program by which Owen intended to become internationally competitive and outclass his European rivals; the summary took the form of a letter by Owen to Buckland written to assist the latter in arguing Owen's case with Peel for a civil list pension.[34] It was Owen's ambition to produce a series of monographs on the vertebrates, each concerned with a particular type of organ or organ system. With these he intended to improve upon, or at least emulate, the greatest Continental authors on comparative anatomy. No contentious issue of methodology entered into his plan.

Did Owen succeed in bringing out the promised series of books on the comparative anatomy of the major organ systems? To what extent, too, did he manage to measure up to European competition and become one of the international greats of his subject? The first of the specifically cited monographs, the *Odontography; or, a Treatise on the Comparative Anatomy of the Teeth* was already in process of being published and was delivered during the period 1840–45. The seed of this two-volume book was contained in Owen's BAAS report of 1838 "On the Structure of Teeth." The *Odontography* fully vindicated the confidence shown in Owen by his patrons. It represented Owen's finest book-sized study and, in the priority it accorded to dental characteristics, provided him with the key to his many later successes in vertebrate paleontology—this key being "that the teeth, by their microscopic structure, as well as their more obvious characters, form important, if not essential aids to the classification of existing, and the determination of extinct species of vertebrated animals."[35]

The subject matter of teeth was topical, particularly because of its applicability to problems of classification, and Owen was by no means the only one to write a dental monograph; witness such Continental studies as Frederic Cuvier's *Des dents des mammiféres* (1825) and Michael Erdl's "Untersuchungen über den Bau der Zähne bei den Wirbelthieren, insbesondere den Nagern" (1843).

When in 1839 the Microscopical Society was instituted, Owen became its founding president; and during the society's first meeting he demonstrated the usefulness of the microscope by applying it to the study of fossil teeth.[36] Such work was made possible by recent improvements in the optical and mechanical properties of microscopes and was encouraged locally by J. S. Bowerbank and internationally by C. G. Ehrenberg, whose classic two-volume opus *Die Infusionsthierchen als vollkommene Organismen* (1838) was dramatic proof of the possibilities of the technique. Owen examined teeth from one extreme of the vertebrate order to the other, and his *Odontography* was a fine example of how, from systematic, anatomical comparisons, physiological issues of growth could be settled. This issue of how teeth grow was just then being addressed by such leading Continental physiologists as J. E. Purkinje, A. A. Retzius, and T. A. H. Schwann, even though—Owen reminded his readers—some presumably novel observations on the microscopic structure of the teeth had been made long ago by Leeuwenhoek and Malpighi. The then widely accepted theory of the growth of teeth was that the pulp, like a gland, secretes the growing tooth. Owen, however, supported the rival theory that the dentine—a term he coined—grows by a process of centripetal calcification of the pulp.[37] He showed a full command of the latest techniques and issues in comparative anatomy and proved himself on a par not just with the generation of Continental greats who had recently passed away but also with the younger masters of the subject, such as Purkinje and Schwann. Moreover, the *Odontography* was lucidly written and suffered less than later publications from Owen's convoluted, evasive style, which seemed intended to hide his meaning rather than clarify it. Included in this monograph was his most spirited priority dispute over certain observations on the microscopic structure and growth of teeth, conducted against the dental surgeon Alexander Nasmyth, who had just published his own *Researches on the Development, Structure, and Diseases of the Teeth* (1839), and whom Owen accused of blatant plagiarism of Schwann.[38] The acknowledgment of debt to various colleagues in the introduction to the *Odontography* was warmer than most of Owen's later expressions of indebtedness. The book itself was beautifully produced in two volumes, the first with text, the second with striking illustrations. Its lasting value is indicated by the fact that even today a specialist may find it a treasure trove of unexpected, valuable observations.[39]

But none of the other monographs promised in Owen's letter to Buckland-cum-Peel—on osteology, neurology, and on the generative system of the vertebrates—ever appeared. One simple reason for this was that the Hunterian approach to comparative anatomy, based on physiological systems, went out of fashion. In his lecture courses of 1837–42, Owen still divided his material into the digestive, excretionary, generative, and other systems (see table 2). A similar arrangement governed the division into chapters of the first generation of textbooks, ranging from Cuvier's *Leçons* to Grant's *Outlines of Comparative Anatomy*. The *Odontography* had been a spin-off from the Hunterian lectures of 1837–39, which among other things included the teeth as part of the digestive system. From 1843 on, however, Owen switched to another approach and used a taxonomic criterion for subdividing the contents of his lectures, and began lecturing on "invertebrates" and "vertebrates." This marked a change of approach from "organs" to "organization" and thus from "function" to "form," or *Bauplan*.[40]

The new emphasis on form also symbolized the emancipation of comparative anatomy. It was no longer just a handmaiden to physiology or an auxiliary subject of medical education but was a subject in its own right. Until this time it had been customary to introduce a textbook of comparative anatomy with the justification that animal structure elucidates physiology, in both animals and humans. This now changed, too, and such apologies became rare. This emancipation was an international trend. In the middle of the 1830s, Rudolph Wagner's *Lehrbuch der vergleichenden Anatomie* (1834–35), for example, was still broken up into chapters dealing with such physiological systems as digestion, circulation, respiration, excretion, and generation. When Wagner's *Lehrbuch der Zootomie* appeared in 1847, the table of contents was fundamentally changed, showing a subdivision of materials by major taxonomic groups.

With the new approach, Owen's lectures became publishable, and the *Lectures on the Comparative Anatomy and Physiology of the Invertebrate Animals* of 1843 was the first result. Little in these lectures was based on Owen's own research papers, however, and they were not of the supreme quality of his *Odontography*, nor did the *Lectures* measure up to the international competition. The *Lehrbuch der vergleichenden Anatomie*, by C. T. von Siebold and H. Stannius, published in 1846, was judged the more authoritative of the two texts. In particular, Owen's second edition of 1855 was not properly updated, and Huxley's savage review of it, though politically motivated, was well founded.[41]

Admittedly, Owen's real expertise lay in the area of vertebrate morphology. Yet his *Lectures on the Comparative Anatomy and Physiology of the Vertebrate Animals*, which appeared in 1846, covered only the fishes, which constituted the class of vertebrates with which he was least concerned. The lectures on the other vertebrate classes took a long time in getting published; not until a decade after Owen's resignation as Hunterian professor did his three-volume *On the Anatomy of Vertebrates* (1866–68) appear (which included fishes). It was hailed as the first and only book on the subject in Britain, a massive undertaking full of information, especially about groups of animals such as the nonplacental mammals, on which Owen had worked for decades. One reviewer called it "a treatise calculated to redeem the reputation of British science."[42] Another praised it as one of "the most important systematic treatises on Comparative Anatomy of the present time."[43] But more telling than these laudatory reactions was the fact that the total number of reviews was small and that Owen's single-biggest opus was ignored by the larger part of the periodical press.

The most likely reason for this was that *On the Anatomy of Vertebrates* represented simply a major compilation rather than a synthesis. To the extent that it was the latter, it was outdated by the time it appeared, in that it provided nothing new to the innovations of the 1840s. It was neither pervaded by the newly formulated principles of the evolutionary thinking of the 1860s nor informed by the most recent German or French work in cytology, histology, experimental physiology, and so on. The introductory and concluding chapters did contain Owen's reactions to disputes with Darwin and Huxley and an outline of his own theory of evolution, but this theory did not function as the yeast that made the text rise to the quality of Darwin's *Origin*; nor did it have the influence which Johannes Müller's earlier *Handbuch der Physiology* had exerted on a whole generation of disciples, or which Carl Gegenbaur's *Grundzüge der vergleichenden Anatomie* was to have.[44]

It could be argued that Owen's international stature should not be measured by his published lectures, not even the ones on vertebrates, but by his contributions to transcendental morphology, which did bring him lasting fame. To the extent that Owen ever wrote a theoretical synthesis, it was the French edition of his major BAAS report in which he mapped out the unity of skeletal composition among the vertebrates. This masterly piece of work, in which Owen placed the philosophical anatomy of Goethe and Geoffroy on a more Baconian basis, stripping it of Okenian excesses, will be discussed below. In Julius Victor Carus's generous assessment of Owen, it placed him on a par with Müller; whereas the latter opened up new avenues of physiological research, Owen brought the notion of "unity in variety" to its fullest completion.[45] Yet in this lay the tragedy of Owen's scientific career, looked at in the international arena. He was less a major innovator of biological theory than an executor of the work left incomplete by great predecessors. As a conservator writing catalogues he completed the work of Hunter. As a paleontologist he rounded off much of what Buckland and Cuvier had begun. And as a comparative anatomist he carried to a level of perfection, not attained on the Continent, the transcendental anatomy of Geoffroy, Oken, and other older colleagues. By the time Owen had finished this task, however, the approach had become antiquated in its place of origin, on the European mainland, where biology was increasingly guided no longer by *Naturphilosophie* but by the positivism of Comte, Feuerbach, and others. In 1850 Owen's friend Jan van der Hoeven at Leiden was already bemoaning the fact that philosophical anatomy was "now nearly abandoned on the Continent."[46] Owen did measure up to the all-time greats of his subject, but only by comparison with predecessors, not contemporaries; he was neither a Darwin nor even a Müller.

All the same, Owen's oeuvre was impressively voluminous and constituted a major advance in comparative anatomy, especially as applied to vertebrate paleontology. He was Britain's most accomplished naturalist of the Victorian period, and the most significant one during the decades before Darwin's *Origin*. Yet the real importance of much of Owen's theoretical work was not its pan-European currency but its local, metropolitan effect. While on the Continent transcendentalism was losing its prevalence, it was new in London and contributed to the cultural ferment of the time. Like Buckland's diluvialism of the 1820s—seemingly antiquated from a Continental point of view—it served as a vehicle for significant issues of reform.[47]

The Lure of *Naturphilosophie*

All along, Owen had aspired to unifying the two philosophies of anatomy represented by the functionalist school of Cuvier and the German school of *Naturphilosophie* led in France by Cuvier's colleague Geoffroy. As early as 1837 he confided to Whewell that he intended to combine into a single, harmonious theory the teleological and the transcendental approaches.[48] In his lecture notes for 1841 Owen again expressed this syncretistic aim; to Bell all was function, to Geoffroy all was form, but, Owen scribbled: "An organ never thoroughly known till both teleological and morphological relations fully known."[49] A few years later, in 1846, when he had outlined his thoughts on the subject in a treatise, Owen sent a complimentary copy of this to Silliman with the following cover letter:

> You may remember the condition in which [the] philosophical department of anatomy was left by the great Cuvier and Geoffroy, and the discussions which unhappily tended to sever those estimable men in the latter period of their lives. The result was the formation of two schools or parties in the French world of anatomy, and subsequently the facts and arguments bearing upon these transcendental questions have been viewed in Paris through the prism of such party feeling.
>
> The chief and most cherished labor and reflections of many past years have been devoted by me to the acquisition of such truth as might lie at the bottom of the well into which the philosophy of anatomy seemed to have sunk after the departure of the great luminaries of the Jardin des Plantes. With what success I have drawn from the deep and obscure source, I leave to the impartial students of my little book.[50]

A decade on, and Owen added iconography to prose in his self-presentation as the mediating unifier. A rather fatuous illustration to a popular rendition of his system of comparative osteology depicted Owen's bust enshrined in a frame of various animal parts, flanked by Cuvier's head on his right and Oken's on his left (fig. 12).

A major challenge of Owen scholarship has been to unravel in his work the intertwined threads of the two intellectual traditions. In *The Darwinian Revolution*, Michael Ruse concurred with Owen's self-assessment, stating that "Owen produced a synthesis between the ideas of two great French biologists, Georges Cuvier and Étienne Geoffroy Saint-Hilaire."[51] Dov Ospovat, in his

FIGURE 12. Owen flanked by Georges Cuvier and Lorenz Oken. This juxtaposition gave visual expression to Owen's claim that he was forging a synthesis of functionalist with transcendentalist anatomy. In reality, he came to assert the superiority of form over function, and his publications for the most part fell into two groups, one guided by the Cuvierian, the other by the Okenian, epistemology. (Owen, *Principal Forms of the Skeleton*, reprint ed., unnumbered figure on first page of text.)

Development of Darwin's Theory, does not go this far but sees Owen as a judicious eclectic who selected the best from both Cuvier and Geoffroy and who demonstrated that functionalism and transcendentalism are not mutually exclusive.[52] Others have expressed a similar view, in particular Phillip Sloan, who in a study of Owen's first seven Hunterian lectures shows how Continental philosophy of anatomy entered the Royal College of Surgeons and became part of Owen's work.[53] Sloan argues that Owen did not predictably follow the precepts of any one intellectual tradition but creatively interwove both German and French lines of research. Underlying these assessments is the assumption that Owen's position was intellectually cohesive and can be satisfactorily explained within the cognitive context of scientific discourse. By contrast, Adrian Desmond reduces Owen's stance to a sociopolitical purpose, arguing, as we have seen, that the adoption of transcendental morphology was part of a strategy to defend the College of Surgeons against the threat of transmutation-espousing reform-minded radicals.[54]

These interpretations of Owen's work are enlightening with respect to some of its aspects, but none gives the requisite prominence to the enduring epistemological duality of his oeuvre. The reality behind the façade of Owen's self-presentation showed no synthesis but two largely segregated bodies of work, one functionalist, the other cast in a transcendentalist mold. Moreover, whenever the two approaches came together, "form" was used as the primary context of explanation, and "function" merely as an incidental one; and in order not to cast adrift the entire functionalist design argument, Owen was forced to redefine the very meaning of teleology. Before addressing the question of how Owen came to move closer to Geoffroy and away from Cuvier, we will need to summarize his contributions to transcendental morphology. The publication history of these is a characteristic illustration of how Owen developed and publicized his ideas.

By his own account, Owen began working systematically on problems of transcendental morphology in 1841, as part of his curatorial task to arrange the osteological collection of the Hunterian Museum.[55] The osteological catalogue was not published until 1853, but in the intervening years various spin-offs of this basic museum work appeared in print. First, Owen's Hunterian lectures of 1841 and also the series covering the years 1844–48 significantly benefited from his cataloguing labor. The part that dealt with fishes was published as the *Lectures on the Comparative Anatomy and Physiology of the Vertebrate Animals* (1846). Second, and more importantly, the catalogue work served Owen in composing his comprehensive account of transcendental osteology, presented to the BAAS in the form of a major report (1846). It was enormously detailed, densely packed with specifics, loaded with technical terms, and tedious to read. This report, with some additions, was published in book form under the title *On the Archetype and Homologies of the Vertebrate Skeleton* (1848). The following year, 1849, Owen expanded upon some parts of his BAAS report in a lecture at the Royal Institution, published as *On the Nature of Limbs.* It was less overloaded with anatomical detail and nomenclature than his report and more accessible to a wider audience. The most popular rendition of this work appeared in 1854 as part of W. S. Orr's *Circle of the Sciences*; it went through at least a dozen different British and American printings, most of which carried the title *The Principle Forms of the Skeleton and of the Teeth.*

The fullest and best-organized rendition of Owen's philosophy of anatomy did not appear in English but in the language of Cuvier and Geoffroy: a translated revision of his *On the Archetype*, brought out under the auspices of Owen's long-term friend and colleague Henri Milne-Edwards. Entitled

Principes d'ostéologie comparée ou recherches sur l'archétype et les homologies du squelette vertébré (1855), its appearance coincided with the award to Owen of a knighthood in the Legion of Honour and his participation in the Universal Exhibition in Paris. Probably the most lucid and concise summary of Owen's transcendentalist epistemology came out that same year in the form of an appendix to Baden Powell's *Essays* (1855) that Owen cowrote.[56]

One immediate and uncontroversial accomplishment of Owen's systematic BAAS report was a simplification and standardization of the terminology of comparative and human osteology. For this, Owen was much praised, among others by Whewell in the latter's *Novum Organon Renovatum* (1858).[57] The process of tracing in detail the many individual bones of the human skeleton to their equivalents in other mammals and in birds, reptiles, and fishes was a task of Herculean proportions, made especially arduous by the terminological "confusion of tongues." The nomenclature for humans differed from that for animals, and in different languages one and the same bone could have different names, some of which were long, descriptive phrases. In particular, Owen standardized the nomenclature of the bones of the head, composing a table of the names proposed by himself, along with the synonyms used by several of the leading French and German authorities.[58]

Moreover, he gave precise definitions to the often synonymously used terms "homology" and "analogy." He defined a homologue as: "The same organ in different animals under every variety of form and function"; and an analogue as: "A part or organ in one animal which has the same function as another part or organ in a different animal."[59] To illustrate the difference, Owen often cited the example of a little lizard, the "flying" reptile *Draco volans*, in which five pairs of ribs are significantly elongated to support a membrane which can be spread out to allow gliding. The little dragon's front legs are homologous with the wings of a bird, whereas its "parachute" is analogous to wings but homologous with ribs. Conversely, the pectoral fin of the flying fish is not only analogous to the wing of a bird but, unlike the flying dragon's "flaps," also homologous with it.[60]

The lucidity of Owen's definitions can be fully appreciated only if we compare them with an effort made earlier by William MacLeay, and repeated by other contemporary naturalists, to distinguish "analogy" from "affinity." "Affinity" was used to express the taxonomic relatedness of two animals. Thus, the affinity between two species of bird is greater than that between a bird and a bat; [61] but in Owen's definition of "affinity," that is, of "homology," the bones in the wings of birds and bats are identical.

With respect to the relative validity of "form" and "function," Owen habitually illustrated the inadequacy of the teleological method by citing the development of the skull, because it had been the interpretation of the skull with which transcendental anatomy had been most publicly, and in certain circles most ignominiously, associated. The skull of the human fetus, at the time of birth, consists of some twenty-eight separate pieces which ultimately unite into an unyielding whole along a series of rigid sutures. The final cause or functional purpose of the loose, dissembled nature of the fetal skull was believed to be that it facilitates childbirth by making possible a change of shape when the head passes through the vagina. However, this function is present only in placental mammals; in various other vertebrates no supple and adjustable cranium is required by the process of birth, and yet their crania are ossified from the same number of points as occur in the human embryo. For example, the tiny, "prematurely" born kangaroo, even though at the time of its birth it is as minuscule as a thimble, nevertheless exhibits uncoalesced skull bones, which do not serve the function of making parturition safer as in higher mammals. The chick when it breaks through its eggshell also has a composite cranium, not yet coalesced into a single, solid whole but made up of a variety of movable pieces which serve no such function as in the human fetus. Owen concluded: "These and a hundred such facts force upon the contemplative anatomist the inadequacy of the teleological hypothesis."[62] Thus, teleological, final causes were barren, like the proverbial vestal virgins cited by Bacon.

A more comprehensive view had to be taken of the fact that in vertebrates, as a rule, the skull is composed of the same number of pieces arranged in the same general way. The answer—Owen believed—is provided by the vertebrate theory of the skull. It had become a commonplace that the sacrum can be seen as a series of "metamorphosed" vertebrae. Much more controversial was the notion that, at the other end, the skull, too, was no more than a number of metamorphosed vertebrae (fig. 13). Lorenz Oken's theory that the head must be interpreted as a recapitulation of the rest of the body was to a large extent intuitively produced and based on limited and selected examples.[63] By contrast, Owen's theory was based on an extensive comparative study of available vertebrate material, and he presented his theory as the end-product of a process of inductive labor.

Owen's first step was to define the ideal vertebra, that is the vertebra in its simplest form yet showing all the elements from which the real modifications of any skeleton can be derived. Such a generalized vertebra is composed of a center from which a number of "apophyses" (bony excrescences) radiate. On the dorsal side two of these touch ends to form the "neural arch," a channel for the central nervous system; while on the ventral, or costal, side two apophyses combine to form a usually larger arch, the "haemal arch," which protects the cardiovascular system. Owen believed "that I have satisfactorily demonstrated

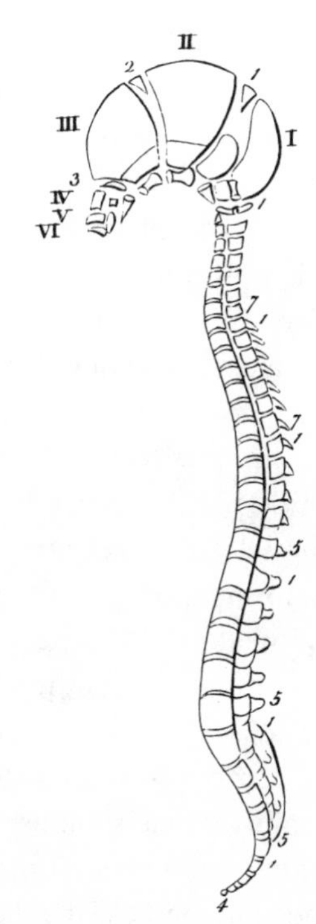

FIGURE 13. Carl Gustav Carus's diagrammatic illustration of the vertebral theory of the skull, in which the skull is interpreted as a composite of metamorphosed vertebrae, similar to the fused vertebrae of the coccyx at the caudal end of the vertebral column. Note his adoption of six cephalic vertebrae, three cranial and three facial ones. (Carus, *Von den Ur-Theilen*, table 5, fig. 15.)

that a vertebra is a natural group of bones, that it may be recognised as a primary division or segment of the endoskeleton, and that the parts of that group are definable and recognizable under all their teleological modifications, their essential relations and characters appearing through every adaptive mask."[64]

Detailed comparative study from fish to human showed that some haemal arches can become disconnected from the vertebrae, especially in the higher classes of vertebrates, and that appendages may develop on certain arches. In this way the bones of the pelvis and posterior extremities were shown to be derived from haemal arches and appendages of the sacral vertebrae. Equally, the shoulder blade, collarbone, and anterior extremities were interpreted as the displaced haemal arch and appendages of the occipital part of the skull, the so-called occipital vertebra. The reason for this was that, whereas the occipital vertebra in humans, for example, appears to be without a costal arch and appendages, in fishes and especially in *Lepidosiren* the homologous part does have an arch which includes the scapula and clavicle; in the series from fish to reptiles to mammals this arch becomes increasingly displaced. A similar process of displacement can be seen by following out the successive stages of embryonal development in the higher vertebrates. Owen concluded that "the human hands and arms are part of the head—diverging appendages of the costal or haemal arch of the occipital segment of the skull."[65]

Like the backbone, the skull is segmented, and four vertebrae with their respective arches can be identified, most clearly in osseous fishes. Owen stated "that in osseous fishes the endoskeletal bones of the head are arranged, like those of the trunk, in segments; that these are four in number, and that they closely conform to the character of the typical vertebra."[66] With considerable effort and in much detail, their exact homologies were traced to humans. Owen dissociated himself from the extreme form in which the cranio-vertebral theory had been formulated by Oken and in particular from Oken's doctrine of cranial recapitulation.[67] In other words, Owen did not subscribe to the view that the head recapitulates the body but believed that the cranial bones can be interpreted as a modified extension of the vertebral column. The crucial Oken-Owen contrast is well shown by the difference in interpretation of the upper extremities. The shoulder blades, to Oken, are repeated in the skull, in particular in the tympanic bones of the middle ear. To Owen, this is fundamentally impossible, because the shoulder blades and the rest of the arms are already part of the head, representing a displaced part of the occipital arch. The scapula and the tympanic bone are merely homotypes, both being modified "pleurapophyses": "The head is, therefore, in no sense a summary or repetition of all the rest of the body: the skull is a province of the whole skeleton, consisting of a series of segments or 'vertebrae' essentially similar to those of which the rest of the skeleton is constituted."[68]

Thus, the vertebrate skeleton could be seen as a series of ideal vertebrae or, rather, "a series of segments succeeding each other in the axis of the body," most closely represented by the relatively simple skeleton of fishes.[69] This fishlike concatenation of virtually undifferentiated ideal vertebrae Owen called the "vertebrate archetype." It represented the vertebrate skeleton in its most elementary form, with only a hint of the modifications that occur in real vertebrates (fig. 14). Also the skeleton of humans, furthest removed from the archetype, can be traced back to the single plan underlying all vertebrate variety, and Owen concluded: "General anatomical science reveals the unity which pervades the diversity, and demonstrates the whole skeleton of man to be the harmonized sum of a series of essentially similar segments, although each segment differs from the other, and all vary from their archetype."[70]

In front of riveted audiences, Owen demonstrated that the human skeleton can be traced from top to bottom, throughout its extremities, and in its most

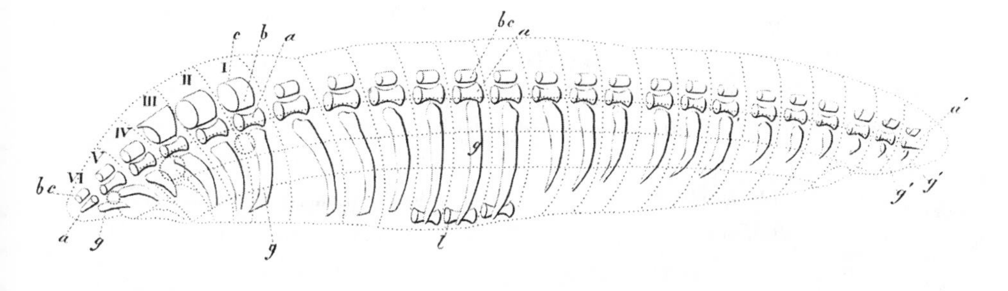

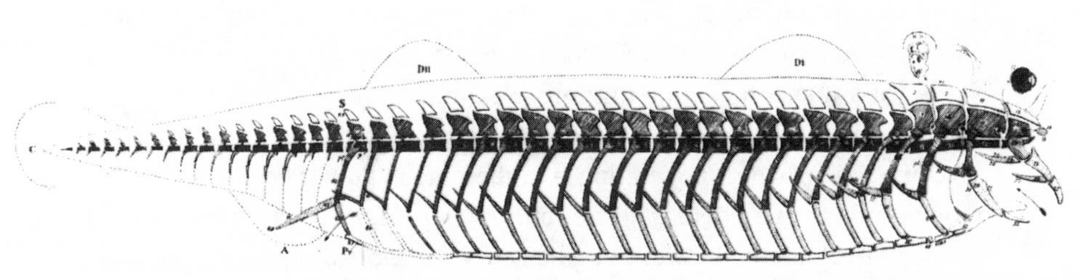

FIGURE 14. A comparison of Carus's vertebrate schema of 1828 (top) with Owen's archetype of 1848 (bottom). Owen may have been more indebted to Carus than he acknowledged. The vertebrate archetype replaced the vertebrate theory of the skull as the shibboleth of transcendental anatomy. (Carus, *Von den Ur-Theilen* table 2, fig. 1; Owen, *On the Archetype*, pl. 2, fig. 1.)

complex and minute components, to the bony frames, not just of other mammals, but also of birds and reptiles, and even to the fish bones of the lowly salmon on our plate. All vertebrates appeared connected, bone for bone, by invisible threads.

In the form of Owen's homological program, transcendental morphology acquired a basis of systematic, rather than incidental, fact. "It is no mere transcendental dream, but true knowledge and legitimate fruit of inductive research," Owen repeatedly boasted.[71] Moreover, Oken's vertebrate theory of the skull was replaced by Owen's vertebrate archetype as the shibboleth of the morphological school. As Hugh Miller stated while tracing the "footprints of the Creator" (1849) throughout geological history, there is a place for Owen's vertebrate examplar, but not for Oken's "vertebrae-developed skull."[72]

The meaning of an organ therefore did not derive from a specific function but from its homological relations, that is, from its place in the makeup of the whole organism. If in this context one wanted to retain the notion of design, it could no longer be defined in terms of the special functions of organic structures but had to be something general at the abstract level of similarities between organisms and of their sequence through geological time. The purpose of an organ, in Owen's transcendental morphology, was to serve us humans as a signpost to direct our minds to the fact of a unity of type and a common ground plan. "For it is certain," Owen stated, "that in the instances where that analogy [of a machine] fails to explain the structure of an organ, such structure does not exist in vain, if its truer comprehension lead rational and responsible beings to a better conception of their own origin and Creator." And William Broderip, to whom Owen dedicated his book *On the Archetype and Homologies of the Vertebrate Skeleton*, concluded: "He [Owen] thus develops a teleology of a higher order than that of Cuvier."[73]

Form rather than function leads us to the conclusion of design in nature. The pectoral fin of the dugong, used for swimming, the forelimb of the mole, employed as a trowel, the wing of a bat, which makes flight possible, or the foreleg of a horse, made for running, all contain the same set of bony pieces. This fact could not be explained by function. Take, for example, human machines, each made for operating in a different medium. They share no common ground plan. "There is no community of plan or structure between the boat and the balloon, between Stephenson's locomotive engine and Brunel's tunneling machinery: a very remote analogy, if any, can be traced between the instruments devised by man to travel in the air and on the sea, through the earth or along its surface."[74] The presence of such a common plan, as in the forelimbs of all mammals, carries our thoughts beyond functional adaptations to a "deep and pregnant principle in philosophy," namely "some archetypal exemplar on which it has pleased the Creator to frame certain of his living creatures."[75]

This line of reasoning seemed validated by the existence of rudimentary or vestigial organs. Darwin later extensively used such atrophied and supposedly functionless structures to argue for evolutionary descent, and with the ascendancy of his theory the two adjectives "rudimentary" and "vestigial" have become synonymous, even though only "vestigial" (*vestigium*, meaning "footprint" or "trace") is correct. After all, "rudimentary" indicates that the structures are imperfect beginnings (*rudimenta*), and this was exactly what several transcendental morphologists, including Agassiz,[76] believed that the apparently nonfunctional parts were. They argued that rudimentary organs are present to complete the architectural design of an organism. To put the distinction somewhat differently: a rudiment, in a purely idealist interpretation, foreshadows what is to come, namely a fully developed and functional organ in a higher organism; whereas in an evolutionary view, the same rudiment is a vestige, a degenerated remnant of a formerly functional part of the body.

Curiously, although Owen did mention rudimentary organs, he made little use of them. Why should this have been so? Did such features not provide splendid illustrations of his belief that form rather than function determined organic architecture? Many examples were known: splint bones in the feet of horses, traces of pelvic and of limb bones in some whales and also in some snakes, and the reduced wing bones in flightless birds. The answer is provided—I should like to suggest–by Owen's preoccupation with the last of these examples. At about the same time that he was working on the homologies of the vertebrate skeleton, he was also studying the flightless birds from New Zealand, both living and extinct. Instances like these made Owen speculate that nonfunctional organs are indications of degeneration and of species change, and as such not reliable as signposts to a divine architect.[77] To him these were not rudimentary but vestigial organs.

Owen distinguished three types of homology: special, general, and serial. An example of special homology is the correspondence of a human arm to a bird's wing. General homology expresses the relationship of the human arm or the bird's wing to its place within the archetype. Serial homology, already described long before Owen by the remarkable Vicq d'Azyr, refers to the repetition of homologous elements within a single skeleton, for example, of successive vertebrae in a vertebral column. For such repetitive elements Owen suggested the term "homotype." Thus, arms are homotypes of legs, and the left extremities are homotypes of the ones on the right, and vice versa. This kind of repetition seemed similar to that of leaves along a branch, and Owen referred to it as "vegetative" or "irrelative." The archetype could therefore be described as an irrelatively repeated series of ideal vertebrae. Such repetitions Owen ascribed to a "polarizing" force similar to the force involved in the process of crystallization: "The repetition of similar segments in a vertebral column, and of similar elements in a vertebral segment, is analogous to the repetition of similar crystals as the result of polarizing force in the growth of an inorganic body."[78] When we descend the scale of animal life, the phenomenon of vegetative repetition becomes more and more pronounced, as it is for example in centipedes, and so does the effect of the polarizing force, which approaches more and more purely geometrical figures—as exhibited, for example, by the external skeleton of the starfish. This demonstrated the concurrence in organic development of an all-pervading polarizing force, interacting with "the adaptive or special organizing force in the development of an animal body."[79]

An Edinburgh Diaspora

Desmond, Jacyna, Ospovat, Rehbock, E. Richards, Sloan, and, more recently, R. Richards have highlighted the importance of transcendentalism in

British biology during the thirty or so years that preceded the appearance of Darwin's *Origin of Species*.[80] Transcendental morphology and physiology helped weave together several disparate strands of research and in addition provided an intellectual stepping-stone for Darwin. Owen may have been the greatest exponent of transcendental morphology in Britain, but he was by no means the only one nor the first. Knox, Grant, Barry, Carpenter, Green, and several lesser figures all sought to reform the biomedical sciences by borrowing the Romantic, or transcendental, program from Continental sources. Significantly, for my argument, these figures were not Oxbridge educated but shared an Edinburgh background.

As Philip Rehbock has shown, during the immediately pre-Darwinian decades of 1830–60 British naturalists brought about a subtle yet far-reaching transformation of the study of life. They looked across the Channel to France and Germany for new ideas and borrowed the idealist approach to comparative anatomy, embryology, and paleontology. The search for generalizations and laws, and the belief that organic form manifests ideal patterns, transformed natural history into a modern science. This transformation—Rehbock argues—was initiated as part of the extracurricular teaching of anatomy at Edinburgh and was brought to London by an Edinburgh diaspora. Robert Knox was "the earliest and most outspoken proponent of idealism in British natural history."[81] He never met Goethe or any of the other German founders of transcendental anatomy, but in 1821 he did visit Paris, where he became acquainted with, among others, Cuvier and Geoffroy. When in 1825 John Barclay retired, Knox took over the teaching of comparative anatomy, and in a series of lectures delivered in 1825–27 he "fully explained" the principles of transcendental anatomy.[82]

Knox's effectiveness was seriously impaired when during the second half of the 1820s he became embroiled in scandal, having bought human cadavers for teaching purposes, not only from body-snatching grave robbers but, unknowingly, also from two notorious murderers who killed to meet his needs. Knox primarily advanced the doctrine of transcendentalism through fervent preaching rather than by original research. In footnotes to his translation of Blainville's lectures on comparative osteology, published in the *Lancet* (1839–40), Knox denounced what he referred to as "the Cuvierian mania and party." He supported "the undoubted fact of the cranium being merely a prolongation, as it were, of the spinal column" and illustrated this with the cranium of a fetal horse.[83] Knox was more a lecturer than a writer, but toward the end of his life he did put his thoughts into writing, for example, in *Great Artists and Great Anatomists* (1852) and in a series of rather-rambling and repetitive papers, again published in the *Lancet* (1855–56). He repeatedly recounted his early conversion to transcendental principles, heaped praise on Goethe, Geoffroy, and even on Cuvier, and scathingly described his fellow countrymen as opportunistic turncoats:

> The pseudoscientific cliques of Britain made, at first, a determined stand against transcendentalism in anatomy and the doctrines of unity of the organization. Their position as educational employees and officials necessitated this. This resistance to science, however, could not continue, and some of the party gradually slid into an unobtrusive low-transcendentalism, in hopes of deprecating the scrutiny of "the powers that be" yet claiming for themselves a wish rather expressed than understood, not to be thought some hundred years behind continental science.[84]

To Adrian Desmond we owe a much improved understanding of the importance of one of Knox's pupils, Robert E. Grant.[85] Like Knox, he visited Paris (1815–20) following his Edinburgh education and gave lectures on comparative anatomy for Barclay (1824). In 1827, he was appointed professor of comparative anatomy at the University of London (the later University College), and in his lectures, published in the *Lancet* (1833–34), he also preached the doctrines of Geoffroy on the unity of type. Notwithstanding the endless diversity of form in the animal kingdom, "we everywhere observe great simplicity and uniformity in the laws which regulate their forms, a perfect unity of plan in their composition and structure, and great regularity and harmony in the order of their development."[86]

A lucid account of von Baer's embryological work was given by another Edinburgh-educated man, Martin Barry, who had studied not only in Paris but also in Heidelberg and elsewhere in Germany.[87] In two *Edinburgh New Philosophical Journal* articles, Barry, himself an accomplished embryologist-to-be, claimed no priority for either function or form—"harmonizing, as they always do"—but did offer some illustrations of the unity of structure in the animal kingdom.[88] He explained the embryological law of von Baer as formulated in the latter's *Entwickelungsgeschichte der Thiere* (1828), "that a heterogeneous or special structure, shall arise only out of one more homogeneous or general; and this by a gradual change." All animals, from infusorians to humans, start at the level of a simple germinal vesicle, and all organic form is a modification of one and the same fundamental form. He emphasized that higher animals do not repeat or pass through stages of the lower animals but follow a diverging, parallel development:

> Strictly speaking, therefore, no animal absolutely *repeats* in its development, the structure of any part of any other animal; and not only is the human embryo at all periods of its existence a human embryo, but the human heart

> and brain, closely as they resemble corresponding organs in other Vertebrata at certain periods of development, are never anything else than the heart and brain of Man.[89]

Rudimentary organs, in this view, are no more than indications of the law "requiring that a fundamental or general type shall uniformly manifest itself before the appearance of one subordinate thereto, and special."[90] He ended by recommending embryology as an aid to the study of comparative anatomy, a view that Owen, during his first course of lectures that year, echoed.

In 1836, William Sharpey joined the diaspora of Edinburgh alumni in London as professor of anatomy and physiology at University College. In his introductory lecture of 1840, published in the *Lancet*, he advocated the unity of type and put forward a typological explanation for such by now standard phenomena as rudimentary organs and embryonal recapitulation. He also subscribed to the vertebral theory of the skull: "strange as it may appear, there is good reason for admitting this opinion, that the head consists of a series of peculiarly-developed vertebrae."[91]

Phillip Sloan has drawn our attention to the transcendentalist lobby inside the Royal College of Surgeons and has pinpointed Joseph Green and his Hunterian lecture course of 1824–28 as the fountainhead from which the early trickle of philosophical anatomy within the college sprang.[92] In 1817 Green befriended the German Romantic poet Johann Ludwig Tieck and also the Germanophile Coleridge. On Tieck's invitation, Green went to Berlin for a summer course in philosophy, and upon his return to London he joined Coleridge's circle and became a close associate. Green's connections with the college were strong and lasting. He was appointed a life member of the council (1835), a trustee of the Hunterian Museum (1841), Hunterian lecturer (twice, 1840, 1847), and president (twice, 1849–50, 1858–59).

Owen attended Green's lectures in 1826, and he admiringly commented: "For the first time in England the comparative Anatomy of the whole Animal Kingdom was described, and illustrated by such a series of enlarged and coloured diagrams as had never before been seen. The vast array of facts was linked by reference to the underlying Unity, as it had been advocated and illustrated by Oken and Carus."[93] Owen, in other words, showed a sympathetic interest in German transcendentalism from the very start of his London career. It is likely that he had become fully familiar with Continental thought even before that, during his short period of study at Edinburgh, through contact with Barclay and his circle.[94] Like the others, Owen established a link with Paris, in 1831, when he not only met the contending parties but was shown around by Grant, who happened to be in France at the same time.[95]

Although Grant and Owen grew apart during the 1830s, Owen was on good terms with several other members of the Edinburgh diaspora. One of these was Barry; the *Life of Professor Owen* refers to the fact that in 1842 the two men met in person.[96] With Edward Forbes, too, whose importance as Knox's main disciple has been stressed by Rehbock,[97] Owen was on friendly terms. During the BAAS meeting of 1844 Forbes presented a paper in which he suggested that there exists a resemblance of zoophytes to plants.[98] Owen "regarded this paper as a beautiful application of the principles of transcendental anatomy."[99] Not long after, Owen used the same national forum to present his own transcendentalist ideas.

Owen's change of emphasis from "function" to "form" was never a change of heart but reflected primarily a change of opportunity. There is no evidence of an intellectual conversion, let alone a sudden one. He knew about and was sympathetic to both positions from the start of his career. The degree to which he pursued one rather than the other approach corresponded—I believe—to the opportunities that each provided for broadening his institutional power base. By the mid-1840s, when Owen began preparing his report for the BAAS on the vertebrate archetype, he was riding a rising tide of metropolitan interest in the transcendental approach. In physics and chemistry, it had all along had a place at the Royal Institution. Now it had also entered the University of London. Even within the College of Surgeons its early representative had risen to a position of power. Moreover, at the annual meetings of the BAAS, philosophical anatomy began to find a voice.

The impact of Owen's homological research program was felt in ever-widening circles. Several presidents of the BAAS lauded his work in their annual addresses. In 1846, the year when Owen delivered his report, the then president Roderick Murchison departed from the custom of reviewing past work and drew attention to the forthcoming presentation by Owen, which showed him catching up with Continental work in the field.[100] The BAAS continued to be a platform for Owen's homological program. The Duke of Argyll proclaimed in his presidential address of 1855 that Owen's work was "leading us up to the very threshold of the deepest mysteries of Nature."[101] When Daubeny took over the chair from Argyll, for 1856, he also interspersed his address with Owenian references to archetypes and homologies.[102]

In his essay "Owen's Position in the History of Anatomical Science" Huxley much later denied that any "of these speculations and determinations . . . were ever widely accepted."[103] This was inaccurate. The appearance of Owen's homological work met with considerable enthusiasm, and "homologizing" was all the rage for a decade or so following Owen's report. In addition to the favorable BAAS publicity, applause was given to Owen's homological work by the professional press. In particular, the *British and Foreign Medico-Chirurgical Review*, which had as a contributor and, from 1847, as its editor the London- and Edinburgh-educated physiologist W. B. Carpenter, pressed the Owenian cause. The terms "archetype" and "homology" came to be the catchphrases of a program of research which defined issues, set goals, and provided a context of explanation that went beyond the by then worn Cuvierian method of function. In this context Owen played a very different role from that of the "British Cuvier." As the grand wizard of homology he had become the "Owen of England," which really meant that his homological work served an indigenous culture of metropolitan science. Not Paleyites but Coleridgeans cheered him on when he created the vertebrate archetype.

In the Shadow of the Archetype

The vertebrate archetype was one of the most fascinating constructs of the "morphological period."[104] It represented the most complete visual expression of a belief in the fundamental relatedness, if not of all organisms, at least of all animals with endoskeletons. More specifically, it formed the centerpiece to Owen's homological research program, embodying his synthesis of biological theory in which organic structures were explained primarily by their morphological relationships and only secondarily by their functions. To many of us today, the concept of an archetype has echoes that range from Plato's theory of ideas to Jung's notion of pervasive cultural symbols in our collective unconscious. During the late nineteenth century, a theory of archetype was introduced into Old Testament philology, too, by the Göttingen theologian Paul Anton de Lagarde, who maintained that all manuscripts of the Hebrew Bible go back to a single, authoritative text from the early part of the second century AD.

In mid-Victorian times, however, the archetype notion had above all a connotation of vertebrate morphology and was closely associated with Owen's book *On the Archetype*. The vertebrate archetype has been Owen's most enduring and most widely acknowledged claim to fame. There is no evidence that Owen used the word "archetype" in the actual delivery of his two-part report to the BAAS, presented on September 17 and 21, 1846, the first part entitled "On the Homologies of the Bones Collectively Called 'Temporal' in Human Anatomy," and the second "On the Vertebrate Structure of the Skull."[105] Used in the sense of an abstract anatomical plan, the term appeared for the first time in Owen's published version of the two-part BAAS report, "On the Archetype and Homologies of the Vertebrate Skeleton" (1847).[106] Until then, for example, in his *Lectures on Comparative Anatomy and Physiology of Vertebrate Animals*, some of which were delivered as late as 1846, he used the expressions "general type" and "fundamental type."[107] The BAAS report of 1847 did not yet contain the actual sketch of the vertebrate archetype. This appeared for the first time in 1848 when the report was published in book form, *On the Archetype*, with additional plates and thirty pages of new text. The philosophical interpretation of the archetype as a Platonic idea was put forward by Owen in his Royal Institution lecture of 1849, *On the Nature of Limbs*, and most explicitly in the *Principes d'ostéologie comparée* of 1855.

Not long after Owen's 1848 book appeared, a priority dispute developed over the osteological application of the term "archetype." Joseph Maclise, a graduate of the University of London and a Geoffroy partisan, pointed out in his entry on the "skeleton" for Todd's *Cyclopaedia of Anatomy and Physiology* that the word had been "first introduced by me in the study of comparative osteology."[108] In an article in the *Lancet*, "On the Nomenclature of Anatomy (Addressed to Professors Owen and Grant)," which appeared on March 14, 1846, Maclise had indeed discussed the possibility of "an original or archetype structure, from which the endo-skeleton designs are struck, and to which they are comparable."[109]

In his response to Maclise, Owen did not counter-claim priority, and this in itself is reason to believe that the introduction in transcendental morphology of the term "archetype," defined in relation to the vertebrate skeleton, bears the date of March 14, 1846, and the name of Joseph Maclise. The *Oxford English Dictionary*, by attributing first usage to Roderick Murchison in 1849, is wrong in several ways. Yet Owen refused to honor Maclise's priority, for various reasons. One was that the term did not constitute a neologism but could be found "in Johnson's and other dictionaries, as the original or pattern of which any resemblance is made."[110] Owen had used "archetype" as early as 1832, although in a different sense, in his *Memoir on the Pearly Nautilus*, where he characterized the nautilus as the living archetype of its extinct relatives, employing the word in the sense of "prototype."[111] In both the British and the Continental literature on anatomy, too, the word "archetype" had been in use, although in a nontechnical and loosely defined, Platonist sense.[112]

Another reason for Owen's refusal to acknowledge the priority of Maclise's use of the term was that the two men defined it differently. In his *Comparative Osteology and the Archetype Skeleton* of 1847, for example, Maclise interpreted the archetype as "unity." By this he meant the sum total of all skeletal modifications from which each particular species could be derived by subtraction of certain complexities. His somewhat convoluted definition ran as follows: "Unity, or the archetype, is a name which may be applied to characterise that whole structure which is capable of undergoing metamorphosis or subtraction through all degrees of quantity severally equal to all those proportional forms which stand in series with itself."[113] Thus, according to Maclise's definition, the more highly developed a skeleton, the closer it is to the archetype. Accordingly, the skeletons of humans and of the higher vertebrates were thought to come nearest to the archetype, and it was on these groups that his *Comparative Osteology* focused.

This was the opposite of Owen's definition, in which the archetype bore a resemblance to the lowest vertebrate class, namely fishes. In fact, Owen's figure of it looked suspiciously like the *Amphioxus lanceolatus*, the "protovertebrate" lancelet, which had just been described by John Goodsir.[114] In his *On the Archetype*, Owen did not offer a precise verbal definition of the morphological "archetype," but in a later dictionary entry he described it as "that ideal original or fundamental pattern on which a natural group of animals or system of organs has been constructed, and to modifications of which the various forms of such animals or organs may be referred." And he added: "The archetypal figure has been most clearly recognised in the study of the modifications of the skeleton of the vertebrate animals."[115]

Elsewhere, and constituting a possible third reason for Owen's repudiation of Maclise's priority, Owen drew attention to the unacknowledged adoption by Maclise of such terms as "serial homology" and "neural arch"; he wryly commented: "It is with pleasure that I see any of the new terms proposed in my 'Lectures on the Vertebrata' (1846) and 'Report on the Archetype and Homologies of the Vertebrate Skeleton' . . . sanctioned by an original author like that of the 'Comparative Osteology.'"[116]

A related reason for Owen's refusal to acknowledge Maclise may well have been Owen's belief that credit belongs to the person who is the first to execute a piece of work properly rather than to someone who has the notion but fails to perfect it.[117] In this case then, priority belonged to whoever gave the archetype its correct and definitive anatomical outline and definition. Maclise had failed to do this. Owen, by contrast, produced the beautiful diagram of the "Ideal pattern or archetype of the vertebrate endoskeleton, as shown in a side view of the series of typical segments or 'vertebrae' of which it is composed."[118] The two crucial stages in the development of this schema were the determination of a typical vertebra and the homological reduction of the entire skeleton to a series of such vertebrae, whether of the sacrum on one end or the skull on the other, or whether in fish, reptile, bird, or mammal. At the anterior and posterior extremities of the archetype figure, Owen indicated the first steps of those modifications that, depending on kind and degree, give the archetype the characters of a class, order, genus, and species (fig. 14). In the same plate he presented figures of the full modifications of the archetype that characterize fish, reptile, bird, mammal, and human being, each the typical skeleton of its respective class: "Here the *Vertebrate Archetype*, so often accepted for the mere verbal and vague indication of a more or less inchoate abstraction—is placed bodily before our eyes in the same 'picture language' as that by which the type-skeletons of the fish, the reptile, the bird, and the beast are distinctly represented in one comprehensive field."[119]

Was the morphological definition and representation of the archetype wholly Owen's work or was the figure, like the term, an improvement upon the less perfect work of a predecessor? The book of 1848 included no mention of any possible source for the archetypal image and suggested that Owen had come upon it in the Baconian course of strict inductive research. This is not to say that Owen failed to refer to relevant literature by predecessors in transcendental morphology. On the contrary, he extensively reviewed the publications by his forerunners, in particular the classics of idealist morphology by Oken, Spix, Bojanus, Geoffroy, and Carus, chronologically listed.[120] In a later autobiographical essay, Owen actually complimented himself on the fact that he "thus generously refers to the services of others."[121] It is therefore the more astonishing that he failed to reveal the source of his archetypal figure, which was Carus's *Von den Ur-Theilen des Knochen- und Schalengerüstes*, published in 1828, a hefty volume, repeatedly referred to by Owen.

In several instances when Owen reviewed the work of predecessors, he did not so much give them credit as use them to present a contrast with his own work, which was, in the words of the Broderip-Owen duo, "new rather than revived, new at least in the best sense as being the result of strict induction."[122] The German transcendentalists, in particular Carus, were described, not as having led the way for Owen, but as having tried and failed in their attempts at transcendental morphology.[123]

Unlike several of his Edinburgh-educated colleagues in London who had come closest to the sources of philosophical anatomy by visiting Paris, Owen, in addition to a Parisian experience, also knew the German scene and extensively tapped its sources of relevant literature. He was especially well acquainted with Oken and Carus, whom he knew personally and with whom he conducted an infrequent but steady correspondence. He met Oken in 1838 during the meeting in Freiburg of the Gesellschaft deutscher Naturforscher und Ärzte. This was a glorious event for Owen, because in spite of his relative juniority, he was treated as an international celebrity.[124] His meeting with Carus took place in London and proved to be equally flattering. In 1844 the king of Saxony visited England and Scotland, accompanied by Carus, the physician-in-ordinary. No sooner had they arrived in the metropolis than Carus hastened to the Hunterian Museum to see Owen. "Owen pleases me greatly," Carus wrote in his travelogue, "a sensible, able man, deeply versed in what is old, and ready for the reception of what is new."[125] Carus seized the opportunity of a further visit to measure the head of the "British Cuvier" and add its measurements to his craniological tables. They met again at a dinner given by Robert Peel in honor of the visiting king, during which Owen addressed the guests and spoke of the high value he placed upon Carus's work.

In the secondary literature, Oken is often singled out as the main representative of German *Naturphilosophie* to have influenced Owen.[126] This pinpointing of Oken is based not on a comparison of Oken's work with that of Owen but largely on the prominence that Owen himself gave to his German colleague, most conspicuously by means of the 1847 translation into English of Oken's *Lehrbuch der Naturphilosophie*, instigated by Owen, and the entry "Oken" in the eighth edition of the *Encyclopaedia Britannica*, written by Owen and published separately in 1860 as the *Life of Lorenz Oken*.[127] A comparison of the oeuvres of the two men shows, however, that Oken's impact on Owen was more general than particular. Oken had a philosophical, highly speculative, and deductive mind; Owen's was none of these. They shared a commitment to transcendental morphology but not to specific research areas. Although Owen particularly praised Oken's inaugural address "Über die Bedeutung der Schädelknochen," he rejected Oken's version of the theory of the skull and adopted instead the more moderate one developed by Goethe and Carus. As Owen later commented: "The brilliant light thrown by Oken on a wider or higher Law of Correspondences, in his 'Discourse on the Cranial Vertebrae' was obscured by his further illustrations of Schelling's transcendental idea of 'the repetition of the whole in every part.'"[128] The vertebrate archetype does not occur in any of Oken's works, while it does appear in those of Goethe and Carus.

Owen's promoting of Oken was self-serving in three ways. Admittedly, Owen was genuinely fond of the tempestuous colleague who had honored him at Freiburg and had translated into German his *Memoir on the Pearly Nautilus*. The fondness shines through in the unusual passion with which Owen defended Oken's priority over Goethe in the "discovery" of the vertebral nature of the skull. To paraphrase Hugh Trevor-Roper's remark about the forger Edmund Backhouse ("The mere fact that Backhouse said something does not necessarily make it untrue"),[129] the mere fact that Owen praised a colleague did not necessarily imply self-glorification. Yet publicizing Oken primarily served Owen's own cause. Oken's *Lehrbuch* could be interpreted as advocating a vague theory of evolution to which Owen was sympathetic. Owen used the English translation of Oken's views like a hat on a stick, testing whether it would draw enemy fire. When Sedgwick and others duly reacted,

Owen decided to keep his head well down. Moreover, by presenting Oken as the prime example of Continental transcendental morphology, Owen's work acquired, by contrast, a look of sound induction that made his claim to Baconian orthodoxy seem all the more solidly founded. At the same time, by placing Oken in the foreground as his main Continental counterpart and predecessor, Owen laid a false trail—whether intentionally or not—away from Goethe and Carus, to whom he was indebted for the idea of an archetypal schema.

Goethe and Carus were far less speculative than Oken and they were the ones who had developed the notion and image of an archetype, although the concept of an organic prototype or an *Urbild* can be traced back to Denis Diderot, J. B. R. Robinet, and, in Goethe's immediate vicinity, to Johann Gottfried Herder.[130] Goethe's famous *Urpflanze*, delineated in his 1790 treatise on the metamorphosis of plants, is to some extent the botanical counterpart of the vertebrate archetype. A significant difference existed, however, in that the archetypal plant was not only composed of its most fundamental element, that is, the leaf (comparable to the vertebra as the elemental building block), but exhibited all its basic modifications such as a stem and all the essential parts of a flower. Goethe did not actually sketch the *Urpflanze* or an equivalent diagram for the animal kingdom, but in 1790–91 he did point to the desirability and possibility of such a visual representation for animals.[131]

It fell to Carus to act upon the programmatic essays by Goethe, about whom he wrote several books. Carus was an interesting figure who made major contributions to biology, psychology, philosophy, painting, and art history. He studied medicine and inaugurated the teaching of comparative anatomy as an independent subject, receiving international recognition for his discovery of the circulation of the blood in insects. In his capacity as a physician, Carus made a name as a gynecologist. He was cofounder of the Gesellschaft deutscher Naturforscher und Ärzte. A member of the Dresden school of Romantic painters, he was particularly close to Caspar David Friedrich, to whose paintings his own bear a resemblance. As an art critic, Carus was most of all interested in landscape painting. In 1829–30 he gave a series of lectures on psychology in which, among other things, he developed the notion of the "unconscious." This led to his most lasting reputation as a precursor of psychoanalysis. In later life, he became increasingly preoccupied with metaphysical issues. Carus's philosophy, which he called "entheism," was essentially Aristotelian in that he interpreted the world as an unfolding unity or developing multiplicity called God. The unknown divine is revealed in nature through organization and organic unity. Real understanding is a function of the universal unconscious, the unknown divine, becoming conscious in us. In the twentieth century, Carus has proved of interest to the anthroposophists.[132]

FIGURE 15. Carl Gustav Carus in 1844, the year he visited England. The book is his 1828 study of transcendental anatomy, *Von den Ur-Theilen des Knochen- und Schalengerüstes.* Portrait in oils by Julius Hübner, 1844 (courtesy Frankfurter Goethe-Museum, Freies Deutsches Hochstift, Frankfurt am Main; used by permission of Frankfurter Goethe-Museum).

In all this, what was probably his proudest accomplishment has drifted into the background, namely his massive work on comparative osteology, the *Von den Ur-Theilen des Knochen- und Schalengerüstes* (1828), on which his arm rests in the well-known painting of 1844, the year he visited Britain (fig. 15). Carus also wrote a more elementary, three-volume textbook on comparative anatomy and physiology.[133] In 1822 he had met Goethe, and they discovered in each other kindred spirits who agreed on the vertebral theory of the skull and even on the disputed number of cephalic vertebrae. This reinforced Carus's determination to carry out Goethe's osteological program; and he fully acknowledged Goethe in the preface to his tome. Carus attributed the concept of a skeletal metamorphosis—the notion that the many differently shaped endoskeletons are modifications of a single fundamental type—to Goethe, and he also praised Oken for having enunciated the more specific belief that a skeleton is in essence a series of repeated vertebrae.[134] Carus systematically worked out this theory for the skeletal features of both vertebrates and invertebrates, following a strict mathematico-deductive route. He pictured the primitive vertebra as a ring and interpreted this as the basic building block of which, through various numerical repetitions, vertebrates are constructed.

Although Owen was not a follower of Carus in this methodological, mathematico-deductive approach, the beautiful and detailed plates of *Von den Ur-Theilen* must be seen as preparatory to the plates in Owen's book on the vertebrate archetype. The typical skeletal arrangement of each different class of vertebrates was lucidly depicted, the simplest archetypal arrangement indicated (fig. 14), and the segmentation of the skull in birds, mammals, and man delineated with precision.[135] There is nothing in the anatomical literature before Owen that comes closer to the famous plate 2 of *On the Archetype* than Carus's plates, and the conclusion may be drawn that Owen's archetype represented the culmination of less complete attempts by Continental predecessors. The line of influence went from Goethe to Carus to Owen, despite the publicity given to Oken.

The similarity between Owen's and Carus's archetype was too obvious to escape notice. Across the Atlantic, Silliman commented on the resemblance between the two schemata.[136] Owen did give slightly more prominence to the importance of Goethe and Carus in the French translation of *On the Archetype* than he had in the English original, but grudgingly, presenting his own work as necessitated by the failure of theirs; and he continued his silence on Carus's archetype figure. Not until the third volume of *On the Anatomy of Vertebrates*, published in 1868, did Owen finally compare his own vertebrate archetype to Goethe's *Urpflanze*, but by this time the comparison was of little more than historical interest.[137]

Not a Platonic Idea

Even though Owen laid claim to superior inductivism, his philosophical interpretation of the archetype was less down-to-earth than that of Goethe and Carus. As G. H. Lewes commented in his *Life of Goethe*, the German's notion of type must not be confounded with a Platonic idea: "It was no metaphysical entity, it was simply a scientific artifice."[138] Similarly, Carus went no further than to describe his vertebrate archetype as the "simplest schema."[139] Early on, to Owen, too, it was no more than a "generalization"–what we might prefer to call a "model"–of vertebrate variability. By the late 1840s, however, he had turned the archetype into a metaphysical, preexistent entity. Various authors have stated that Owen construed the archetype as a Platonic idea.[140] The full story is more interesting than that. Owen initially defined his vertebrate archetype in terms consistent with the speculations of a form of pantheistic *Naturphilosophie* and in contradistinction to a Platonic idea. Yet in due course he came to accept and to promulgate a Platonist connotation, even though his archetype differed significantly from a Platonic idea *sensu stricto*.

A brief reminder of what was understood by a Platonic "idea" may help clarify the contrast with Owen's conception of an archetype. In Plato's view, the notions that we have of values ("good," "truth," and "beauty"), of mathematics (triangle, etc.), and of natural phenomena (fire, etc.) are not products of our mind but stem from ideas of these concepts that have an independent, metaphysical reality. Above the material world, which we know through our senses, there exists an immaterial world of noncorporeal ideas that are perfect, immutable, and everlasting. The idea is the "*Urbild*" of which, in the sublunary reality, the "*Abbild*" exists. What we perceive through the senses may resemble the idea but never fully expresses this. A beautiful object never entirely embodies the idea of beauty; the sketch of a triangle approximates the pure triangle but always has imperfections; no particular fire has the set of properties of all possible fires.

We therefore do not arrive at a knowledge of ideas by the observation of physical reality but by thought processes that tap the memory of our souls, which existed before inhabiting our bodies. In its preexistence in the immaterial world, the soul has seen the ideas of all things on earth. The senses are therefore not the source of our knowledge but merely the inducement to intuitive perception. In the *Republic*, Plato's famous image of the cave helped to clarify this point. People incarcerated in this cave and chained to a wall are able to observe only the wall in front of them. Above and behind them light enters the cave and casts shadows of objects from the outside world on the opposite wall. To the captives inside, the only reality is the shadows in front of them. Similarly, our eyes perceive merely the shadows of the reality of the ideal world. Plato's theory of ideas was modified by the later Neoplatonists and by Christian theists such as Augustine who interpreted Plato's perfect originals or ideas as divine thoughts by means of which the world had been created.[141]

Was Owen's archetype a Platonic idea, the shadows of which we perceive in the skeletal types of fishes, reptiles, birds, and mammals? The answer is a definite "no." Whereas a Platonic idea is the highest, most perfect reality, the vertebrate archetype represented the opposite, namely the simplest and least perfected conception of a vertebrate. In one sense, Owen's archetype was all potentiality and as such his position more Aristotelian than Platonist, close to the "entheism" of Carus. Add to this Owen's belief in "innate tendencies" as the driving force of organic evolution, and the Aristotelian nature of his stance will appear clearer yet.[142] Thus, real vertebrates are produced from the archetype, not by subtraction, but by addition. This was the very point on which Owen differed from Maclise, whose concept of an archetype, defined as the totality of all skeletal modifications, most closely approximated by the human skeleton, was indeed Platonist. In the *Timaeus*, Plato describes the organic world as having been formed by the degradation of man.[143] Owen's archetype was the opposite extreme, showing the greatest degree of vegetative repetition, that is, the simplest, invertebrate-like, plantlike, and even crystalline development. "Man, whose organization is regarded as the highest, departs most from the vertebrate archetype," Owen stated.[144] It is a rather startling fact that so many historians who have read, cited, or even quoted the relevant passages in Owen's work have concluded that Owen presented the vertebrate archetype as a Platonic idea. Ruse, as one of a few exceptions, noticed that this was not quite so; he believes that Owen became confused and did not know precisely what he was talking about.[145]

Owen, however, sharply distinguished "Platonic idea" from "archetype" in his BAAS report of 1847 and in its book form of 1848. An organic body develops by the interaction of two antithetical forces, he stated, one "the adaptive or special organizing force," the other the "general and all-pervading polarizing force."[146] The first he interpreted as a Platonic idea, as a force or model that produces specific modification of organization and adaptation. By contrast, the archetype was a reflection of the second force, which produces repetition of parts and thus unity of organization. The Platonic idea, by counteracting the force represented by the archetype, produces species and is the cause of organic diversity. Having attributed the repetitive simplicity of the archetype to a "general polarizing force," Owen continued: "The platonic idea or specific organizing principle or force would seem to be in antagonism with the general polarizing force, and to subdue and mould it in subserviency to the exigencies of the resulting specific forms."[147] The force that produces the sequential arrangement of ideal vertebrae of the archetype is the same force that causes crystallization. In essence, the vertebrate archetype was something similar to a model of crystal symmetry.

In 1847 and 1848, Owen explicitly stated that the archetype force acted "in antagonism" with the Platonic vital force. Yet not long after he had thus placed his archetype and Plato's ideas at opposite ends of the spectrum of influences that shape an organism, Owen made, without so much as a hint of an explanation, an about-face, equating his archetype with an idea "in the Platonic cosmogony." He did this in his Royal Institution lecture "On the Nature of Limbs," delivered in early February 1849, adding that the archetype represented a "predetermined pattern."[148] He based his belief that the archetype represented a preexistent pattern on the fact that its possible modifications exceed the actual ones. Different and higher ones may exist on other planets or arise in future times. Equally, in the distant geological past only the simplest modifications of the vertebrate archetype existed, yet the later, more complex possibilities were foreshadowed and thus were cognitively preexistent.

In his *Ostéologie comparée* Owen expanded upon this notion of preexistence by arguing that our recognition of an archetype as the basis for the organization of all vertebrate animals proves that a knowledge of the human species, as one of many possible modifications of the archetype, must have existed prior to our appearance in the flesh: "all the parts and organs of man had been sketched out in anticipation, so to speak, in the inferior animals; and the recognition of an ideal exemplar in the vertebrated animals proves, that the knowledge of such a being as man must have existed before man appeared."[149] That this did not match Plato's notion or that of the Christian Platonists was left unaddressed. Rather than lose himself in a maze of philosophical niceties, Owen simply asserted that his archetype helped to clarify what a Platonic idea really was: "J'ose aussi me flatter que ces essais jeteront quelque jour sur la philosophic platonique, et faciliteront l'intelligence de l'idée dont le divin Philosophe traça la première esquisse."[150]

Thus, the archetype no longer reflected a material force but became a divine forethought, a blueprint of design for the formation of animal life. Owen adopted the archetype figure as a crest, elucidating a wax impression of this crest as follows in a letter to his sister Maria:

> It represents the Archetype, or primal pattern—what Plato would have called the "Divine Idea"–on which the osseous frame of all vertebrate animals, i.e., all animals that have bones, has been constructed. The motto is 'The *One* in the *Manifold*,' expressive of the unity of plan which may be traced through all the modifications of that 'pattern,' by which it is adapted to the varied habits and modes of life of Fishes, Reptiles, Birds, Beasts and Human Kind. Many have been the attempts to discover the Vertebrate Archetype, and it seems now generally felt that it has been found.[151]

Yet—to reiterate this critical point—Owen's vertebrate archetype was not a Platonic idea. First, it represented the simplest, rather than the most complete, conception of a vertebrate, and although the conception of a simple blueprint could be interpreted as a Neoplatonic idea in the divine mind, it did not, strictly speaking, match a Platonic notion, in particular not as understood by

Owen's contemporaries; Henry Wentworth Acland, for example, commented that Owen's archetype was not akin to a Platonic idea.[152] Second, it was initially defined as the reflection of an immanent force, in contradistinction to a transcendent entity. The question therefore arises: "Why this parading of the archetype as a Platonic idea?"

The traditional view is that Platonism and Kantian philosophy were formative intellectual influences on Owen.[153] I strongly doubt this; the extent to which Owen intellectually came to accept a Platonized archetype is difficult to establish, but it is certain that the Platonization took place retrospectively and was imposed on Owen by his social circle and forced on him by the constraints of patronage. It is erroneous to believe that Owen was directly influenced by any philosopher at all, whether it be Plato, Spinoza, Kant, Fichte, Schelling, Hegel, or Schopenhauer, even though Schopenhauer cited him scatteredly in *Die Welt als Wille und Vorstellung* (third and later editions) and elsewhere.[154] It is true that in his *On the Archetype* Owen cited, as instances of the use of the word "archetype" prior to Maclise, two philosophical treatises, namely the *Scepsis Scientifica* (1665) by Joseph Glanvill, an Oxford sympathizer with the Cambridge Platonists, and the *Logick* (1725) by Isaac Watts. It is very unlikely, however, that Owen had read even a single page from these treatises; he misspelled the names of both authors and gave no page numbers for his citations; more damningly, these are the two sources quoted in Samuel Johnson's *Dictionary of the English Language* to illustrate the meaning of the word "archetype." Most likely, Owen's familiarity with the two philosophers went no further than Johnson's quotations.[155] Owen simply did not have a philosophical turn of mind. To try and make a coherent system of philosophy out of Owen's various theories would be an unjustified and futile undertaking.

Of course, one could argue that Owen was indirectly influenced by "philosophy" via the work of his more philosophically literate colleagues. The indirect impact of Schelling was apparent in the transcendentalist approach that he adopted from Carus and others. Specifically with respect to a Platonist interpretation of the vertebrate archetype, one might point to two possible scientific sources which Sloan has identified as significant influences on Owen: in Germany Johannes Müller and, at Owen's own College of Surgeons, Joseph Green.[156] Müller equated archetypes with "the *eternal ideas* of Plato" as put forward in the *Timaeus*.[157] Green, too, used "archetypes" as synonymous with Plato's "ideas."[158] Neither man, it should be added, gave a well-defined morphological definition to the archetype concept.

Although Owen may have borrowed from these colleagues when, in 1849, he Platonized his vertebrate archetype, the reason for his about-face is not likely to have been his undoubted familiarity with their work, which, after all, had been in print for some time when, in 1847, he first defined his archetype very differently from the way his colleagues had done. Müller had written on the subject in both the first and second volumes of his *Handbuch der Physiologie* (1833–40), the English translation of which appeared during 1838–42 under the title *Elements of Physiology*. Two years earlier, Green had attempted "to rescue the speculations of Plato" in his Hunterian Oration *Vital Dynamics* (1840).[159] To Müller, the archetype is a creative, vital force which maintains a living being's organization and determines the species to which it belongs. Green spoke of the archetype as an artist's mental ideal, embodying the highest, most perfect development of, for example, the female form. Thus, both used the term "archetype" for what Owen referred to as the Platonic "adaptive or special organizing force," which produces diversity, while Owen's archetype was a reflection of the diametrically opposed "polarizing force," which engenders unity of organization.

The reason Owen made an about-turn and Platonized his archetype is not hard to find. It was—I believe—to get himself out of the quagmire of pantheism into which the promotion of his chosen version of transcendental morphology was slowly letting him sink. The danger of being tainted by pantheism was very real, given the hold this philosophy had taken among the Germans from whom Owen and his Germanizing circle borrowed ideas and inspiration. One of the many contemporary authors who recognized pantheism as a quintessential part of German Romanticism was Heinrich Heine, who in his marvelously readable *Zur Geschichte der Religion und Philosophie in Deutschland* (1834) argued extensively that Goethe was influenced by Spinoza, that the *Naturphilosophen* were pantheists, and that the Romantics in general acted from a pantheistic instinct.[160]

By promoting German transcendentalism in London, especially a version significantly indebted to the Schelling-influenced Jena school, a door seemed opened for pantheism to enter England. Owen's initial definition of his archetype must have added fuel to this fear. By defining the archetype as a reflection of the all-pervading polarizing force, he implicitly attributed the origin of species, constructed upon the archetypal plan, to natural causes and made it part of the theory of the pantheistic, self-developing energy of nature of which Oken spoke in his *Lehrbuch*, which had just been translated into English at Owen's behest. Carus's "entheism" was no safer, merely being a different name for a particular variety of pantheism. Such theories appeared to amount to a form of pantheist organic evolution, in essence the same as the materialist organic evolution of *Vestiges*.

The popular press raised the alarm, accusing Owen of introducing a pernicious pantheism from across the Channel.[161] More undermining of Owen's position, his Cambridge friends Sedgwick and Whewell put up their hackles. It was worrying enough that Owen had bolted the Cuvierian stable of functionalist epistemology. To see him now run away with the transcendentalist bit in his mouth caused real consternation. In the material added to the fifth edition of his *Discourse on the Studies of the University*, Sedgwick expanded his attack on the materialism of *Vestiges* to include Oken's idealism. Owen's friend Oken was made as much the butt of Sedgwick's fierce condemnation as the Vestigiarian, and just as Nieuwentyt had opposed Spinoza with an appeal to empiricism, so, too, Sedgwick countered German idealism with an appeal to healthy Baconian principles. That this represented a warning to Owen was left in no doubt when Sedgwick strongly condemned the translation of Oken's *Naturphilosophie* and filled no fewer than twenty-four pages with extracts to demonstrate "the depths of mysticism, pantheistic profanity, and arrant nonsense, into which a very clever, inventive, and well-informed physiologist may sink, when he deserts the track so nobly delineated by Bacon, and so gloriously trodden by men like Galileo, Newton, La Place, and Cuvier."[162] Oken's pantheistic evolution enraged Sedgwick, who scolded his German colleague for being "the great modern high-priest of the theory of development" and "the great hierophant of transmutation."[163] "May God for ever save the University of Cambridge from this base, degrading, demoralizing creed!"[164]

Whewell's prose was free of the invective that weighs down the language of his Trinity College colleague, and Whewell more gently steered Owen away from the direction he had taken. In a paper before the Cambridge Philosophical Society, on November 13, 1848, some three months before Owen began the public Platonization of his archetype—a lecture warmly referred to by Sedgwick[165]—Whewell warned of the excesses of idealist philosophy.[166] He followed up with a series of lectures in which he favorably discussed various aspects of Plato's philosophy, in particular the theory of ideas.[167] This represented a continuation of his early interest in Platonism, apparent ever since the *History of the Inductive Sciences*, in which he presented the tenet of Platonism ("The One in the Many") as the opposite of that of pantheism ("All Things Are One"). In fact, Owen's motto, "The One in the Manifold," may well have been taken from Whewell's *History of the Inductive Sciences*.[168]

Whewell's interest, it should be added, was not an isolated instance of philo-Platonism but part of a larger revival of Platonist philosophy. Its wide popularity was indicated by encyclopedia articles, which ranged from the major entry in the eighth edition of the *Encyclopaedia Britannica* to another relatively large one, in the *Dictionary of Science, Literature, and Art*. Various editions of Plato's works were produced, including Whewell's *Platonic Dialogues for English Readers* (1860) and the magisterial *Plato and the Other*

Companions of Socrates (1865), edited by George Grote. The importance of Platonism as an ingredient of the Romantic movement has been recognized for some time.[169] More specifically, Richard Yeo has argued that Whewell's Platonism was an attempt to affirm the assumptions of natural theology. It is indeed true that Owen's archetype, reinterpreted as a Platonic idea and a blueprint of creation in the divine mind, could be integrated into the design argument.[170]

Owen actually received explicit and specific advice as to how he should reinterpret the archetype from a mere product of the polarizing force to a Platonic idea in the sense of a divine forethought, analogous to the plan in the mind of an instrument maker. This advice came from the old Christ Church *confrère* of Buckland, Conybeare, who suggested a way to Christianize and Platonize the vertebrate archetype. Given the precise match of his suggestion with Owen's reinterpretation of the archetype, Conybeare's advice may have been a crucial factor in the Platonization process. He wrote to Owen: "The true sense in which I believe Plato to have used such terms as 'ideas' . . . is one which entirely approves itself to my mind—he meant the archetype forms of things, as they existed in the creative mind, and by their participation in which the innumerable individual existences of actual nature possessed a true specific unity."[171]

Other friends of Owen contributed as well to this Platonist Christianization of the vertebrate archetype. They lavished praise on "the greatest of living comparative anatomists" for having discovered in the archetype "the true Ariadne thread by which he is guided in the midst of what, without this thought, must have been but a tangled maze."[172] Yet they cheerfully ignored, whenever convenient, the Owenian definition of the archetype as a reductionist construct and, like Maclise, used its properly Platonist definition as the fullest, highest, and most complex of types. Parallels were drawn with biblical history: Adam was interpreted as the archetype of mankind, man as the archetype of the natural kingdom, and Christ as that of the spiritual kingdom. Oxford's Charles Daubeny gave his imprimatur to this when as president of the BAAS he approvingly cited *Typical Forms and Special Ends in Creation* (1857), by James McCosh and George Dickie, who attempted to rescue teleology by assigning "purpose" to the archetypal plan of creation.[173] One of Owen's correspondents, Thomas Worsley, the master of Downing and at the time Cambridge University's Christian Advocate, noticing the similarity of the archetype to a fish, extended the comparison to the fish symbol of Christianity and implicitly to Christ: "The best way of overthrowing the Dragonworshippers with their idols is to set before them the true ichthus," he wrote in a tone of Christian camaraderie. "Do you comprehend this?" Owen inquired of Acland, adding dryly: "I suppose the evidence of an Archetype must be congenial to the mind of my accomplished friend, as illustrating the One in the Manifold."[174]

It was to hide its pantheist birthmarks—I believe—that Owen covered his limpid archetype with a coat of Platonist fabric. This explanation differs from the one offered by Desmond, who interprets Owen's archetype as a Platonist construct produced in reaction to the materialist threat of transmutation. In Desmond's view, Owen represented the establishment, and the reason he put forward the notion of an archetype was that it legitimized the social pretensions of the upper classes, whose members were gratified to trace their pedigree to a pure, Providential, and ideal source. By contrast, those who did not believe in Platonist archetypes but in evolutionary ancestors were the metropolitan radicals whose plebeian backgrounds predisposed them to champion a descent from a common, down-to-earth origin. In his *Archetypes and Ancestors*, Desmond has developed this thesis in a study of the contrast between Owen and "Darwin's bulldog" Huxley; in his *Politics of Evolution*, he has gone further by placing Owen in opposition to Grant, known for his advocacy of the transmutation theory of Geoffroy Saint-Hilaire.[175]

Desmond's thesis is predicated, however, on the assumption that Owen's vertebrate archetype was conceived of as a Platonist idea. Given the fact that, in origin, the archetype was not a Platonist but an Aristotelian, vaguely pantheistic construct, Desmond's interpretation, however engaging, is inaccurate. Conybeare's letter and the other Oxbridge reactions of alarm, especially that of Sedgwick, who lumped materialism and pantheism together as a single dastardly attack on Christian belief, show moreover that the Platonization of the archetype was an effort to safeguard natural theology and to defuse the threat that Owen and the Goethe-inspired Germanizers of the metropolis represented to Oxbridge, not whatever threat Grant may or may not have been to Owen. This is not to deny that, as Desmond believes, Geoffroy's transmutationist materialism was a threat, nor does it invalidate Desmond's general thesis that a Platonized archetype could function as an antimaterialist, antiradical concept; but the Platonist archetype was not formulated in response to the Grantian radicals. Rather, the Platonization was a response to the nonfunctionalist and vaguely evolutionary transcendentalism of the Schelling-inspired *Naturphilosophen* and was intended to reintegrate morphology into a traditional, teleological epistemology.

Owen went along with the Platonization of his archetype because it placed him in a religiously safe corner and gave social legitimacy to his transcendental morphology. To the French translation of his book *On the Archetype and Homologies of the Vertebrate Skeleton* he added two title-page quotations:

one, from Bacon, may be seen as an affirmation of his empiricist orthodoxy in answer to Sedgwick; the other, from Plato's *Timaeus*, as an acquiescence to Whewell's Platonization effort. Thus, to the extent that the vertebrate archetype seemed a Platonic idea, it served to placate the powerful Oxford and Cambridge faction among Owen's supporters.

In summary, Owen's most famous accomplishment, the vertebrate archetype, was an incongruous construct composed of borrowed parts. The term came from Maclise, the morphological schema from Carus, and the Platonist interpretation from Owen's Oxbridge patrons. Exclusively Owen's was all the hard work.

From Archetype to Ancestor

The Platonist interpretation of the vertebrate archetype may have protected Owen on his right flank, yet the very same interpretation opened up a serious gap in his defenses on the left. Specifically, the Platonist coloring of the vertebrate blueprint, intended to invalidate accusations of pantheism, presented an irresistible target to the followers of Comtean positivism. In his major, six-volume *Cours de philosophie positive*, issued during the first dozen years of Owen's career as a publishing naturalist (1830–42), Comte argued that the development of science passes through three successive stages, from the theological, through the metaphysical, to the positive. The interpretation of the vertebrate archetype as a metaphysical entity placed Owen's morphological work in the "backward" metaphysical stage, soon to be left behind in the march of science toward the adoption of natural laws and an abandonment of the search for first or final causes.

When Comtean positivism began to take hold in Britain, the homological bandwagon ran into a ditch and Owen's Carlylean hero status became tarnished. Among the people who turned against the Owenian notion of a vertebrate archetype were Huxley, Herbert Spencer, and Lewes, the three musketeers of positivism, who attacked and seriously damaged Owen's archetype construct. The person initiating the attack was Huxley, who—although no Comtean in every sense of the word—through the 1850s had already chosen a curious mixture of imitation of Owen and systematic disagreement with him as a method of self-advancement. By the late-1850s Huxley was actively demythologizing the "English Cuvier" and opposing Owen's museum plans and was about to start his major campaign to discredit both Owen's scientific preeminence and his integrity. Whether fair and accurate or not, Huxley was clever and effective in his role as a controversialist. In 1858 he was invited to present the Croonian lecture, and on June 17 of that year he delivered a speech before the Royal Society entitled "On the Theory of the Vertebrate Skull." He thus chose the best-known and most controversial feature of transcendental morphology and undercut it from the side which had been neglected by Owen, namely the embryological.

In his first course of Hunterian lectures of 1837, Owen had criticized the a priori approach to the interpretation of the skull bones, adopted by Oken and Geoffroy, because it lacked the empirical evidence of embryology; Geoffroy's hypothesis that the vertebrate skull is composed of seven vertebrae "is wholly unsupported by an examination into the primary formation of the cranial bones, to determine how many are actually developed from the circumference of a gelatinous *Chorda dorsalis*; the only true embryonal condition of a vertebra."[176] In spite of this programmatic critique, Owen himself relied little on the criterion of embryological differentiation but based his conclusions on a comparative study of the osteological collection in the Hunterian Museum, which contained for the most part postnatal specimens. In this way the museum, while providing him with his peculiar strength, also imposed limitations.

Huxley perceptively focused on these by arguing for the primacy of the developmental method. Cannibalizing various German sources, he denied the validity of the vertebral theory of the skull on the grounds that the skull had no relation with vertebrae developmentally. If the skull were made up of modified vertebrae, its vertebrate character should be clearest in its earliest, least modified stages of development. Yet these embryonal stages showed no resemblance to vertebrae at all. As soon as the backbone begins to look like one, the part that is to become the skull starts to resemble something very different. In particular, in the early membranous and cartilaginous stages of development, the segmentation of the vertebrate body cannot be traced into the cranial region. When completely ossified, this region does indeed consist of four definite segments, as Owen had maintained, but Huxley contended that there was no ground for interpreting these segments as homologous to vertebrae. As Huxley later recalled, the developmental method removed the venue of the interpretation of skeletal morphology "from the court of comparative anatomy to that of embryology," that is, out of the court of Owen's expertise.[177]

Huxley made no secret of the fact that by undercutting Owen's skull theory he was opening a door to let the winds of positivist change blow away the cobwebs of what he called "an obsolete scholastic realism."[178] The transcendental conception of skull morphology was based on the same misconception—Huxley maintained—as the recapitulation theory of the *Naturphilosophen* before von Baer; just as a mammal in its embryonal development never is a fish but simply shares with the fish a common point of origin, so "the spinal column and the skull start from the same primitive condition . . . whence they immediately begin to diverge."[179] Huxley did not deny the unity of plan of all vertebrate skulls, nor did he principally object to putting into a "diagrammatic form" such a plan or even calling it an "archetype"; "but I prefer to avoid a word whose connotation is so fundamentally opposed to the spirit of modern science."[180] Elsewhere, Huxley explained that the "common plans" must be regarded simply as devices to express the fact that the various parts of a living body are bound together according to definite laws.[181]

No sooner had Huxley made his incisive positivist attack on Owen than Spencer followed suit. Spencer had attended Owen's classic course on comparative osteology given in 1851, when many a foreign visitor to the International Exhibition had also come to sit at the Hunterian professor's feet; but, as Spencer recalled in his *Autobiography*, "my scepticism respecting his theory of the archetypal skeleton and archetypal vertebra grew gradually stronger, until at the close of the course it ended in complete disbelief." He failed to muster the courage to express this publicly, but very shortly after Huxley had given his Croonian lecture, Spencer arranged with Carpenter, then editor of the *Medico-Chirurgical Review*, which before had thrice given excessive praise to Owen's theory, "to write a criticism on the several works in which Prof. Owen had embodied it."[182] There is reason to doubt the accuracy of Spencer's autobiographical recollections; in 1857, in an anonymous article for the *National Review* entitled "The Ultimate Laws of Physiology," Spencer showed no skepticism whatsoever about the vertebral theory of the skull. He approvingly used it to illustrate the "longitudinal integration" of organisms: "The coalescence of four vertebrae to form the skull is one instance of it. It is further illustrated in the *os coccygis*, which results from the fusion of a number of caudal vertebrae."[183] The following year, however, in the wake of Huxley's attack, Spencer did come out of the closet to present "A Criticism on Prof. Owen's Theory of the Vertebrate Skeleton."

Spencer's critique was rather amateurish and far less skilled than Huxley's, but it was equally anti-Platonist and antitheist. He denied the vertebral composition of the skull and also the fundamental notion that an ideal or generalized vertebra can be deduced from the actual ones, using as his main, rather weak line of argument that real vertebrae are so variable that they cannot be subsumed under a single model; furthermore, the disagreement among the experts as to the number of cranial vertebrae proved that the very notion was nonsense: "Does not the fact that different comparative anatomists have arranged the same group of bones into *one*, *three*, *four*, *six*, and *seven* vertebral segments, show that the mode of determination is arbitrary, and the conclusions arrived at fanciful?"[184] Spencer implicitly rejected the very concept of serial homology, believing that the segmented nature of the spinal column was not primary and genetic but functionally determined and acquired through inheritance of individual adaptations. Whereas the notochord in the primitive chordate *Amphioxus* is continuous, in higher, more developed vertebrates the central axis had to be firmer, and as a result a vertebral column had developed. To preserve the necessary flexibility, however, it needed segmentation: "Hence, increasing density of the central axis necessarily went hand in hand with its segmentation: for strength, ossification was required; for flexibility, division into parts."[185] Thus, the segmentation of the endoskeleton was produced mechanically by interaction of organisms with the environment in a Lamarckian process of "*the superposition of adaptations upon adaptations.*"[186] "The average community of form which vertebrae display, is explicable as necessarily resulting from natural causes."[187]

Of the three anti-Owenites to attack the vertebrate archetype, Lewes was the most philosophically trenchant. His criticism was to a certain extent an about-turn. During the heyday of homologizing he had praised the program of transcendental morphology and idolized Goethe, Geoffroy, and Owen alike. In the second edition of his *Life of Goethe*, however, published in 1864, Lewes backtracked by cautiously adding the conclusions of Huxley's "remarkable" Croonian lecture: "Although Huxley insists, perhaps, too much upon the *differences*, in his impatience at the too great emphasis which has been laid on the *resemblances*, his criticism seems to me conclusive against the vertebral theory as generally understood."[188] Lewes had been more or less a follower of Comte ever since his *Biographical History of Philosophy* (1845–46). It was not until its third edition, under the title *The History of Philosophy from Thales to Comte* (1867), published some twenty-one years later, that Lewes, in the added "Prolegomena," presented an incisive Comtean criticism of transcendental morphology, without naming Owen, however, who at that time was still his friend.

In his *History of Philosophy*, under the inflammatory heading "Some Infirmities of Thought," Lewes described instances of the tendency to give objective reality to one's mental abstractions. A classic example was the notion of a vital principle. Life—Lewes stated—is manifest as the interlinkage of organic processes and thus as a complex whole of various particulars. Because each part cannot exist without its participation in the whole, or connexus, some have argued that the connexus is prior to the parts, "the whole *generating* the parts, instead of being a *generalisation* from the parts."[189] Thus, the result of the connexus of the parts, namely life, is turned into a necessary antecedent, and an independent entity has been created out of a relationship.[190]

The metaphysician had replaced the "vital principle" with the notion of a "plan" that produces organic unity. This is another instance of "putting the cart before the horse"; "the resultant is transformed into the cause." We see an architect arranging a plan for a house, and a builder arranging the materials in accordance with this plan. Finding in an organism a certain adjustment of parts that may be reduced to a plan, we are easily led to conceive that this plan was made before the parts and that the adjustment was determined by the plan. The difference is, however, that, whereas building materials do not have a spontaneous tendency to come together into houses, "what we know of organic materials is that they *have* this spontaneous tendency to arrange themselves in definite forms; precisely as we see chemical substances arranging themselves in definite forms, without the intervention of any extra-chemical agency."[191] A plan simply evolves, and the parts are not constructed after a plan. "From an observed *nexus* men rashly infer a *nisus*, from an actual conjunction a previous intention."[192] The plan is no more than the generalized expression of the facts of a nexus. For the purpose of scientific thought such a plan is indispensable, but "men are led by an infirmity of thought to realise the concept; and having first used it only as a convenient expression, they grow into a belief of this nexus being *also* a nisus."[193] To counter that the plan preexists, not as a fact but as a potentiality, is mere "jugglery of thought." "Nothing exists before it exists."[194]

Lewes did not explicitly refer to Owen's vertebrate archetype in the "Prolegomena" of 1867, but that he had Owen in mind was made clear when the following year he reviewed for the *Fortnightly Review* the Duke of Argyll's *Reign of Law*, in which Argyll towed an Owenite line. Lewes again criticized the concept that a plan determines phenomena rather than summarizes them. This review contained Lewes's most concise Comtean critique of the notion of a transcendental archetype:

> All that Science discovers in the investigation of Nature's processes is a *nexus formativus*—the law of causation. By a well-known infirmity [of the mind] we transform this into a *nisus formativus*, and the law becomes a plan. Thus because the *nexus formativus* of a vast animal group can be abstracted as the concept of a Vertebral Type, *i.e.* a Plan according to which each part may be exhibited as correlated, there have been philosophers, from Plato downwards, who believed that this Type existed before animals were created, and that when animals were created they were constructed after this model.[195]

While the Comteans were firing at a Platonized Owen, Darwin quietly expropriated Owen's extensive labor on the archetype and homologies of the skeleton of vertebrates. In the middle of the 1840s, he had followed Owen's homological work with close attention. Upon the appearance of the *Anatomy of Fishes*, Darwin wrote: "I have just read your first chapter and have been delighted with it. Those vertebrae are awfully difficult to understand."[196] When Darwin had seen an early version of the BAAS report of 1846, he informed Owen: "I have lately read with *very great* interest all the parts which I could follow in your Report on Archetypes, etc."[197] And he penciled in his copy of Owen's *Nature of Limbs*: "I follow him that there is a created archetype, the parent of its class."[198] It had of course been obvious to Chambers and to Owen himself that unity of type was an indication of genetic relationship and thus evidence of descent and organic evolution; Darwin went one step further and connected unity of plan to the process of natural selection while turning the archetype with a click of his fingers into a primitive ancestor of flesh and blood.[199]

As historians from Russell to Ospovat have shown, Darwin effortlessly transferred the bulk of Owen's vast labor on the comparative anatomy of vertebrates to his grand synthesis of evolution by means of natural selection.[200] In consequence of Darwin's reinterpretation, homology not only became evidence of ancestry but, conversely, ancestry was turned into the criterion of homology. As a result, evolutionists have been accused of founding their theory on a circular argument, but although this may be true for the post-*Origin* period, it does not apply to the 1840s and 1850s.[201] By incorporating Owen's work, Darwin used a body of homological inferences that had not been based on the notion of descent, not even much on embryological development. Owen's criterion of homology had almost entirely been that of relative position as revealed by comparative anatomy, providing Darwin with a straight, noncircular argument for his evolution theory.

The extent to which the Platonist interpretation had lost its currency in the early 1860s is apparent from an article, "Vertebral Patterns," in the *Medical Times and Gazette*. The magazine had been supportive of Owen, and this time, too, assented to his theory of special, general, and serial homology, praising him as the inductive demonstrator of the ideal archetypal pattern on which vertebrates are constructed; but the article's author also speculated that the archetype might be conceived of not only as a divine forethought but also as an ancestor of flesh and blood, adding: "We rather prefer to believe that the archetypal vertebrate skeleton, whatever exact form it might have possessed, was once manifested in the flesh, as an objective entity, than conceive it merely as a process of the Creator's thought."[202]

The article in the *Medical Times* demonstrated that it was possible to think of Owen's vertebrate archetype as both an ideal blueprint and a real ancestor, or—to translate this into terms of personalities—that one could bring Owen and Darwin together, at least on this crucial point. There were various nonscientific reasons, however, why such peacemaking proved no easy matter.

First, Huxley's systematic assault on Owen, particularly in the Croonian lecture, had injected much poisonous hostility into the dispute. Second, Owen had grown to expect a respectful acknowledgment of his supremacy and was not about to give way to any pretender. Third, and more profound, the Darwinian reinterpretation of the archetype was couched in terms of Comtean positivism, which clashed not merely with the Platonist exposition of the archetype but also with Owen's genuine and deeply felt theism, which additionally did not harmonize with the chance nature of Darwinian natural selection. Moreover, at the time that the interpretation of the vertebrate archetype as a divine blueprint was being torn to shreds, an attempt was made to inflict similar damage on Owen's blueprint for a national museum of natural history, and this by the same people.

Darwin's accusation that Owen was unfair in his assessment of *The Origin of Species* has become part of the staple food for students of the Victorian controversy over evolution. To the well-founded charge of unfairness should be added, however, that not only Huxley but also Darwin was perfectly capable of uttering an untruth about Owen. One example is his well-known line to Asa Gray that "no fact tells so strongly against Owen, considering his former position at the College of Surgeons, as that he has never reared one pupil or follower."[203] This was nonsense. Owen had a wide following, in the museum movement, as a Cuvierian and as a transcendentalist. Darwin himself owed more to Owen than he was willing to admit.

5

Eclipsed by Darwin

By changing from functional to transcendental anatomy, Owen placed the argument from design on a new footing. With his notion of archetypes he shifted the evidence for the existence of a Supreme Designer from concrete adaptations to an abstract plan or, to use Dov Ospovat's terminology, from special to general teleology.[1] Divine contrivance was to be recognized, not so much anymore in the characteristics of separate species but in their common ground plan. God was no longer the Supreme Watchmaker but the Supreme Architect, who had personally conceived the blueprint of nature, yet employed natural laws for the actual construction work.

Thus, in Owen's transcendentalist work the argument from design did not require a belief in the miraculous creation of species, and Owen cautiously began to formulate a theory of theistic evolution, arguing that individual species had come into existence by a preordained process of natural, or secondary, laws. Having demonstrated the existence of an ideal type among vertebrate species, the step was inevitable, as Owen later stated, "to the conception of the operation of a secondary cause of the entire series of species, whether of plants, or vertebrates, or other groups of organisms, such cause being the servant of predetermining intelligent Will."[2]

Over a period of some four decades, from the mid-1840s to the mid-1880s, Owen explicitly and repeatedly expressed in articles, monographs, a textbook, and letters his belief in a natural origin of species. It is astonishing that this extensive body of primary evidence has been systematically ignored by all but those who in recent years have begun to reexamine Owen's role in the debate about evolution. The belief that Owen was a creationist, opposed to any sort of evolution, has been widely held and occasionally still continues to be expressed. The one-sided and distorted portrayal of Owen as given by

Darwin in the expanded "Historical Introduction" to the fourth (1866) and later editions of his *Origin of Species* and as given also in the *Life and Letters* of Huxley (1900), Darwin (1901), and Hooker (1918) and in *More Letters of Charles Darwin* (1903) has been uncritically reproduced by many historians.[3] It is a measure of the extent to which the historiography of "Darwin's century" has served the partisan purpose of vindicating Darwinism. However, over the past four decades or so, increasing numbers of historians have distanced themselves from a Whiggish interpretation of the Darwinian conflict. In the process, Owen's participation in it and his significance for the development of nineteenth-century natural history have become the subject of substantial, revisionist scholarship.[4]

Toward a Natural Origin of Species

Admittedly, throughout the 1830s Owen had been an outspoken and well-known advocate of the doctrine of the creation of species. Having followed the famous debate between Cuvier and Geoffroy in 1830, he initially had been left unconvinced by Geoffroy's transmutationist ideas, siding instead with Cuvier. The last time, however, that Owen publicly and unambiguously stated his belief in special creation was in 1841 at the conclusion of his major BAAS "Report on British Fossil Reptiles." Summing up, Owen asked: "Does the hypothesis of the transmutation of species, by a march of development occasioning a progressive ascent in the organic scale, afford any explanation of these surprising phaenomena?"[5] He answered this question in the negative on the basis of a closely reasoned survey of the stratigraphic distribution and anatomical characteristics of the reptiles of his "Report." Reptilian species appear suddenly, and only at a very general level do they exhibit a progressive trend through time. The most primitive, fishlike organization in reptiles was not the earliest. The nearest reptilian approximation to fishes, the ichthyosaurus, appeared relatively late in the stratigraphic record. The most advanced reptilian organization, found in the dinosaurs, did not appear or stay latest. Moreover, in such morphological lineages connecting "ichthyosaurus to plesiosaurus to teleosaurus," the different genera did not evolve one from the other, because they showed up simultaneously in the record—an antitransformist argument that Buckland a few years before had put forward in his Bridgewater Treatise. Owen concluded:

> Thus, though a general progression may be discerned, the interruptions and faults, to use a geological phrase, negative the notion that the progression has been the result of self-developing energies adequate to a transmutation

> of specific characters; but, on the contrary, support the conclusion that the modifications of osteological structure which characterize the extinct Reptiles, were originally impressed upon them at their creation, and have been neither derived from improvement of a lower, nor lost by progressive development into a higher type.[6]

This was a highly competent refutation of the French theories of transmutation, especially Geoffroy's, and of the home-grown variety promulgated by Robert Grant. Robert Jameson printed the entire concluding section of Owen's report in his *Edinburgh New Philosophical Journal*;[7] but Owen would never again explicitly state that the anatomical characteristics of animal species "were originally impressed upon them at their creation."

An early indication that Owen had abandoned Cuvierian fixism was his tolerant, partly appreciative reaction to Robert Chambers's anonymously published *Vestiges* (1844). Owen's friends and patrons, especially the Cantabrigians among them, strongly condemned the evolutionary ideas of the book and were deeply worried about its popularity. Roderick Murchison urged Owen to consider writing a damning review for Lockhart's *Quarterly*; "a real man *in armour* is required, and if you would undertake the concern you would do infinite service to *true* science and sincerely oblige your friends." Adding to the flattery of his appeal, Murchison continued: "I cannot say how much you would gratify your friends and admirers by this effort, which would entitle you to another niche in the temple of good works in which you already occupy so high a place."[8] But Murchison failed to move Owen, who left the public condemnation of *Vestiges* to Adam Sedgwick and William Whewell. The latter issued some selections from both his history and his philosophy of the inductive sciences under the title *Indications of the Creator* (1845), reaffirming the validity of the creationist argument from design. Sedgwick went further and wrote a deeply emotive and bellowing attack for the *Edinburgh Review* (1845).[9]

Owen kept his distance, although he drew the attention of his Trinity College supporters to a number of substantial inaccuracies in Chambers's amateur account, asking Whewell at the same time not to use his—Owen's—private comments in any public refutation of *Vestiges*.[10] Owen also brought a number of errata to the attention of the anonymous author himself. In addition to "a few mistakes," "easily rectified in your second edition," and a need for the Vestigiarian to look up Owen's published observations on embryological recapitulation, two other points had to be made. The first was that there existed no evidence for the spontaneous generation of simple forms of life (Owen later insisted that this process takes place all the same), and the second was that, although "the development of the Hottentot from the chimpanzee" was an unfounded hypothesis, he—Owen—did not hold "the slightest prejudice" against the possibility of such a relationship.[11]

This open-minded attitude toward the possibility of human descent from animal ancestry was not shown, however, in a further letter to Whewell about *Vestiges* in which Owen wrote: "The man who is willing to believe with the writer of that book, that his great-great-great-etc. grand father was a Baboon, and his great-great-etc. grand mother a Chimpanzee, will not be converted by whatever manifestation of the *mens divinior* may shine in a refutation of such an opinion."[12] Given the ambiguity of Owen's overall reaction to *Vestiges*, the question of his real belief has been the subject of controversy. John Brooke has argued that Owen's constructive and friendly response to the Vestigiarian was merely "the minimum that was commensurate with the etiquette of the situation" and "that Owen's enigmatic letter ought to be construed, not as a straightforward letter of mild encouragement, but as a deft rebuke."[13] Others, such as Michael Ruse, Adrian Desmond, and, in most detail, Evelleen Richards, maintain that Owen did indeed sympathize with the notion of a natural origin of species.[14]

This, too, is my reading of the affair. At the time, Owen was crucially dependent on the patronage of his Oxbridge friends, whereas he owed no allegiance to the author of *Vestiges*. He may not even have known who the author was. Desmond speculates that Owen probably mistook the phrenologist George Combe for the author of the book.[15] In spite of the conflicting signals he sent, Owen is likely to have been genuine when he concurred with the Vestigiarian that, given adequate evidence, "the discovery of the general secondary causes concerned in the production of organised beings upon this planet would not only be received with pleasure, but is probably the chief end which the best anatomists and physiologists have in view."[16] Also, even if common courtesy had been the reason for this concurrence, there would still be the fact that Owen declined to write a critical review, resisting his friends' entreaties.

So it would appear that by the mid-1840s Owen had no quarrel with the fact of a natural, noncreationist origin of species and, in the words of his grandson, "had a certain leaning towards the theories enunciated by Robert Chambers."[17] What was it that had changed since his public affirmation, earlier in the decade, of the doctrine of special creation? To repeat, Owen's adoption of an idealist position had lifted the burden of design evidence from specific adaptations to homological similarities and ultimately to the ideal type. But this change merely facilitated a belief in organic evolution; it did not necessitate such a major shift in position of theory. His transcendentalist idealism may well have been the underlying cause, but it was not the triggering insight.

What then had come to Owen's attention to convince him of the fact of species evolution? The answer to this question may be contained in a review he wrote for the *Quarterly* (1847) of the *Zoological Recreations* by his loyal supporter Broderip, a book that was dedicated to Owen. A substantial part of this review essay was not about Broderip's charming sketches of natural history but about Owen's own recent studies of past and present zoogeographical provinces. In 1844 Owen had presented a report to the BAAS entitled "On the Extinct Mammals of Australia," primarily to describe in new detail the diprotodon and two species of a related genus of pachydermoid marsupials. Owen ended this report with a preliminary formulation of a "law of geographic distribution of extinct Mammalia" which stated that during the most recent geological past—the Pliocene and Pleistocene—mammals were spread across the globe according to distinctive provinces, just as today's mammals are, and that both past and present provinces were approximately the same, characterized by the same mammalian types: "with extinct as with existing Mammalia, particular forms were assigned to particular provinces, and, what is still more interesting and suggestive, that *the same forms were restricted to the same provinces at a former geological period as they are at the present day.*"[18]

Owen regarded the Old World—the Euro-Asiatic and African continental masses—as a single province, marked by such animals as elephants, rhinos, hippos, hyenas, and various feline carnivores, both fossil and living. The New World, or at least South America, represented a second province, with its extinct megatherium and glyptodon and today's allied species of sloth and armadillo. Australia had proved to be a third distinct province, inhabited in the past by megatherium-sized marsupials, such as the diprotodon, and known today for its kangaroos, wombats, and other pouched mammals. Owen added that New Zealand was and is characterized by its flightless birds, such as the extinct giant dinornis and the diminutive wingless *Apteryx*, or kiwi, alive today.

It is true that Robert Jameson had already recognized the existence "at a former period" of a characteristically Australian faunal province.[19] In fact, very soon after Mitchell's fossils from the Wellington Valley caves had arrived in London and Paris, Joseph Pentland had concluded that in Australia, as in Europe, the Pleistocene mammals are "referable" to the same genera to which today's species belong.[20] Darwin, too, in his *Journal* (1839) of the *Beagle* voyage, had described for South America the "law of the succession of types," that is, that within a particular province of distribution the fossil mammals bore a close resemblance to today's living ones. Yet Owen had been a participant in this Australian and South American work virtually from its mid-1830s start. A decade later, in the mid-1840s, Owen took the lead in discussing the subject comprehensively, in his BAAS report of 1844, in an 1846 lecture at the Royal Institution entitled "On the Geographical Distribution of Extinct Mammalia," and, later that year, in a BAAS lecture on the same subject which, in the words of the *Literary Gazette*, "raised him still higher than he stood before (and that was no easy task) in the ranks of scientific fame."[21]

At this time, the excitement about biogeographic provinces and their continuity from the Pleistocene to the present was fueled by the related issue of the origin of species. How to account for the fact that continental masses and islands, isolated from one another by stretches of water, each have—and had—their own, characteristic assemblage of species? A widely held theory of the period was that the biogeographic provinces reflected "specific centers." or "centers of creation." Through the 1840s, this theory grew into a doctrine of which Edward Forbes was a noted advocate.[22] Although Forbes did not elaborate on the means by which the first individual or individuals of a new species had come into being, it was assumed by some, predominantly from the English- and French-speaking countries, that this had happened by divine acts of miraculous creation; others, a majority of them Germans and a few living in France, postulated that discrete floral and faunal provinces not only prove that species have no shared, single place of origin (Paradise or Mount Ararat); but, more to the point, that the autochthonous—restricted to their own provinces—distribution of species across the world can best be explained in terms of a natural origin by means of a spontaneous aggregation of their first germs in situ, spontaneously generated under conditions that varied from province to province. Thus, species were autochthons and could be either fixed or variable within certain limits, but they had not originated by means of descent from a single, common ancestor.[23] Forbes, too, stated that the fact of unrelated centers of distribution "is necessarily opposed to the hypothesis of *the evolution* of all species from one first form."[24]

Owen, too, adopted the notion of specific centers of distribution. Moreover, in his—customarily anonymous—essay in the *Quarterly Review* of 1847 he rejected the biblical notion of a single Asiatic center of postdiluvial dispersal from Noah's ark. Such a single center was unimaginable, given the fact that neither Australia's characteristic mammals nor New Zealand's flightless birds would have been able to wander across. The alternative possibility, put forward by the anthropologist James Cowles Prichard in his authoritative *Researches into the Physical History of Mankind* (1813 and later editions),[25] was also rejected, namely that in addition to an Asiatic center, there may have been new, postdiluvial creations in Australia and elsewhere. This was implausible—Owen argued—in view of the discovery that the antediluvial faunal distribution was similar to today's. Why should new creations geographically have matched old ones in a world that had undergone a radical change in physical geography?

How then to interpret these distributions? Owen left the reader with no other option but to conclude that the characteristic fauna of today's provinces had originated naturally within these provinces. He did not spell this out but alluded to the need for a revolutionary revision of our ideas about the origin of species, suggesting that the proper solution to the problem of animal distribution in space and time would equal the Copernican revolution in intellectual importance. "Let any one reflect on the limited powers of locomotion" of such animals as the burrowing wombat, the diving duck-billed platypus, the climbing sloth, and the flightless birds—Owen wrote—"and say whether Zoology has not presented a problem which, when rightly solved, will effect as great a revolution in men's ideas of the time and mode of the dispersion of animal life over the earth's surface as the Copernican system did in those regarding the relations of our planet to the sun."[26] In later years, he explicitly and recurrently referred to the Copernican revolution as the proper historical analogy to the theory of organic evolution.

Modes of Evolution

When Owen spoke out against transmutation, as he continued to do, this may well have given people the impression that he opposed the notion of an origin of species by natural means, but this impression would have been false. "Antitransmutation" was not a synonym for "miraculous creation," although Owen made little, if any, effort to enlighten his audiences about this. How could he have opposed transmutation and yet rejected a supernatural genesis? The answer lies in the fact that to Owen transmutation meant only one of several modes of the natural origin of species. At the time, three different mechanisms were put forward: transmutation, envisaged as for the most part gradual change from one species into another due to an organism's exertions in coping with environmental pressures; autogeny, believed to involve an origin of species not from preceding forms of life but directly from primordial germs, spontaneously aggregated from lifeless matter under favorable chemical and geological conditions; and heterogeny, by many of its advocates combined with a limited degree of autogeny, whereby lower species were thought to have originated spontaneously but higher ones by major mutations of embryonal and other germs—eggs, seeds, spores, or other units of reproduction.

Owen abandoned creationism to become, in the main, a heterogenist. This helps explain the apparent ambiguities in his pronouncements on organic evolution. He could denounce French transmutationism and later also

Darwinism while genuinely adhering to an evolutionary theory of the origin of species. Owen could, moreover, show abhorrence of the notion that humans had emerged by a gradual transformation of apes, yet at the same time truly believe that human existence proceeded from apes. Such a derivation could be visualized as a process of heterogeny, whereby the human form had come into existence suddenly as a result of major, mutational changes—a kind of autogenous generation of humanoid forms in the uteri of anthropoid females. Arthur Schopenhauer, an admirer of Owen as we saw above, referred to this mode of evolution as *generatio aequivoca in utero heterogeneo*, attributing the idea to *Vestiges*. Owen's sympathetic reaction to, and Darwin's later criticism of, the Vestigiarian were due to his book's slant toward heterogeny.[27]

In private correspondence with the publisher John Chapman, Owen mentioned that he knew of no fewer than half a dozen possible ways in which species could originate naturally:

> Transmutation of species in the ascending course is one of six possible secondary causes of species apprehended by me, and the least probable of the six. When I remarked to the (reputed) author of "Vestiges," the last time he visited the museum, how servilely the old idea had been followed by . . . Lamarck, and the author of "Vestiges"–viz. of "progressive development"—and that there were five more likely ways of introducing a new species, he asked suddenly and eagerly, "What are they?" I declined to give him the information, but shortly after brought prominently under his notice the facts that might have suggested one, at least, of the more likely ways. He saw nothing of their bearing, and I shall refrain from publishing my ideas on this matter till I get more evidence.[28]

Owen, who seldom missed an opportunity to refer back to one of his earlier publications, reminded Chapman that his negative opinion about the notion of the transmutation of species "continues the same as that expressed in the concluding summary of my second report on British Fossil Reptiles." This was at best a half-truth, because, although his objections to transmutation remained, his belief in direct creation had been irrevocably replaced by a belief in the efficacy of "secondary causes." It has been alleged that Owen never revealed which "six possible secondary causes of species" he had in mind,[29] but in places he did refer to various modes of organic change. For example, in a later encyclopedia entry on "species" he cited atrophy and hypertrophy, invoked by Lamarck's transformism; sudden congenital malformations or monstrosities, in which Geoffroy had shown such interest; premature birth and prolonged gestation, to which the Vestigiarian had drawn attention; and in particular the phenomenon of parthenogenesis or, more specifically, "*Generationswechsel*," which Owen himself believed to be of relevance to the question of the origin of species.[30] This last "secondary cause of species" requires explanation.

In 1745 Charles Bonnet had described reproduction without sexual union in aphides, or plant lice. The fertilized eggs of an aphid produce a wingless, asexual (larval) generation existing of only one form. Without copulation, this form is capable of producing similarly asexual offspring. A series of asexually reproducing, viviparous generations unfolds until some individuals develop into winged males and oviparous females. In the latter, the ova develop after copulation with the males. The fertilized eggs again develop into a wingless, asexual generation. Thus, an alternation of oviparous sexual and viviparous asexual generations takes place. Throughout the first half of the nineteenth century, there was a growing interest in similar phenomena.[31] From the middle of the 1840s, and for a period of some fifteen years, Owen was much taken by the phenomenon of asexual reproduction. In 1845 the Ray Society published an English translation of *Ueber den Generationswechsel* (1842), in which the Danish zoologist J. J. S. Steenstrup described the many instances of "alternation of generations" (a phenomenon that had been described by the eighteenth-century Swiss naturalist Charles Bonnet), especially in medusae, polyps, salpae, and trematodous fluke worms. The *Medico-Chirurgical Review* published a glowing account of this "deeply scientific and profound work," singling out for special praise the section on trematodous entozoa.[32]

In 1849 Owen himself devoted the introductory two lectures of his Hunterian course to the phenomenon of asexual reproduction, for which he coined the term "parthenogenesis." The two lectures were published separately under the title *On Parthenogenesis, or the Successive Production of Procreating Individuals from a Single Ovum* (1849). In trying to explain this phenomenon, Owen suggested that the primary impregnated "germ-cell" divides into "derivative impregnated germ-cells" constituting a "germ-mass," and that not all of this germ-mass is used up in the growth of an individual but that part is preserved to produce the next generation.[33] When August Weismann's *Essays upon Heredity* (1889) appeared with its theory of a "germ plasm," St. George Mivart pointed out that the by then "aged and illustrious" Owen "in many respects actually anticipated the ideas of the Freiburg Professor."[34]

The German zoologist Carl von Siebold, in his *Ueber wahre Parthenogenesis* (1856), distinguished the phenomenon of viviparous reproduction by asexual larval creatures from that of "true parthenogenesis," which he limited to reproduction by means of true eggs produced from the ovaries of true females, capable of development without fertilization, as in the case of bees. Owen revised and annotated the English translation of Siebold's book, *On a True Parthenogenesis in Moths and Bees* (1857), and he managed to generate considerable excitement about these phenomena in his circle of friends. During a Royal Academy dinner in 1849, Lord Brougham approached him "and plunged at once into the mysteries of 'Parthenogenesis,' about which the world is beginning to talk, as the subject of my 'Lectures' oozes out in conversation."[35] Owen tried to get his and Siebold's books on parthenogenesis reviewed in the *Quarterly* but he failed because, to his chagrin, the unavoidable, sexual terminology led to censorship.[36]

A twinned review of Owen and Siebold did appear but in the medical periodical literature,[37] not in the *Quarterly* or in the *Edinburgh Review*. Although Owen made no major contributions to the puzzling phenomenon of heredity, which remained unsolved until the discovery of chromosomes much later in the century and the rediscovery of Mendel's work in 1900, he did apply his work on parthenogenesis to the question of the origin of species. Characteristic of the alternation of sexual and asexual generations is that they can differ from one another as much as different species—or even genera, families, and orders—do. This change of form Owen referred to as "metagenesis," which he defined as: "The changes of form which the representative of a species undergoes in passing, by a series of successively generated individuals, from the egg to the perfect or imago state."[38] It is distinguished from "metamorphosis" in that, in the latter instance, one and the same individual undergoes the changes of form.

A good example of metagenetic change is the reproductive cycle of the medusa, or jellyfish. Its eggs develop into polyps, which by gemmation produce other polyps, each of which breaks up into disk-shaped parts, which in turn grow to become medusae, some of which are male, others female. Another example is that of entozoan parasites such as the Trematoda, or fluke worms. The eggs hatch into free-swimming, infusorial larvae, which may bore into water snails, where, via a wormlike stage, they multiply into innumerable tadpole-like animalcules. These bore their way out of the snail, disperse in the water, and may ultimately end up in the intestines of, for example, waterfowl. There they change into sexually differentiated fluke worms, which produce fertilized eggs. This entozoan example was particularly close to Owen's heart, because he had nursed an interest in intestinal parasites ever since his description in 1835 of the existence in humans of the entozoon *Trichina spiralis*.[39]

Thus, the trematodous metagenetic cycle consisted of a succession of three distinct forms, infusorial, wormlike, and tadpole-like, connecting the egg and the mature fluke worm. One could imagine that under particular circumstances the cycle could be broken and that the separate stages would go on reproducing. In this way wholly new genera or even orders might originate.

Admittedly, no such breakup had ever been observed. Some years after the publication of his *Parthenogenesis*, Owen commented: "The first acquaintance with these marvels excited the hope that we are about to penetrate the mystery of the origin of different species of animals; but as far as observation has yet extended, the cycle of changes is definitely closed."[40]

Yet, as Desmond concludes, "Owen showed, if nothing else, that natural causes need not mean transmutation."[41] Moreover, I believe that the phenomenon of metagenesis provided Owen not only with one of his six modes of the origin of particular species but also with an analogy, a visualizing aid, to the process of evolution in general. Whereas embryological development was used by the transmutationists as an analogy for imagining the gradual change of one species into another, the metagenetic sequence of, for example, the aphids evoked a very different picture of how a species, after generations of unchanging reproduction, might suddenly give rise to a wholly new species, and how such a process of descent might be thought of as a process of unfolding, driven by an inherent tendency to change, not by external causes.

Reactions to Owen's *Parthenogenesis* differed. Huxley was critical and suggested that all the different stages of a metagenetic cycle, taken together, form a single individual, and that what Owen had considered discrete individuals were merely free-floating organs.[42] In his *Antiquity of Man* (1863), Lyell discussed the phenomenon of "alternate generation" as a possible mode of transmutation, and although he ignored Owen's work, he came to the same conclusion that Owen had drawn: the cycle of change "returns to the exact point from which it set out, and no new form or species is thereby introduced into the world."[43] Others, such as John Duns, Scottish divine and editor of the *North British Review*, misinterpreted Owen's work and saw it as an answer to the ungodly belief in the spontaneous generation of fluke worms or other primitive forms of life.[44] Yet others felt confused because of Owen's "appalling terminology" and the mystifying, Germanic flavor of his ideas. After a lecture by Owen entitled "On Metamorphosis and Metagenesis,"[45] the *Lancet* joked:

> "Twixt Owen and Oken what difference I pray?"
> "Simply one spells with *w*, t'other with *k*;"
> "But which is most mystical?"—"No one can say,
> For their myths and conundrums all tend the same way,
> Transcendentally leading the judgment astray."
> Oh! I now comprehend the gist of the thesis,
> On metamorphosis and metagenesis;
> It is, that all science and wisdom worth knowing
> Are homologies 'gendered by *Oken* and *Owen!*[46]

It therefore comes as no surprise that apart from Owen the main biologists to bring the phenomenon of the alternation of generations to bear on the origin of species were from the German-speaking world. One of them was the Swiss-born botanist Carl Wilhelm von Nägeli, who as early as 1856 considered the possibility of an origin of species by means of metagenetic jumps.[47] Another was the Swiss zoologist Rudolf Albert von Kölliker, who, in *Ueber die Darwinsche Schöpfungstheorie* (1864), criticized the mechanism of natural selection because of what he believed was its overemphasis on the phenomenon of adaptation. Kölliker offered an alternative "theory of heterogeneous generation," arguing that variations are more apt to appear suddenly than gradually, in the way that widely differing forms succeed one another in metagenetic sequences.[48] This theory was effectively identical to Owen's "derivative hypothesis" and foreshadowed—one might argue—the mutation theory.[49]

Evolution in Disguise

The fact that, not long after *Vestiges* had been published, Owen engineered the translation into English of Oken's old *Lehrbuch der Naturphilosophie* (1809–11) and its publication by the Ray Society may be interpreted as part of his strategy for a cautious furthering of the idea of evolution. The outdated handbook certainly had no value with respect to Owen's studies of comparative anatomy and paleontology, but its Germanic, rather enigmatic, imagery of heterogeneous development was suggestive of evolution, yet less directly so than *Vestiges.* An example is the passage where Oken states that the first plants and animals emerged from the sea and that "*Man also is a child of the warm and shallow parts of the sea in the neighbourhood of the land.*"[50]

Owen himself understood Oken in an evolutionary sense. We have direct evidence for this in the form of an abstract of the *Lehrbuch* which Owen prepared as part of his entry "Oken" for the eighth edition of the *Encyclopaedia Britannica* (1858). We must remember that the developmental language of German *Naturphilosophie* did not necessarily imply an actual, physical evolution in time but could equally well indicate a logical progression, as in a numeric or geometric sequence; moreover, specifically evolutionary utterances in Oken's textbook are rare and require attentive scrutiny of the metaphor-ridden text. Yet Owen left no doubt about his evolutionary understanding of it, quoting Oken on how "by self-evolution into higher and manifold forms they [the elements] separated into minerals, became finally organic, and in man attained self-consciousness." Owen further cited Oken's belief that God created only infusorial-sized organisms: "Whatever is larger has not been created but developed."[51] The same passage was seized upon by

Hugh Miller in a fuming condemnation of Oken. Edward Hitchcock, Samuel Wilberforce, and others, too, censured Oken's textbook for its evolutionary speculations "carried to their logical and legitimate conclusion."[52]

Owen's trial balloon—if that is what his promotion of Oken's *Lehrbuch* represented—was promptly shot down in an uproar within the Ray Society. The society was one of twenty-two such printing clubs established in the first half of the nineteenth century. One of its purposes was "the promotion of Natural History, by the printing of . . . translations and reprints of foreign works which are generally inaccessible from the language in which they are written, or from the manner in which they have been published."[53] Soon after the foundation of the Ray Society in 1844, Owen became a member of its council. The publication of the English translation of Oken's book as one of the society's official volumes, under the title *Elements of Physiophilosophy* (1847), was instigated by Owen and took place without much consultation. A report on the book, commissioned by the council, warned of its evolutionary flavor; "details resemble a little the 'Vestiges of Creation.'" The report also drew attention to "paragraphs of a questionable nature," such as the one expressing the view that "Man has not been *created*, but *developed*," and concluded "that the Ray Society has not been happy in its choice of the work for one of its authorized publications."[54] Two council members, J. H. Balfour and R. K. Grenville, both Scottish botanists of a traditional, Paleyite allegiance, withdrew their names. Another councillor and close friend of the other two Scotsmen, William Jardine, an editor of the *Annals and Magazine of Natural History*, also expressed concern. The society's minutes record: "Notice of motion by the President [Thomas Bell] that on account of the extremely objectionable character of the translation of Oken's work lately published by the Society the Council do immediately take into consideration whether any and what means can be taken to obviate the injury which that publication is calculated to do to the Society."[55]

Apparently, none of the by then well over seven hundred regular members left the society, and some parsons actually joined in order to obtain a copy of the condemned book. The council decided, however, that in future no work was to be published without consultation with "some competent person or persons," and that a report on the work "be sent to each country member of council"; also, that no manuscript be sent to the printer without the imprimatur of the secretary of the society.[56] Owen had failed; evolutionary allusions, whether or not disguised in the mystifying language of *Naturphilosophie*, were decisively censured, even though John Ray himself had been anything but orthodox.

On February 9, 1849, Owen made another cautious attempt to bring his belief in the natural origin of species to the public. He had already indicated this belief in his BAAS report of 1846, "Archetype and Homologies of the Vertebrate Skeleton," and in the subsequent book based on the report, stating that it is a legitimate course of inquiry to study "the law which has governed the successive introduction of specific forms of living beings into this planet."[57] This loaded sentence had remained buried, however, under the report's mass of descriptive data until Owen now resurrected it in an attention-drawing way. At the conclusion of his lecture "On the Nature of Limbs," delivered before a large audience at the Royal Institution, he melodramatically stated concerning the origin of life on earth:

> To what natural laws or secondary causes the orderly succession and progression of such organic phenomena may have been committed we as yet are ignorant. But if, without derogation of the Divine power, we may conceive the existence of such ministers, and personify them by the term "Nature," we learn from the past history of our globe that she has advanced with slow and stately steps, guided by the archetypal light, amidst the wreck of worlds, from the first embodiment of the Vertebrate idea under its old Ichthyic vestment, until it became arrayed in the glorious garb of the Human form.[58]

That the natural laws had worked in any other way than by descent was inconceivable, Owen maintained in a later essay.[59] The evolutionary implications of this rather-mystifying paragraph may have escaped the audience, for whom the lecture of an hour and a half was "too technical" and "too transcendental" anyway.[60] Yet its figurative language did not pull the wool over the watchful eyes of the *Manchester Spectator*, a moderately liberal weekly that, although interested in the advancement of science, was "also solicitous for the promotion of the interests of true Religion." It accused Owen of importing into England the scientific pantheism of German naturalists such as Oken and supporting the Vestigian theory of development "that God has not peopled the globe by successive creations, but by the operation of general laws."[61]

The *Spectator*'s explication of Owen's opaque concluding paragraph, although simplified, was accurate. Owen, seeing his second trial balloon shot down by the Manchester periodical, sought safety behind a change of emphasis and terminology. The facts were—he stated in a letter to the editor—that, at successive periods, animals of progressively higher organization had appeared on this planet and that they were constructed on a common plan. "Of the nature of the creative acts by which such successive races of animals were called into being, I have never presumed to offer an opinion, save in refutation of some inadequate hypotheses, more especially of the stale, but lately revived, one of Development and Transmutation." The evidence of the unity of plan, Owen added, was evidence of the oneness of the Creator, "as the modifications of the plan for different modes of life illustrate the beneficence of the Designer"; in general, his work on the vertebrate archetype was an argument in refutation of the old atheists and pantheists.[62]

This answer was vintage Owen. There was nothing untrue in what he wrote, yet he skirted the real issue and the reader was left to draw his own, possibly wrong, conclusion. Owen indeed had objected to a Lamarckian mechanism of transmutation but not to an origin of species by natural laws. He disguised his belief in this by wrapping it in the orthodox expression "creative acts." The *Manchester Spectator* withdrew its accusation that Owen promoted pantheism and Vestigiarian ideas and flatteringly explained: "The high authority of Professor Owen in the scientific world renders every deliberate opinion pronounced by him a matter of importance." The withdrawal was only partial, however, and the *Spectator* continued to suspect that Owen's true belief was no longer creationist: the first sentence of Owen's concluding paragraph was quoted with the offending words "or secondary causes" underlined.[63]

No one jumped to Owen's defense. Neither friends nor colleagues supported him in his cautious furtherance of the idea of the origin of species by natural laws. On the contrary; Sedgwick issued a warning to Owen in the fifth edition of his *Discourse on the Studies of the University* when he commented as follows on the concluding paragraph of *On the Nature of Limbs*: "Had I not known the opinions of this great comparative anatomist, as they are expressed in many of his recent works, I should, perhaps, have thought that in this passage he meant to indicate some theoretical law of generative development from one animal type to another along the whole ascending scale of Nature."[64]

No major reviews of *On the Nature of Limbs* appeared, not even in the *Westminster Review*. Darwin remained silent, and such later Darwinians as Huxley and Lyell were at this time still firmly opposed to the evolution hypothesis, even to Owen's paleontological progressivism (although they may have sympathized with German autogeny thinking);[65] and Owen's ever-loyal supporter Broderip gently nudged him away from the dangerous precipice of outspoken evolutionism. In his analysis of Owen's "Generalizations of Comparative Anatomy," written—one should be reminded—with Owen's participation, he denounced *Vestiges* ("the only real merit of which lies in its clever literary composition") and presented Owen's position as a creationist refutation of transmutation.[66]

A change of emphasis and the replacement of "secondary laws" by the ambiguous "creative acts," which could refer to divine, as well as to natural, creativity, remained Owen's shelter for a number of years. The effect of the Ray Society uproar and the *Manchester Spectator* incident was obvious in the popular rendition of his osteological work, issued in *Orr's Circle of the Sciences*

(1854). The hypothesis that a reptile was transmuted into a mammal or an ape into a Negro was unsubstantiated—Owen declared. No known cause of organic change, he maintained, was capable of altering the premaxillary bones, which so remarkably distinguish the gorilla, the highest of all apes, from the lowest race of mankind. Such differences contravene the hypothesis of transmutation, until—he added tantalizingly—the modifiability of such characters can be demonstrated. This time he ended not with an affirmation of his belief in "secondary causes" but with an almost exact quotation from his opaque response to the *Manchester Spectator*: "Of the nature of the creative acts by which the successive races of animals were called into being, we are ignorant. But this we know, that as the evidence of unity of plan testifies to the oneness of the Creator, so the modifications of the plan for different modes of existence illustrate the beneficence of the Designer."[67]

While keeping away from public controversy, Owen continued to ponder the zoogeographic evidence for a natural emergence and also disappearance of species. In one of his many contributions on the dinornis read before the Zoological Society (1850) he discussed, and rejected, the hypothesis that the large extinct mammals of South America and Australia, and the extinct giant birds of New Zealand, could have given rise to the smaller, allied species that populate these regions today. Such a hypothesis might find support in the law of continuity of zoogeographic provinces. But Owen discounted the notion that today's comparatively diminutive forms had originated by a process of degeneration in size from any gigantic fossil precursors. This idea of degeneration, which had been suggested by Buffon in his study of the Elephantidae and which many Romantics had enthusiastically embraced, seemed confirmed by the geographic and taxonomic proximity of the fossil giant sloth and the very much smaller living sloth. It seemed further supported in the course of the 1830s and 1840s by the discovery of the gigantic glyptodon from South America, much larger than today's comparatively tiny armadillo; of the huge diprotodon from Australia, also much larger than any present-day marsupial; and of the dinornis from New Zealand, which towered over the still-living kiwi.

However, the small species in question were not of a recent origin—Owen pointed out—but their fossil remains were found associated with those of the extinct monsters. In England, too, today's mole, water vole, hare, weasel, stoat, badger, and fox had coexisted with the large hyenas, bears, and other beasts from the Pleistocene that cave studies had brought to light. Owen concluded that there was "no particle of evidence" to suggest that the extinct giants had been modified and become today's dwarf species. The Goliaths of the geological past had simply died out; and the fact that their taxonomic relatives were still living in the same region indicated that the process of extinction had not been the result of a universal deluge, "no general sweeping away of the peculiar aboriginal land animals of those continents or islands," but that "the smaller and feebler animals have bent and accommodated themselves to changes which have destroyed the larger species."[68]

In other words, Owen believed that extinction may have been a function of the size of an animal. The larger a species is, the less adaptable it will be to a major change in environmental conditions. A drought, a diminished water and food supply, or the introduction of new enemies will affect larger species more than smaller ones. Also, the latter tend to be more numerous. It is the "contest . . . against the surrounding agencies" which explains the extinction of species. This view bears more than a superficial resemblance to Darwin's "the struggle for life," and in order to solve a priority dispute between the two men that flared up in the early 1860s we need to quote Owen's actual words:

> In proportion to its bulk is the difficulty of the contest which, as a living organized whole, the individual of such species has to maintain against the surrounding agencies that are ever tending to dissolve the vital bond, and subjugate the living matter to the ordinary chemical and physical forces. Any changes, therefore, in such external agencies as a species may have been originally adapted to exist in, will militate against that existence in a degree proportionate, perhaps in a geometrical ratio, to the bulk of the species. If a dry season be gradually prolonged, the large Mammal will suffer from the drought sooner than the small one: if such alteration of climate affect the quantity of vegetable food, the bulky Herbivore will first feel the effects of stinted nourishment: if new enemies are introduced, the large and conspicuous quadruped or bird will fall a prey, whilst the smaller species conceal themselves and escape. Smaller animals are usually, also, more prolific than larger ones.[69]

The notion of a struggle for existence, both between different species and between individual members of a single species, was already a familiar one during the first half of the nineteenth century and could be found, for example, in Lyell's *Principles of Geology*.[70] Moreover, in the third volume of his magnum opus Lyell had discussed extinction in terms of physical conditions, especially climate. Yet to formulate something akin to a law of extinction, as was done by Owen, was still a pioneering effort. In the same year that Owen put forward the "contest against the surrounding agencies" as a cause of extinction, his close and competitive colleague Mantell maintained that "we are as utterly ignorant" of the causes of extinction of species as we are of their origin; "both are veiled in inscrutable mystery,—the results only are within the scope of our finite comprehension."[71]

The BAAS Address of 1858

As president of the BAAS in 1858 Owen had the opportunity to speak to a national audience of colleagues on issues of his choosing. His address to the plenary session of the Leeds meeting, on September 22, is a fascinating document. In this Owen pays some attention to progress in astronomy and physics, but the larger part of his lengthy address, which may not have been read *in toto*, concerns his own interests. Thus, he presented an outline of his homological research program and also discussed the museum question in addition to the larger issue of the relationship of science and government. More specifically, he returned to the geographic distribution of animals, today and in the geological past. To the rule of continuity from the past to the present, exemplified most dramatically by Australia's marsupial fossil and living species, Owen added another one, namely that the further back in time we go, the less applicable the present geographic distribution of animals becomes[72] because of major changes throughout geological history in the relative positions of land and sea that provided links and bridges where today there are none.

This additional rule was an elaboration on Edward Forbes's internationally acclaimed paper of 1846 on the geological changes that have contributed to the present-day distribution of British flora and fauna.[73] Now—to paraphrase Owen—such changes complicated and weakened the geographical evidence for postdiluvial creations of species in situ as postulated by Prichard.[74] Moreover, "creation" was a concept to which biologists would in due course be able to give a naturalistic meaning. Owen then showed a glimpse of the German theory of autogenesis when he tortuously stated that, even if those scientists were right who supposed that species were indigenous "creations" that had spontaneously come into being within particular provinces of distribution, there would still be reason to believe in a Creator who had decreed those natural origins.

> These phenomena [land bridges and species migrations] shake our confidence in the conclusion that the Apteryx of New Zealand and the Red-grouse of England were distinct creations in and for those islands respectively. Always, also, it may be well to bear in mind that by the word "creation," the zoologist means "a process he knows not what." Science has not yet ascertained the secondary causes that operated when "the earth brought forth grass and herb yielding seed after its kind" and when "the waters brought forth abundantly the moving creature that hath life." And supposing both the fact and the whole process of the so-called "spontaneous generation" of a fruit-bearing tree, or of a fish, were scientifically demonstrated, we should still retain as strongly the idea, which is the chief of the "mode" or "group of ideas" we call "creation"

> viz. that the process was ordained by and had originated from an all-wise and powerful First Cause of all things.[75]

Between the lines of obfuscating biblical quotations one can read that Owen had been led away by his zoogeographical studies from the doctrine of special creation toward a belief in natural causes, whatever these may have been; and in front of the gathered multitude he declared that the appearance of reptiles, birds, and beasts had to be explained on the basis of "[t]he axiom of the continuous operation of Creative power, or of the ordained becoming of living things."[76] Although his language was ornately traditional, his message was indubitably noncreationist. As before, he declared himself ignorant of precisely what laws of nature had been at work in the "creating" of new species. He may have intended to hint at his preference for heterogeny by adding to his address an outline of his own and other people's work on parthenogenesis.

More explicitly, he referred favorably to the ideas that Darwin and Wallace had only just put forward during the historic meeting of the Linnean Society on July 1 of that year, less than three months before the BAAS meeting. Owen cited environmental changes as the cause of extinction; Darwin and Wallace had proposed the same changes as a mechanism for the origin of varieties. Owen quoted Darwin's example of an island on which canines such as dogs or foxes preyed on rabbits and to a lesser extent on hares. A slight "plasticity" of the dogs or foxes, combined with a decrease in the number of rabbits and an increase in the hare population, would favor the canines "with the lightest forms, longest limbs, and best eyesight." In the course of a thousand generations, the dog or fox would become adapted to the catching of hares instead of rabbits, just as the speed of greyhounds can be improved by selection and breeding. Owen cautioned that the degree of plasticity or variability needed to be established by observation of animals in the wild, and that the fossil record should be scrutinized to discover the "ante-types," the "more generalized structures," from which varieties might have originated, "so as to give rise ultimately to such extreme forms as the Giraffe for example."[77]

Owen was one of the few naturalists, if not the only one, who some time before the publication of *The Origin of Species* publicly alerted the scientific community to the importance of the novel ideas of Darwin and Wallace;[78] and while not losing sight of empirical obstacles to a theory of descent, he hinted that his studies of the fossil record might reveal the steps that had led from a more general form to such extreme adaptations as the giraffe's long neck. Only a few months later, however, in his Fullerian lectures of January and February 1859, just weeks after Darwin and his cronies had come out against Owen's plans to keep the British Museum's natural history collections together and provide spacious accommodation for their exhibition and storage, Owen retreated to an agnostic position with respect to the origin of species. He reiterated his belief that species had become extinct as a result of changes in their environment, not because it was "inherent in their own nature." As to "the more mysterious subject of their coming into being," Owen confessed that he had nothing to say that was "profitable or to the purpose." One could speculate that new species had come into being as a result of particular varieties having been selected by changing conditions, as suggested by Darwin and Wallace; "but to what purpose?" Owen asked. The fact was that in the case of the mammalian class—the subject of Owen's Fullerian course—successive extinctions of Mesozoic forms had been followed by the introduction of more numerous, more varied, and more highly organized forms during the Tertiary. It is just as unlikely that an ichthyosaurus will be found swimming in the Pacific as that the fossil bones of a whale will turn up in the Liassic. With respect to extinctions, Owen did not subscribe to catastrophist explanations:

> Not that the extinction of such forms or species was sudden or simultaneous: the evidences so interpreted have been but local: over the wider field of life at any given epoch, the change has been gradual; and, as it would seem, obedient to some general, but as yet, ill-comprehended law. In regard to animal life, and its assigned work on this planet, there has, however, plainly been "an ascent and progress in the main."[79]

This Fullerian lecture, "On the Extinction and Transmutation of Species," was printed as an appendix to Owen's Rede lecture *On the Classification and Geographical Distribution of the Mammalia* (1859). Before its publication, a period of fifteen years had elapsed during which Owen had repeatedly drawn attention to the difficulties which the existence of past and present zoogeographical provinces present to the doctrine of special creation and had highlighted their importance in speculating about a naturalistic origin of species, cautiously drawing attention to Continental theories of autogenesis. He had played a major role in the discovery and delineation of the provinces to which some Late Tertiary and Pleistocene mammalian fossils belonged. He had been the leading authority on the issue of a progressive succession of fossil types in opposition to the Lyellian doctrine of nonprogression. He had advanced a noncatastrophist, ecological theory of extinction. And although, with respect to the mechanism of the origin of species, he had consistently preached caution, it was no secret that Owen had nailed his colors firmly to the mast of an origin of species by natural law, notwithstanding the fact that at times he obscured this by using a reverential, biblical phraseology.

Clash with Darwin

Imagine Owen's feelings when in November 1859 *The Origin of Species* appeared, in which Darwin cast Owen in the role of a leading advocate of the immutability of species, lumping him together with outspoken antievolutionary creationists such as Agassiz and Sedgwick. Owen was outraged, and whereas he had passed up the opportunity to review *Vestiges* all those fifteen years ago, this time he forsook caution and wrote an essay on *The Origin of Species* for the *Edinburgh Review*, the editor of which was his fellow club member Henry Reeve. It was a condemning review. Some have cited Owen's critique as evidence that he opposed evolution,[80] but it shows nothing of the sort. In fact, the review was most of all a forceful reiteration of Owen's own evolutionary stance, to which was added a critique of natural selection. In addition to Darwin's book, nine other publications were included in this review, among them the French translation of Owen's own work on the archetype, his presidential address to the BAAS, and his *Palaeontology*, which had just come off the press.

In Owen's view, Darwin made the fundamental mistake of confusing the issue of the origin of species by natural law with that of the nature of such a law. Apart from some well-known transmutationists, zoologists had kept aloof from any hypothesis of the origin of species.

> One only [Owen wrote in reference to himself], in connexion with his palaeontological discoveries, with his development of the law of irrelative repetition and of homologies, including the relation of the latter to an archetype, has pronounced in favour of the view of the origin of species by a continuously operative creational law; but he, at the same time, has set forth some of the strongest objections or exceptions to the hypothesis of the nature of that law as a progressively and gradually transmutational one.[81]

Owen quoted from his *Palaeontology* "that perhaps the most important and significant result of palaeontological research has been the establishment of the axiom of *the continuous operation of the ordained becoming of living things.*" With respect to the mode of operation, various ideas had been put forward by the French (Buffon, Demaillet, Lamarck, Geoffroy) and by the author of *Vestiges*; but—asked Owen rhetorically—is the entire series of phenomena of parthenogenesis or "alternation of generations," on which he and others had worked, of no relevance with respect to the introduction of new species on this planet? This was an unambiguous indication that Owen had visualized the evolutionary process as saltational heterogeny, analogous to the cycle of metagenesis. As this and other possibilities had by no means been exhaustively explored, why consent to Darwin's additional speculation? "The natural phenomena already possessed by science are far from being exhausted on which hypotheses, other than transmutative, of the production of species by law might be based, and on a foundation at least as broad as that which Mr. Darwin has exposed in this Essay."[82]

It was not that Owen supported any rival hypothesis to that of natural selection; "we merely affirm that this at present rests on as purely a conjectural basis." If the struggle for life indeed represented a mechanism of natural selection, it would have to be demonstrable in such organs of struggle as the antlers of deer. Even though antlers are the most variable appendages of the quadruped, millennia of combat have not produced changes beyond the specific limit. Individual variability obviously exists; but

> [w]e have searched in vain, from Demaillet to Darwin, for the evidence or the proof, that it is only necessary for one individual to vary, be it ever so little, in order to [validate] the conclusion that the variability is progressive and unlimited, so as, in the course of generations, to change the species, the genus, the order, or the class. We have no objection to this result of "natural selection" in the abstract; but we desire to have reason for our faith. What we do object to is, that science should be compromised through the assumption of its true character by mere hypotheses, the logical consequences of which are of such deep importance.[83]

Some of Owen's clergyman-naturalist friends concurred. Wilberforce, who reviewed Darwin's *Origin of Species* for the *Quarterly Review*, emphasized that neither in nature nor even under domestication do species vary beyond fixed limits, and that the discovery of mummified animals in the graves of Egypt's pharaohs had shown "that there has been no beginning of transmutation in the species of our most familiar domesticated animals."[84] Wilberforce has been ridiculed as an ecclesiastical ignoramus in matters of natural history,[85] but he was in fact one of Copleston's Oriel fellows who had sat attentively at Buckland's feet and in later years had also attended Owen's Hunterian course. "I could give the Bishop of Oxford a certificate for most regular attendance," Owen wrote in 1850.[86]

In his presidential address to the BAAS Owen had maintained that the question of spontaneous generation of the simplest life form "cannot be answered until every possible care and pains have been applied to its solution."[87] Darwin, too, left the problem of the primordial origin of life unsolved, wrapping it in a biblical phrase when he stated "that probably all the organic beings which have ever lived on this earth, have descended from some one primordial form, into which life was first breathed."[88] This gave Owen a chance to "out-naturalize Darwin," as Neal Gillespie puts it,[89] and Owen asserted that

Félix-Archimède Pouchet's work *Hétérogénie*, which had just appeared, did now answer the question of spontaneous generation—in the affirmative.

Owen's self-portrayal as the misrepresented and ignored evolutionist went on till the very last page of his review. The assumption that Owen wrote this critique of *The Origin of Species* in subservience to the powers-that-be, or to his own or his patrons' religious belief in Genesis,[90] is incorrect. "We have no sympathy whatever"—Owen declared openly—"with Biblical objectors to creation by law, or with the sacerdotal revilers of those who would explain such law."[91] Why then had Owen changed his reaction to natural selection since the BAAS meeting in Leeds from "could be" to "can't be"? Was it just jealousy, as Darwin himself suggested, and with him many historians since?[92]

It is perfectly possible, even likely, that Owen, who lacked magnanimity, did feel pangs of jealousy when he read Darwin's magnum opus, given its unparalleled breadth and masterly interweaving of hitherto-loose strands of natural history. Most galling of all may have been the fact that Darwin finally succeeded in unifying the two schools of anatomical thought represented by Cuvier and Geoffroy, that is, of functional and transcendental anatomy, using the phenomenon of adaptation in his concept of natural selection and interpreting the "unity of type" as evidence of "common descent." Owen, whose long-term ambition it had been to bring about this unification, had succeeded in no more than demonstrating that the two were not mutually exclusive, that attention to "final causes" did not preclude an interest in "first causes." In his essay "Darwin on the Origin of Species" for the *National Review*, W. B. Carpenter urged respectful consideration for Darwin's work, because "it brings into mutual reconciliation the antagonistic doctrines of two great schools—that of Unity of Type, as put forward by Geoffroy St. Hilaire and his followers of the Morphological School, and that of Adaptation to Conditions of Existence, which has been the leading principle of Cuvier and the Teleologists."[93]

Professional jealousy was by no means the whole story, however. More substantively, Owen fundamentally objected to the concept that a process of chance phenomena is the cause of the origin of life and its history, even though he was willing to assign a limited, subsidiary role to natural selection, just as he had done to the Lamarckian inheritance of acquired characteristics. Furthermore, as we have seen above, shortly after Owen had initiated the drive to establish a separate museum of natural history in his presidential address of 1858, and considerably before the appearance of his essay on Darwin in the *Edinburgh Review*, the Darwinians had attempted to sabotage Owen's plans for what was to become his most monumental accomplishment—the Natural History Museum in South Kensington. A systematic attack on Owen by Huxley, too, had already begun. When Darwin then lumped Owen together in *The*

Origin of Species with a number of vehemently outspoken opponents of evolution, this must have seemed an intentional misrepresentation, part of the anti-Owen strategy. Darwin's correspondence with Hooker, Huxley, Lyell, and a few other friends indeed shows a conspiratorial obsession with Owen and a tendency to suspect Owen's hand in more attacks on the Darwinians than those for which he was actually responsible. A critique of Lyell's *Antiquity of Man*, for example, published in the *Quarterly Review*, was ascribed to Owen,[94] although it was in fact written by John Phillips, by then professor of geology at Oxford.

Darwin later attributed his "misunderstanding" of Owen's position to the difficulty "to understand and to reconcile with each other" Owen's "controversial writings,"[95] but this was a weak excuse. The *Manchester Spectator* had experienced no difficulty in perceiving the implications of Owen's advocacy of the origin of species by "natural laws or secondary causes"; nor had several creationist fellows of the Geological Society. Conybeare, for one, disagreed with Owen as to the extent of the operation of natural laws in producing organic diversity.[96] Sedgwick, too, was worried about Owen's noncreationist recourse to nature's laws.[97]

Conversely, the pro-evolution Baden Powell fully understood the naturalistic implications of Owen's *Nature of Limbs* and paid ample tribute to it in his "Essay on the Philosophy of Creation" (1855).[98] Across the Atlantic, the pro-Darwin Asa Gray admitted, albeit with a well-targeted sarcasm, that he had appreciated the drift of Owen's position, which had prepared him—Gray—to a certain extent, for Darwin's book.[99]

Moreover, the people who attended Owen's Hunterian lecture courses, the last of which was delivered in 1855, had not come away with the impression that Owen was a creationist. The Duke of Argyll recalled: "I had heard many of his lectures, and all these had left on my mind an impression that he himself believed that somehow or another different races of animals had descended from each other by ordinary generation."[100] Mivart also confirmed that his revered teacher had in no way been averse to the idea of evolution; "on the contrary, he was its decided advocate."[101] In his review "Darwin's Descent of Man" for the *Quarterly*, Mivart pointed to another feature of Darwin's writing which may have given Owen just cause for chagrin, namely the panegyrics which Darwin bestowed upon the names and opinions of his followers, whereas colleagues of equal merit, but who did not agree with Darwin, Owen in particular, were quoted for the most part without commendation.[102]

Owen's reviled review of 1860 was not what in Darwinian mythology it has become, namely an obscurantist attack on an honest and persecuted prophet of truth.[103] The self-portrayal by Owen as a misrepresented evolutionist was a true likeness, and his discussion of natural selection correctly exposed a weak link in the chain of Darwin's reasoning, namely the nature of organic variability. All the same, Owen's was not an admirable essay. His reviews in several instances were poorly composed with unnecessary repetitions, and he would rather refer to his own work than the book under review. The essay on *The Origin of Species* was no exception. His self-preoccupation was combined with anonymous self-praise, and this "bad form" undermined Owen's otherwise perfectly tenable position with respect to Darwin and his theory. Owen never owned up to having been the author of the review in question, and his grandson, too, left open the possibility that someone else might have written it, quoting a letter by Sedgwick in which he asked Owen: "Do you know who was the author of the article in the 'Edinburgh' on the subject of Darwin's theory? On the whole, I think it very good. I once suspected that you must have had a hand in it, and I then abandoned that thought."[104]

Post-*Origin* Clarity

Owen's textbook *Palaeontology* appeared very shortly after Darwin had set the cat among the pigeons, but it did little to restore calm. In the introduction, Owen merely restated his "oracular" axiom "of the continuous operation of the ordained becoming of living things," and the book's conclusion of exactly eleven pages of text was no more than a collage of paragraphs from his presidential address of 1858, his concluding Fullerian lecture of 1859, and some remarks he was about to repeat in his essay for the *Edinburgh Review* of 1860. The *Athenaeum*, which had printed Owen's Fullerian lecture, noticed the repetitiveness with disappointment. Darwin had written "boldly"; why should Owen remain "reserved"? "While merely summarizing dentitions and structures, he is hidden in the valley of dry bones."[105] The reviewer ended his essay by admonishing the "accomplished successor of Cuvier" to declare all that is "within the power of his pen." Although the review of Owen's paleontology textbook in *Fraser's Magazine* was not only much longer (forty-two columns) than the *Athenaeum*'s but also more sympathetic, it, too, regretted his "utmost caution" and his "obstinate silence."[106]

Such public admonitions, coming on top of Darwin's misrepresentation, were indications that Owen's cautious strategy no longer worked. He had been content to add to the observational basis of the fact of evolution, obscuring his belief in that fact by wrapping it in Pentateuchal phrases and keeping speculations about the mode of evolution at arm's length. At a minimum, more clarity was required. In 1862, with his answer to the *Times* museum editorial out of the way, Owen addressed the BAAS "On the Characters of the Aye-aye, as a Test of the Lamarckian and Darwinian Hypothesis of the

FIGURE 16. The aye-aye. Owen believed that its attenuated middle finger, used for extracting grubs from their burrows, could not be explained by Darwinian natural selection. Such adaptations had evolved, he thought, as the result of sudden mutations in accordance with a predetermined plan of evolution. (Owen, "On the Aye-aye," pl. 18.)

Transmutation and Origin of Species."[107] He included an expanded version of this talk in his beautiful *Monograph on the Aye-aye* (1863), having first lectured on the subject to the Zoological Society. Via the colonial secretary for Madagascar, Owen had received an adult, male *Chiromys madagascariensis.* It feeds on, among other things, wood-boring grubs by gnawing down to their burrows with its strong incisor teeth and extracting the larvae with the claw of its attenuated middle finger (fig. 16).

This most remarkable of the Malagasy lemurs (some now regard it as separate from lemurs) provided Owen with a unique opportunity to test hypotheses of the origin of its peculiar adaptations and to elucidate his own views. He began by clarifying that the succession of species by a continuously operating law was not a "blind operation" but a process of preconceived progress, guided by divine volition. In a footnote he recalled the 1849 fracas with the *Manchester Spectator*. To the accusation that his *Nature of Limbs* had contributed to "blotting God out of creation" instead of providing an argument from design, he answered: "Could the pride of the heart be reached whence such imputation came, there would be found, unuttered,—'Unless every living thing has come to be in the way required by my system of theology, Deity shall have no share in its creation.'"[108] No wonder that Asa Gray called some of Owen's phrases "oracular" and "sonorous."[109] Fortunately, the fog of verbal obfuscation lifted in the paragraphs that followed. Owen explicitly and lucidly stated that he subscribed to the "derivative hypothesis" of the origin of life by the spontaneous generation of a variety of primordial forms, and the origin of species by organic descent. Such a belief did not imply the adoption of either the Lamarckian or the Darwinian mode of evolution. The following is the first comprehensive summary of Owen's evolutionary beliefs:

> What I have termed the "derivative hypothesis" of organisms, for example, holds that these are coming into being, by aggregation of organic atoms, at all times and in all places, under their simplest unicellular condition; with differences of character as many as are the various circumstances, conditions, and combinations of the causes educing them,—one form appearing in mud at the bottom of the ocean, another in the pond on the heath, a third in the sawdust of the cellar, a fourth on the surface of the mountain rock, etc., but all by combination and arrangement of organic atoms through forces and conditions acting according to predetermined law. The disposition to vary in form and structure, according to variation of surrounding conditions, is greatest in these first-formed beings; and from them, or such as them, are and have been derived all other and higher forms of organisms on this planet. And thus it is that we now find energizing in fair proportions every grade of organization from Man to the Monad. Each organism, as such, also propagates its own form for a time under such similitude as to be called its kind.
>
> Specific characters are those that have been recognized in individuals of successive generations, propagating similar individuals, as far back as observation has reached; and which characters, not being artificially produced, are ascribed to nature. Instead of referring such characters to an originally distinct creation, the derivative hypothesis, whilst admitting their transmissibility and their maintenance for an unknown period through generative powers obstructive of departure from such characters, holds that observation has not yet reached the actual beginning of such species, nor the point at which variation stops.[110]

Species were, however, fixed for the descriptive purpose of the taxonomist; "they will last our time." Furthermore, the "derivative hypothesis" was merely "a kind of vantage-ground artificially raised to expand the view of the outlooker for the road to truth." This road did not run via the transmutational hypothesis, if only because it failed the aye-aye test. Lamarck's notion that the increased use of an organ increases its development through a succession of generations cannot explain the incisors of the *Chiromys.* These teeth already possess their specific size and shape before protruding from the gum. "The great scalpriform front teeth thus appear to be structures fore-ordained—to be predetermined characters of the grub-extracting Lemur; and one can as little conceive the development of these teeth to be the result of external stimulus or effort, as the development of the tail, or as the atrophy of the *digitus medius* of both hands."[111] As far as the Darwinian "preservation of favoured races in the struggle for life" was concerned, there was no evidence of a shortage of food for, and therefore no selective pressures on, the various lemurs. "We know of no changes in progress in the Island of Madagascar, necessitating a special quest of wood-boring larvae by small quadrupeds of the Lemurine or Sciurine types of organization."[112]

Yet Owen did believe that the various Lemuridae were related by descent from a common ancestral form. The evidence for the phenomenon of descent existed, in general, in the fossil succession and "the progressive departure from a general to a special type"; in the analogy of embryonal stages in higher species to the mature forms of lower species; and in the "unity of type" which Owen had demonstrated in his homology work. Nonadaptive organs, too, negated an origin of species by primary creation. He hinted at the saltatory way in which descent may have occurred as indicated by the "the phenomena of parthenogenesis." Owen ended by stating that further work to determine the way in which species originate would prove beneficial to the progress of zoology.

In spite of his general dismissal of Lamarckism, Owen in fact used the French hypothesis for the particular instance of his flightless birds. In 1865 Owen received via the bishop of Mauritius several bones of the *Didus ineptus,* which formed the subject of yet another first-rate monograph, *Memoir on the Dodo* (1866). The dodo had died out because of overkill by Dutch East India sailors, who used Mauritius as a stopover and from whom the birds had possessed no means of escape. They had lost their capacity of flight, Owen speculated, as a consequence of the disuse of their wings and the extra use of their legs, that is, as a result of Lamarckian atrophy and hypertrophy.[113]

This Lamarckian "easy and seductive vein of speculation" was not surprising in Owen, given his interest for many years in the wingless birds of New Zealand. It would have been surprising indeed, however, if he had given assent to Darwinian natural selection—an assent that the *London Review,* in a notice of the first two volumes of Owen's three-volume *On the Anatomy of*

Vertebrates (1866–68), suggested he had in fact given. The preface included a few paragraphs that repeated almost verbatim Owen's views on the extinction and origin of species as expressed in his address of 1858. The reviewer noted "a significant though partial admission which he makes of the truth of the principles of Natural Selection."[114] The *Popular Science Review* also commented that Owen's book "contains a partial concurrence, on the part of the author, in the theory of *natural selection*."[115] In the introduction to the first volume, Owen had repeated his long-standing theory of the extinction of large mammals, enunciated as long ago as 1850 in a paper on the dinornis, namely that they are less well equipped than small species to cope with environmental changes in "the contest . . . against the surrounding agencies." This time he added the expression that the smaller species "have fared better in the 'battle of life.'" In a "courteous and temperate letter" to the editor of the *London Review*, Owen objected to the imputation of his concurrence with the Darwinian view; if he was thought to have followed Darwin's magnum opus of 1859, then the fact that he—Owen—had actually expressed his theory in 1850 would place Darwin in the position of "adopter."[116]

Darwin immediately seized this opportunity, in the fourth edition of *The Origin of Species* (1866), to maneuver his opponent into the embarrassing position of having claimed that he—Owen—had given to the world the theory of natural selection long before the appearance of *The Origin of Species*.[117] Some historians, uncritically taking the Darwinian line, have accused Owen of shameless inconsistency, anonymously attacking Darwin's ideas and publicly claiming priority in having developed them. Even Roy MacLeod maintained in his pioneering 1965 paper on Owen "that Owen had become more convinced of the probable truth of Darwin's concept, and was trying desperately to regain his fallen prestige by accepting and claiming as his own as much of the Darwinian synthesis as he could."[118] A number of leading Darwin scholars, including James Moore and Michael Ruse,[119] have put forward the same interpretation of what MacLeod termed "this blot upon his honor."

There was nothing disingenuous, however, in Owen's claims, and the allegation that Owen was a turncoat with respect to the idea of natural selection was a further instance of misrepresentation by Darwin and his allies. Darwin's complaint has often been cited, namely that Owen in his review of *The Origin of Species* "scandalously misrepresents many parts."[120] Owen's review, as we have seen, was calculated to present an accurate picture, not so much of Darwin, as of Owen himself. Darwin's accusation has never been critically examined. No Darwin scholar has considered the possibility that Darwin was perfectly capable of misrepresenting Owen. The latter had publicly enunciated his notion of the "contest with surrounding agencies" nearly a decade before

Darwin came out of the closet, and in the course of the 1850s, Owen had applied it on several occasions to the extinction of species. During the early 1860s, unwisely perhaps, he changed his original, awkward phrase of "contest against the surrounding agencies" to the Darwinian-sounding "battle of life" and "contest for existence." But this was merely a terminological adoption; at no time did Owen suggest that competition between animals had contributed to the origin of species. On the contrary, he repeatedly maintained that it explained extinctions only.[121]

The *Quarterly Review*, through the pen of J. B. Mozley, regarded Owen's case as just, reiterating that he had put forward "natural selection" as "the cause of *extinction* of species," whereas Darwin's theory "applies not only to the extinction, but also to the *origin* of species."[122] The accusation by the *London Review* and by Darwin irritated Owen, and he returned to the issue in a comparatively long entry on "species" for the new edition of Brande and Cox's *Dictionary of Science, Literature, and Art* (1867). It was Owen's most lucid, concise, and well-written contribution to the controversy on the origin of species, and the one most readily available to a large readership. He summarized the main evidence for, and the main theories of, a natural origin of species, adding that changes in external conditions are the cause of a "*natural rejection*" of species, not of a "*natural selection*," as Darwin maintained. External stimuli, whether those of French or of English transmutation theories, do not add up to an adequate cause of new species, Owen argued. Take, for example, the whales with their gradational and broad range of morphological features, especially of skull and dentition. The wide open oceans simply do not present the required environmental pressures to produce the different whale species, and only an innate, congenital departure from parental type can have produced their known diversity of form.[123]

The Derivative Hypothesis of 1868

The fullest refutation of the Darwinian accusation that Owen had claimed priority in formulating the theory of natural selection appeared as a long footnote to the concluding chapter of the third volume of Owen's *Anatomy of Vertebrates* (1868). This chapter is the longest and most detailed statement of where he stood with respect to the theory of evolution. It was reprinted in its entirety in the *American Journal of Science*[124] and published as a separate, forty-page booklet entitled *Derivative Hypothesis of Life and Species* (1868). Owen reminded his readers of Cuvier's opinion that, if evolution had occurred, one should be able to find intermediary forms in the fossil record, connecting, for example, the genus *Palaeotherium* with today's hoofed quadrupeds.[125]

This condition had now amply been met, Owen maintained, in particular by the discovery of *Hipparion*, a fossil genus of the Equidae, from the Miocene at Eppelsheim and the Sewalik Hills in India. Their molar teeth differed from those of existing horses mainly in the greater extent of separation of the inner column; and the second and fourth digits of their feet, suppressed in the form of splint bones in the existing horse, were retained in a functional state.

In his lectures on fossil mammals at the School of Mines in 1857, Owen had earlier illustrated the consecutive modifications of teeth and feet in the morphological series from *Palaeotherium* to *Hipparion* and eventually to today's horse. He rejected the possibility that these forms were successive, separate creations: "If the alternative—species by miracle or by law?—be applied to *Palaeotherium, Paloplotherium, Anchitherium, Hipparion, Equus*, I accept the latter, without misgiving, and recognise such law as continuously operative throughout Tertiary time."[126] If no direct, creative acts had taken place, how then had *Palaeotherium* ultimately become *Equus*? The development from a three-hoofed to a one-hoofed foot, the latter with splint bones, might be explained on the basis of Lamarckian volition or Darwinian selection, but these hypotheses "are less applicable, less intelligible, in connection with the changes in the structure and proportion of the molar series of teeth." Owen envisaged no gradual transmutation but sudden "mutations." He drew attention to the phenomenon of reverse evolution exemplified by modern horses that are born with supplementary hoofs in which the second and fourth metacarpals, normally present in the form of splint bones, carry phalanges and a terminal hoof. The feet of such polydactyl horses resemble those of *Hipparion*, and the pairing of a tridactyl colt with a tridactyl filly "might restore the race of hipparions."[127]

> Now the fact suggesting such possibility teaches that the change would be sudden and considerable: it opposes the idea that species are transmuted by minute and slow degrees. It also shows that a species might originate independently of the operation of any external influence; that change of structure would precede that of use and habit; that appetency, impulse, ambient medium, fortuitous fitness of surrounding circumstances, or a personified "selecting Nature," would have had no share in the transmutative act.[128]

Thus, Owen continued to prod the Darwinians about the nature of organic variability. Objecting to a chance process of variation, he believed that evolution was a teleological process, a movement toward a preordained goal. Mutations were not randomly useful or useless but a logical embroidering on the archetype. This deterministic interpretation of evolutionary progress had been implicit in Owen's use of metagenesis as an analogy to evolution. He no longer referred to his work on parthenogenesis, however, but he did reiterate his belief in the teleological nature of the history of life. Of all the quadrupeds, he stated, none had proved of more value in war and peace than the horse. The fact that ungulates with equine modifications had originated at about the time of the earliest evidence of the human race was no coincidence; they were meant for each other:

> No one can enter the "saddling ground" at Epsom, before the start for the "Derby," without feeling that the glossy-coated, proudly stepping creatures led out before him are the most perfect and beautiful quadrupeds. As such, I believe the horse to have been predestined and prepared for Man. It may be weakness; but, if so, it is a glorious one, to discern, however dimly, across our finite prison wall, evidence of the "Divinity that shapes our ends," abuse the means as we may.[129]

From among the invertebrates, Owen selected the reef-building corals as a further test of evolutionary hypotheses. The calcareous cups of certain corals show internal, longitudinal partitions, or septa, which in some types are arranged in a pattern of four or its multiples and in others show a six-part symmetry. In the course of earth history, the hexapartite corals superseded the tetrapartite ones. Considering the very many species of Anthozoa that have come and gone in the course of geological time, the miracle of special creation becomes hard to believe, "inconsistent with any worthy conception of an all-seeing, all-provident Omnipotence!" A Lamarckian mode of evolution would not work either, "from the impotency of a coral-polype to exercise volition." Nor would a Darwinian mode—Owen asserted. Just as in the case of the whales, the required external influences do not exist in the oceanic environment. In the Red Sea, for example, a minimum of a hundred types of coral exist under more or less the same conditions:

> So, being unable to accept the volitional hypothesis, or that of impulse from within, or the selective force exerted by outward circumstances, I deem an innate tendency to deviate from parental type, operating through periods of adequate duration, to be the most probable nature, or way of operation, of the secondary law, whereby species have been derived one from the other.[130]

Whereas Darwin's theory left the origin of species to the chance occurrence of individual variations and milieu conditions, Owen's "derivation" of species proposed that the evolution of life followed a preordained course that was determined by an organism's "innate capacity or power of change." Peter Bowler has identified three fin-de-siècle scientific theories intended to undercut Darwin's theory of natural selection: the neo-Lamarckian transmission of induced somatic traits, the theory of orthogenetic progress in the fossil record, and the mutation theory.[131] Here Owen proposed a combined orthogenetic-mutational mechanism, and in specific instances such as that of flightless birds he resorted to a Lamarckian process of atrophy.

The reason Owen no longer used the phenomena of metagenesis to illustrate his concept of mutational and premeditated evolution was that he had abandoned his former belief in the theory of "omnis cellula ex cellulae," most recently represented by Pasteur's panspermist theory. To him, Pasteur's panspermism was akin to Darwin's hypothesis of "pangenesis" and of the evolution of all life from a primordial form "into which life was first breathed." The latter was a belief in "thaumatogeny," a miraculous origin of life, whereas Owen advocated "nomogeny," a coming into existence caused by natural law. Over the years Owen had paid little attention to the issue of spontaneous generation, although in his *Lectures on the Comparative Anatomy and Physiology of the Invertebrate Animals* (as in his letter to the author of *Vestiges*) he had explicitly rejected the hypothesis of "equivocal generation," calling it "gratuitous."[132] But now, in the heat of the Darwinian debate, Owen took the opposite stance, speculating in his *Anatomy of Vertebrates* that life in the form of "protogenal" jelly-specks originated as the result of a favorable concurrence of physicochemical conditions that produced "organic crystallisation."[133]

In a manuscript note on "spontaneous generation," Owen even more explicitly than in his *Anatomy of Vertebrates* contrasted himself with Darwin by means of this issue. Natural selection could work only once life existed, and to get over the hurdle between dead matter and a living creature a different mechanism is required:

> The doctrines of the *generatio spontanea* and of the transmutation of species are intimately connected. Who believes in the one, ought to take the other for granted, both being founded on the faith in the immutability of the laws of nature. If Darwin had adopted the theory of the spontaneous generation, we could not raise that objection against him, viz. that the continuous occurrence of the lowest forms of life is not explained by his doctrine, according to which they would have made way to more perfect forms. If there be spontaneous generation, the process of their creation is constantly going on.[134]

Accordingly, Owen sided with Pouchet in the latter's debate with Pasteur.[135] He admitted that Pasteur's experiments were very accurate but believed that they "do not settle the question, and may be explained in a different way." A fairly lengthy sketch of Owen's views *in casu* had already appeared in 1863 in the form of a review in the *Athenaeum* of Carpenter's magnificent *Introduction to the Study of the Foraminifera*. The anonymous reviewer, recognized by the Darwinians as Owen,[136] noted with disapproval Carpenter's belief that the vast diversity of foraminiferal forms does not reflect primordial differences but is the result of a process of divergence by genetic descent from "a small number of original types"[137] or from an original "father of all Foraminifera"–a notion that reminded the reviewer of Darwin's "primordial form, into which life was first breathed." This, he charged, was invoking an "occult, miraculous, creative cause operating once for all." "The inevitable corollary of the Darwinian hypothesis [i.e., the generation of the first form or forms of life by means of a divine act of creation] seems to us to demonstrate its weakness; and to show by contrast the superiority of the Lamarckian principle of the heterogeneous production of the primitive types of organisms." In the reviewer's opinion, microscopic organisms, including the foraminifera, were and continued to be formed by a spontaneous aggregation of dead matter; and he speculated

> that at the period when life became possible on the earth's surface the conditions were sufficiently varied to permit the conversion of the general polaric into a special organic mode of force to operate under circumstances resulting in a variety of the simplest forms of life, such as "monad," "mucor," "amoeba," "lichen-spore," etc., and that such conditions have continued to operate in the heterogeneous production of "organisms without organs" to the present day.[138]

Darwin, seeing the weakest part of his position prodded, was furious: "Who would ever have thought of the old stupid *Athenaeum* taking to Oken-like transcendental philosophy written in Owenian style!" he exclaimed to Hooker,[139] and a rare public exchange with Owen in the correspondence pages of the *Athenaeum* followed.[140] By trying to be more evolutionary than Darwin, Owen chose the side of the losers in the Parisian debate about spontaneous generation, but it should be added that he never went as far as Huxley, who, also in 1868, not only postulated spontaneous generation but mistakenly identified and named dead, albuminous matter as *Bathybius haeckelii*, believed to represent primordial life generated at the bottom of the oceans.[141]

The concluding chapter of Owen's *Anatomy of Vertebrates* failed to generate much debate. It did not offer the rebuttal of Darwin that his antievolution friends and followers were hoping for, and the Darwinians, by this time, had won the day and felt no appetite for spurious combat. Michael Foster did review Owen's book for the *Quarterly* but avoided any direct argumentation.[142] A rare mention of the "derivative hypothesis" was made by Mivart in his treatise *On the Genesis of Species* (1871), in which he commented that he had arrived "in complete independence" at the same views advocated by Owen.[143]

By the end of the 1860s Owen himself appeared to have grown tired of much further involvement in the evolution question. He fully applied himself to the study of colonial natural history and other topics directly linked to the advancement of his museum plans. Through the 1870s, however, he occasionally repeated some of his views about evolution, illustrating them with examples from new studies. In 1872, in the peroration to a long and detailed presentation at the Linnean Society, "On the Anatomy of the American King-Crab (*Limulus polyphemus*)," Owen speculated that *Limulus* is related to its Paleozoic predecessors by "a preordained plan of derivation by congenital departures from parental form." That the ancient ocean should have afforded the chance conditions of natural selection for the origin of crustaceous sub-classes, orders, genera, and species, Owen regarded as inconceivable: "I hold by the conviction that all forms and grades of *Articulata* are due to 'secondary cause or law' as strongly as when I expressed the same belief in regard to the *Vertebrata*, and defined it as 'the deep and pregnant principle in Philosophy' evolved in the researches on the General Analogies and Archetype of the Vertebrate Skeleton."[144] Moreover, the Darwinian theory was inadequate because it did not allow for imponderables such as aesthetics and ethics.

A new, more specific controversy flared up when Huxley suggested that bipedal dinosaurs had given rise to flightless birds, that is, to the *Struthiones* and *Cursores*, and that these, and not *Archaeopteryx*, were the ancestors of today's winged birds.[145] Owen did not agree. In the "Monograph of a Fossil Dinosaur (*Omosaurus armatus*, Owen) of the Kimmeridge Clay" (1875) and in his major collection *Memoirs on the Extinct Wingless Birds of New Zealand* (1879), he further developed his earlier idea that flightless and wingless birds have lost their ability to fly by process of degeneration, the *modus operandi* having been a Lamarckian atrophy of wings and hypertrophy of legs:

> By long disuse of the wings, continued through successive generations, those organs would become enfeebled, and ultimately atrophied to a degree effecting their capability to raise the body of the bird in the air.
>
> The legs then monopolizing all the functions of locomotion would attain, through the concomitant frequency of exercise, proportional increase of power and size.[146]

As examples, Owen cited the dodo and the solitaire from the Mascarene Islands, and the kiwi, moa, and related extinct birds from New Zealand. Before humans arrived on these islands, these isolated birds had known few, if any, foes, and in the absence of a stimulus to flight—in the sense of both fleeing and flying—wings would have become stunted. Owen supported his case by pointing out that most wingless birds are closely related to birds which do possess the capacity to fly and that a high percentage of wingless birds are found on isolated, "paradise" islands. Anyhow, Owen regarded it as more likely that birds had originated from flying reptiles and that *Archaeopteryx* had emerged from *Ramphorhynchus*. At the same time, however, Owen interpreted the pterosaurs as fully reptilian and therefore envisaged them as cold-blooded, bat-winged gliders—a view which led to a controversy with the King's College geologist H. G. Seeley, who maintained that pterosaurs were birdlike, that is, warm-blooded and active flyers.[147]

Although in this and some other instances Owen resorted to a Lamarckian mode of evolution, he stressed that this mechanism was of limited applicability. Generally speaking, the way in which "the secondary law of the origin of organic species" worked was not yet known. Biology was at the stage where astronomy had been in the days of Copernicus, who knew that the earth had a diurnal rotation around its axis and an annual rotation around the sun but could not explain away various objections. Owen persisted in believing that natural selection in particular was not an adequate explanation of evolution. In a presentation to the Geological Society, "On the Influence of the Advent of a Higher Form of Life in Modifying the Structure of an Older and Lower Form," he speculated that anatomical differences between the Mesozoic and the Neozoic crocodiles were related to changes in the nature of their available prey. The advent in Tertiary times of many species of large mammals, frequenting the banks of rivers and the shores of lakes, made it tempting for crocodiles to rush on to dry land to catch this new source of food. Improved mobility on dry land was made possible in the Neozoic crocodilian species by a procoelian ball-and-socket articulation of the trunk vertebrae, a lighter, dermoskeletal armor, various palatonarial features, and an increase in the size and strength of the forelimbs. When Seeley suggested that this study demonstrated the evolution of crocodiles in a struggle for life under changing conditions of food supply, Owen retorted that his paper was about adaptations, not about evolution, although he was willing to take account of the applicability of Lamarck's theory.[148]

A Copernican Analogy

Owen's last contributions to the evolution question, written around 1880, were autobiographical letters that looked back over the several decades of his involvement in the issue. In them he elaborated on the Copernican comparison that he had used as long ago as 1847 but that had been appropriated by Huxley in his review of *The Origin of Species* for the *Westminster Review* (1860). Here Huxley had suggested that just as the planetary orbits of the

Copernican system turned out to be not circular but elliptical, so too "the orbit of Darwinism" might turn out to be "a little too circular" and some organic phenomena not explicable in terms of natural selection.[149] By contrast, Owen used the Copernican analogy to clarify his distinction between the established fact of evolution and its as yet unknown mechanism. To the anthropologist C. O. Groom Napier he wrote:

> I regard the science of biology to be at the stage in which Copernicus left the science of astronomy when his successors began to explain their discoveries on the "heliocentric hypothesis." The geocentric one, prevalent, prior to Copernicus, was, like thaumatogeny [miraculous creation], in accordance with "scripture." But the accession of facts was such as to render their explanation of the mechanism it required too cumbrous. Copernicus suggested the heliocentric hypothesis, as facilitating explanation. To the obvious objection that all things loose would be whirled off by his supposed rotation of earth on her axis in twenty four hours, neither Copernicus nor his followers could, then, reply.
>
> So with nomogeny [origin by natural law]. The accumulation of facts in embryology and palaeontology is such as to make their conception or rational exposition impossible on the miraculous "creation-of-species-hypothesis" (thaumatogeny). Consequently, every young cultivator of biology, as a rule, is a nomogenist.
>
> But, "How do you get one species out of another; say, a dodo out of a dove, a Bushman out of an ape?"
>
> Lamarck has suggested one way of operation of the secondary law; Darwin has eloquently advocated another way. In the case of the dodo I find the theory of Lamarck applicable.
>
> Remember, that even Lord Bacon had his fling at the Copernicans: "Those carmen that drive the world about." It seemed to involve, indeed, a contradiction to common sense and experience. But, *it was an easier way of explaining* the movements of the heavenly bodies.
>
> By analogy, therefore, although, with Copernicus, I must say to the objecting geocentrist as to loose things sticking to the earth; and to the thaumatogenist as to raising an ape to a Bushman, *I don't know*, yet I believe nomogeny to be true. Since its adoption as an expository hypothesis the progress of biology has been rapid as was that of astronomy after Copernicus. By analogy, biology may expect her Galileo, her Kepler, finally, her Newton, who will *demonstrate* the way of operation of the secondary law. But some generations have still to work.[150]

This letter was written "*currente calamo.*" A more smoothly composed but very similar letter was addressed to the physician George Burrows, who, a few months before, had invited Owen to unveil the statue of William Harvey at

Folkestone, during the International Medical Congress of 1881.[151] In the same year, the Museum of Natural History finally opened, and Owen's career-long dream had become a reality. At this juncture, Owen wrote a paper on how the museum had come about, to be read at the BAAS meeting of 1881. The manuscript original of this paper was followed by a sketch of the steps leading to his belief in the origin of species. Although the second part of this twin paper was never published or even fully completed, it was a concise, well-argued piece in which he elaborated on the fossil evidence for the evolution of the horse, citing recent work by the American paleontologist O. C. Marsh. He once more went over the old ground of his differences with Darwin: environmental changes can cause the extinction but not the origin of species; in spite of annual combat among roebucks, "[t]he specific characters not of antlers only but of every other part of specific value have undergone no change in the individual of *Cervus dama* or *Dama vulgaris* which roamed in Saxon times, centuries before the Conquest".[152]

Owen never lost his conviction that Darwin failed to solve the problem of the origin of species, but he belatedly acknowledged the tremendous breadth of Darwin's work and the inspiration it had given to biology. To Owen, Darwin was "the Copernicus of British biology," not its Newton, who had not yet arisen. In 1882, shortly after Darwin's death, Owen sent the following formal appreciation of his erstwhile adversary to the British Museum trustee Spencer Horatio Walpole. It was Owen's final word on Darwin, and as such may be worth quoting in full:

> The great value of Darwin's series of works, summarizing all the evidences of embryology, palaeontology, and physiology experimentally applied to producing varieties of species, is exemplified in the general acceptance by biologists of the secondary law, by evolution, of the "origin of species." As a rule, additions by summaries and monographs now published in natural history are in the terms of such "law."
>
> In this respect Charles Darwin stands to biology in the relation in which Copernicus stood to astronomy.
>
> The rejection of the origin of species by primary law or miraculous creation, is equivalent to the rejection of the fixity, centricity and supreme magnitude of our earth, i.e. to the substitution for the geocentric of the heliocentric hypothesis. The accelerated progress of natural history under the guidance of "evolution" parallels that of astronomy under the guidance of heliocentricity.
>
> But the adoption of Darwin's hypothesis of the evolutional way of work is not general. Lamarck's hypothesis is found in some cases to be more applicable. So it seems to me that Darwin parallels Copernicus. The latter knew not how the planets revolved around the sun: to know that required the successive

> labours of a Galileo, a Kepler and finally a Newton. Analogy raises a cheerful hope and a confident expectation that the science of living things will also be blessed with its Galileo, its Kepler and finally its Newton. And that the way of operation of the secondary law originating species will then be as firmly established as the "law of gravitation."
>
> Meanwhile, our British Copernicus of biology merits the honour and the gratitude of the Empire, which is manifested by a statue in Westminster Abbey.[153]

The historiographic model of the mid-Victorian controversy over the origin of species, initiated by the Darwinians themselves and followed by historians even after the centenary celebrations of the publication of *The Origin of Species*, has been one in which the development of Darwin's theory is described in terms of enlightened, evolutionary "forerunners" versus dogmatic, creationist "opponents." Because Owen was an opponent, he had to have been an antievolutionist. This role seemed tailor-made for him, because he represented the establishment, and also because his personality, unlike Darwin's, did not invite hero worship. Furthermore, he was the only proper candidate for the role of opponent, because none of the real antievolutionists was suitable. Hugh Miller had committed suicide; Sedgwick was old and dying; Benjamin Brodie was a surgeon; Samuel Wilberforce was a clergyman; Agassiz lived far away in America; and so on. The model, which needed a prominent opponent, virtually dictated the misrepresentation of Owen.

A better historiographic approach, suggested by Jonathan Hodge, would be to study "evolution" as part of one or more explanatory traditions and programs concerned with the phenomenon of organic diversity.[154] One such tradition is that of a Romantic program of autogenesis and homological research. In such a context, Owen can be given due credit for his contributions to the question of the origin of species, without the need to ignore, belittle, dismiss, or distort him. After all, by the middle of the 1840s, he appreciated the importance of zoogeography in its bearing on the riddle of organic diversity, and he led the way in extending today's zoogeographical provinces to the fossil record. During the 1840s and 1850s his work in comparative anatomy and paleontology became the cornerstone of the evidence for evolution. With little more than a flick of the fingers, Owen's archetype could be turned into an ancestor.

It is true that his understanding of the concept of natural selection was naive. His assertion, for example, that the Malagasy jungle is too well stocked a larder, or the ocean too open an environment, to exert selective pressures for a struggle of life to take place shows that he was a poor ecologist. The reason for this is that Owen was a museum man, not a field naturalist. His observations of living animals were restricted to Regent's Park Zoo or to the birds in his garden and the deer in Richmond Park. Darwin, Hooker, Huxley, and Lyell had all been on one or more major navigational expeditions; Owen on none. When he traveled, he tended to visit museums, concert halls, or country houses. His journeys of discovery took place in anatomical theaters and museum galleries, not aboard a *Beagle* or a *Rattlesnake.* But while this provides explanatory background to his ecological naivety, it also explains why he was so acutely aware of the enormity of the anatomical hurdles that separate many species and of the need to resort to sudden and major mutations.

The museum context explains not only Owen's particular expertise, or lack thereof, with respect to the evolution question but also his strategy. Owen—we have seen—was deeply involved in the metropolitan museum movement, which was central to his career. One cannot look at Owen's involvement in the evolution debate without simultaneously considering his efforts to establish a national museum of natural history. The two were linked in the concrete reality of his career. This is illustrated by the fact that his 1881 BAAS paper on the origin of the natural history museum in manuscript form was twinned with one on the development of his anticreationist and evolutionary stance. Owen's efforts were directed toward the two interwoven aims of the advancement of scientific theory and the establishment of its institutional basis. Darwin's constituency was a small circle of intimate friends and fellow naturalists. Owen's was a very much larger and multifarious constituency, and additionally he was more dependent on their votes, quite literally so in the case of his parliamentary supporters. Significantly, all of Owen's closest allies and major patrons were either outspoken creationists or at least deeply distrustful of evolutionary theory. This was true of his supporters at Oxbridge, those on the boards of trustees of the College of Surgeons and of the British Museum, those among the members of "The Club," and his admirers among the country clergy.

It is a measure of Owen's basic sincerity that, in spite of this, he walked a course that led away from creationism and toward the establishment of the fact of an evolutionary origin of species. The plan of action he followed in doing so, particularly before 1859, was calculated to avoid alienating his constituency and was actually borrowed from them in that it was a Peelite strategy of gradual change that at the time was leading to such varied but related results as the reform of the Royal Society and the establishment of Owen's museum of natural history. Owen's trial balloons, his ambiguously phrased early pronouncements on the origin of species, his focus on the fact of evolution rather than its mode, his emphasis on the inadequacies of various evolutionary hypotheses, and his emphasis, too, on humanity's uniqueness—all these fitted a moderate course of action to reform both scientific beliefs and institutions without alienating the mainstream of Anglican England. To characterize this as mere duplicity, as Darwinian historians are wont to do, reflects a tunnel-vision view of mid-Victorian society, encompassing little more than Darwin and his immediate circle. Although Owen expressed his evolutionary opinions with much more clarity after *The Origin of Species* had appeared, this does not mean that he was a Johnny-come-lately. The appearance of Darwin's book destroyed Owen's cautious strategy, forcing him to show his hand. The only fundamental difference between Owen's evolutionary views before and after 1859 was his post-1859 adoption of spontaneous generation.

Of course, more than sociopolitical differences separated Owen from Darwin. Owen never wrote a major treatise on organic evolution; no *Origin of Species* ever flowed from his pen. His views on the matter were invariably expressed in restricted perorations. His talent for generalization operated at a lower, more specific level than Darwin's and was most effective when he alternated wielding scalpel and pen in composing his masterly anatomical monographs. Also, Owen was a deeply religious person and an outspoken theist who was hesitant to accept the full consequences of his evolutionary stance with respect to the origin of humans. It was this reluctance that led to a major clash with Huxley, producing a debilitating crack in his reputation.

6

Cerebral Constructs

Of all the issues raised by the theory of organic evolution, the origin of *Homo sapiens* was the most emotive. The idea that humans are related to apes or monkeys by a process of genetic descent filled many Victorians with a feeling of repugnance. To Adam Sedgwick, for example, it appeared to annul "*all distinction between physical and moral.*"[1] The sensitivity of the issue was such that Charles Darwin himself in his *Origin of Species* chose not to confront this most contentious of questions directly but merely hinted at the possibility that humans, too, had come into being as a result of natural selection. "Light will be thrown on the origin of man and his history," he parsimoniously commented.[2] And Charles Lyell, despite his conversion to Darwin's theory, only reluctantly accepted in his *Antiquity of Man* (1863) that *Homo sapiens* had come about in no more exalted a way than via animal ancestors.[3] Others also treated the question warily.

The person to breach this tender approach to the origin of humankind was T. H. Huxley. With characteristic bluntness he argued in his *Evidence as to Man's Place in Nature* (1863) that humans are intimately related to the rest of the living world and most directly to the anthropoid apes. One reviewer of the book put "the great doctrine of which Professor Huxley, in England, is the chief apostle" in the following worried words: "That man should be absolutely identical . . . with the beast of the field—that his direct ancestors should have been like the howling brute of the Gaboon [the gorilla], and his collateral relation another and more degraded Bornean form [orangutan]."[4] It is well known that *Man's Place in Nature,* arguably Huxley's most influential piece of writing, was inspired by Darwin's *Origin of Species.* Less well known is that Huxley's book was in part also the product of a protracted controversy with Owen over the question of humanity's genealogical relationship to "brutes."

Contrasting the Frames of Apes and Humans

Shortly after the appearance of *Vestiges*, Owen had let its author know that he, Owen, held no preconceived opinion against the possibility of the development of a "Hottentot" from a "chimpanzee."[5] Throughout the second half of the 1840s, however, when Owen developed his cautious strategy of allowing for the fact of evolution while disallowing the hypothesis of gradual species transformation, he simultaneously reinforced this strategy by placing particular emphasis on the anatomical differences that separate humans from the anthropoid apes. In 1848, the year in which Owen wrote his letter to the publisher John Chapman about "six possible secondary causes of species," he ended his Hunterian lectures for the year with "Comparison of the Apes or Anthropoid Quadrumana with Man," maintaining that the transmutation hypothesis was inadequate to account for the origin of the human species.[6] The following year saw the appearance of Owen's *On the Nature of Limbs*, with its public avowal of a natural origin of species, and heralded the start of the series "Osteological Contributions to the Natural History of Anthropoid Apes," in which Owen drew attention to anatomical hurdles that neither Lamarck nor the author of *Vestiges* were, in his view, able to surmount.[7]

From the very start of his Hunterian career, Owen had worked on apes, in particular on the chimpanzee and the orangutan. An early English classic of simian morphology, a study of the chimpanzee by the London physician Edward Tyson, had appeared in 1699, under the title *Orang-outang, sive Homo sylvestris*, a misnomer as it turned out, but an excellent work all the same. In spite of this propitious start, few English contributions to the comparative anatomy of nonhuman primates were made until the 1830s. It was in the course of this decade that the Zoological Society received its first anthropoid apes for public display in Regent's Park Zoo. The first orang arrived in 1830; the first chimp in 1835; and the first gibbon at the end of the decade, in 1839. Because of the primitive conditions of care under which the animals were held captive, they died from a few days to a few years after entering the zoo.[8] To Owen, the cloud of these deaths had a silver lining in that the carcasses provided him with an opportunity to dissect and describe the animals. His first zoological—as distinct from medical—paper was "On the Anatomy of the Orang-outang," presented to the Zoological Society in 1830; and in 1835 the death of the society's first chimpanzee enabled Owen to start his classic series on the comparative osteology of the orangutan and chimpanzee.[9] The apes' crude mimicry of humans, in both appearance and behavior, fascinated the general public and the naturalists alike. Broderip, upon seeing the corpse of a chimpanzee, offered to have a cast made "of this most interesting animal—the next step, though a long one, below man," before the anatomist's hand took the dead body apart.[10]

Owen, too, contributed to the visual representation of the zoo's simians, for example, in the form of a delicate watercolor of a young, living orang.[11] When he was working on the dissection of the muscles of a mature chimpanzee at the Royal College of Surgeons, he justified his work to the Museum Committee by mentioning the proximity of apes to humans.[12] Much excitement was added to Owen's work on human-like apes by the discovery of the largest anthropoid of all, the gorilla, initially also referred to as "the great chimpanzee" or *Troglodytes gorilla* (later renamed *Gorilla gorilla*). Before the end of the 1840s, neither the skeleton nor any of the soft organs of the gorilla had reached the scientific world or had come under the scalpel of any anatomist. In 1846, Thomas Savage, an American Episcopalian missionary in the Gabon region, accidentally discovered a skull of this huge ape. A sketch of that specimen and two skulls of full-grown males from the collection at the Philosophical Institution of Bristol, which had been able to obtain the skulls via a West African trading connection, reached Owen in 1847. In a paper to the Zoological Society, Owen named the gorilla *Troglodytes Savagei*, but in the priority race he had been beaten to the post by Isidore Geoffroy Saint-Hilaire and by the American naturalist Jeffries Wyman, who also had received parts of this anthropoid ape. The first complete skeleton of a gorilla sent to Europe was received at the Hunterian Museum in 1851. A nearly full-grown male preserved in spirits was presented to the British Museum in 1858, and a collection of skins of a full-grown male, a full-grown female, and immature gorillas, killed by the explorer Paul B. Du Chaillu, was purchased by the British Museum in 1861 for mounting and exhibition.[13]

Owen, having moved in 1856 from the Hunterian to the British Museum, was the intended recipient and main beneficiary of these new and thrilling additions to the primate collection, and for a dozen or so years he capitalized on his gorilla discoveries with papers to the Zoological Society, lectures at the Hunterian Museum and at the Royal Institution, especially during his tenure as Fullerian professor, presentations to the BAAS, and various incidental lectures to both popular and educated audiences. During the Liverpool meeting of the BAAS in 1854, Owen's sectional paper "On the Anthropoid Apes" aroused so much interest that at the request of the Council it was delivered at the general anniversary assembly.[14] His work on the chimps and orangs from Regent's Park Zoo, on the Gabon gorillas, and on such prosimians as the aye-aye from Mauritius made Owen one of very few European authorities on primates and the foremost authority on primate osteology.

The interest generated by Owen's gorilla work was caused not only by its novelty but also by the fact that, in Owen's view, these giants were the most anthropoid of all the apes. In other words, on the taxonomic ladder the gorilla had to be placed more closely to *Homo sapiens* than even the chimp or the orang. Although some French and American naturalists demoted the gorilla to a rank below that of the chimpanzee, Owen argued that the following sequence of decreasing resemblance to humans exists: gorilla, chimpanzee, orang, and gibbon. Assessing similarity to humans was no straightforward matter, however, because one anthropoid genus may well come close in certain features but not in others, by reference to which a different genus may appear more anthropoid. Owen solved the problem by weighing the nature and number of similarities to humans. In comparing the orang and the chimpanzee, the former had three characters of closer correspondence to humans, whereas the latter had sixteen; and in juxtaposing the orang to the gorilla, the former had nine, but the latter had no fewer than twenty-four points of closer human likeness in cranial and dental features alone.[15]

The dissection and description of the gorilla thus became of the greatest possible significance in contrasting "man and monkey" or, conversely, in speculating how humans might have descended from apelike ancestors. Owen's many lectures and publications on the anthropoid apes and their relationship to humans almost invariably presented the subject with reference to the Lamarckian theory of transmutation—or, in Owen's own words, "in reference to the hypothesis that specific characters can be so far modified by external influences, operating on successive generations, as to produce a new and higher species of animal, and that thus there had been a gradual progression from the monad up to man."[16]

The anatomical features in which the gorilla resembles humans more closely than the chimpanzee were, Owen pointed out, the greater grinding surface of the teeth, the nasal bones, which show an incipient projection, the greater conformity of the foot, "and many minute anatomical points." Rather than elaborate on the similarities between humans and apes, Owen focused on the differences. These pertained to the cranium, to dentition, and to various characteristics of the skeleton. The gorilla's skull and teeth alone presented as many as thirty "chief differences" from their human equivalents.[17] For example, skull measurements illustrate that the mean capacity of the cranial cavity in humans is much greater than that in gorillas. Owen ended one of his early comparative studies of the cranium of gorillas, chimpanzees, and various human races with the observation "that the Hottentots and Papuans of Australia have the smallest cranial capacity amongst the human races; but

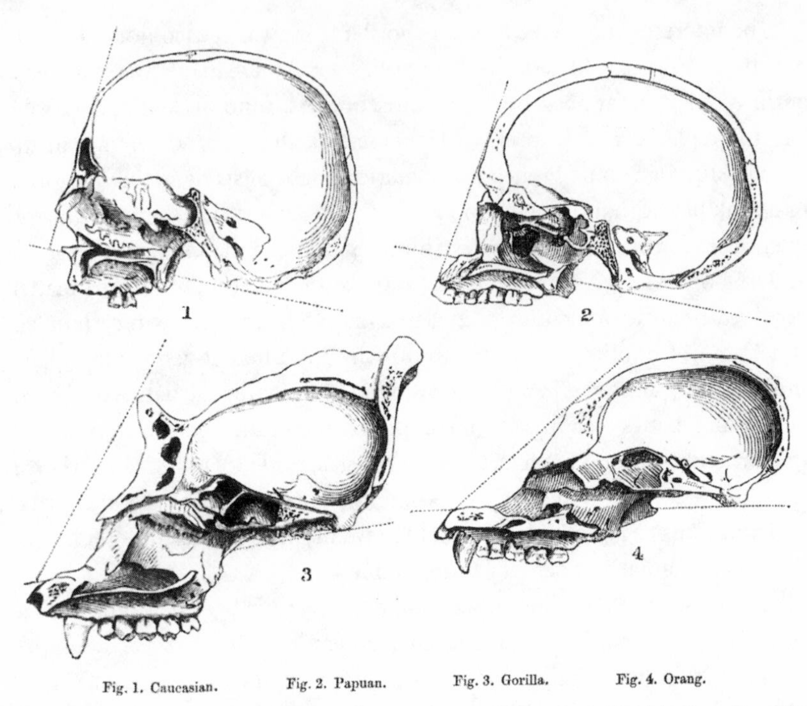

Fig. 1. Caucasian. Fig. 2. Papuan. Fig. 3. Gorilla. Fig. 4. Orang.

FIGURE 17. Cross sections of the skulls of men and apes. Owen, who believed in the unity of the human species, repeatedly made the point that the difference in skull capacity between the human races is small compared with that between humans and the highest anthropoid apes. ("Proceedings of Societies: Zoological Society," *Literary Gazette, and Journal of Science and Art*, no. 1817 [1851]: 777, figs. 1–4.)

that the largest capacity yet observed in the adult male Gorilla is less than one-half the mean capacity in those Aethiopian races" (fig. 17).[18]

This difference was the more pronounced given the greater average body weight of gorillas compared to that of humans. Also, the very pronounced supraorbital, or eyebrow, ridge and the prognathous premaxillary bones containing the incisor teeth distinguish the gorilla from *Homo sapiens.* All apes differ from humans in that they have proportionally larger canine teeth, which necessitates a break in the juxtaposition of the teeth to allow the points of the canines to fit when the mouth is closed. The human dental system is peculiar because of the equal length of the teeth, which are arranged in an uninterrupted series; no sexual dimorphism occurs.

Furthermore, our erect position is reflected in various parts of our skeletons, in particular in the feet. Whereas both the fore- and hind-limbs of the apes have an opposable thumb, or pollux, the human foot is characterized by a hallux, or "great toe." On the basis of this difference, Blumenbach and Cuvier had classified the nonhuman primates as four-handed quadrumana and placed humans in a separate order of two-handed bimana. Others had maintained that the criterion of opposable thumbs was not applicable to all quadrumana, because New World monkeys lack opposable digits on their hind-limbs. Yet Owen concurred with the opinion of the founding fathers of comparative anatomy that the human foot is of decisive taxonomic value: "The great toe which forms the fulcrum in standing or walking is perhaps the most characteristic peculiarity in the human structure; it is that modification which differentiates the foot from the hand, and gives the character to his order (*Bimana*)."[19] John Hunter had expressed a similar view on the difference between humans and monkeys in notes that Owen, at about this time, was in the process of editing.[20]

Owen subscribed to the belief that when muscles are attached to particular bones, these bones may undergo a change in shape as a result of muscular exertion. He cited the example of the changes that occur in the sternum of a young bird when it grows up and exercises its wing muscles. This and other types of variation seemed well known from the breeding of *Canis familiaris*, varieties of which differ in size, color, hair, shape of the head, and protuberances, crests, and ridges for the attachment of muscles. But would similar variations in the higher quadrumana suffice to turn them into even the "lowest" race of *Homo sapiens*? Owen did not think so. The features he had cited to distinguish humans from gorillas were not susceptible to change by external stimuli. The prominent supraorbital ridge, for example, is not the result or even the concomitant of muscular development; no muscles are attached to it that could have stimulated its growth. The same is true of the premaxillary bones. As early as February 1848, when he delivered his first gorilla paper to the Zoological Society, Owen concluded: "No known cause of change productive of varieties of mammalian species could operate in altering the size, the shape, or the connexions of the premaxillary bones, which so remarkably distinguish the great *Troglodites gorilla*, not from man only, but from all other anthropoid apes."[21]

The development of the teeth provides a similar insurmountable barrier. No muscular or other external influences determine the characteristics of teeth such as the ape's canines, which exhibit their typical form even before they break through the gums:

> The whole crown of the great canine is, in fact, calcified before it cuts the gum or displaces its small deciduous predecessor; the weapon is prepared prior to the development of the forces by which it is to be wielded; it is therefore a structure foreordained, a predetermined character of the chimpanzee, by which it is made physically superior to man; and one can as little conceive

> its development to be a result of external stimulus, or as being influenced by the muscular actions, as the development of the stomach, the testes, or the ovaria.[22]

In conclusion, Owen maintained that the vast majority of osteological features in which humans differ from gorillas cannot have been produced by any of the then known, essentially Lamarckian causes of organic variability, and that speculation about the origin of *Homo sapiens* must wait till more became known about how such specific differences are generated.[23]

Part and parcel of Owen's belief that a wide gap separates humans from the great apes was another belief, namely that the various human races are anatomically closely related to each other, forming a single, unitary species—a view associated with such names as Blumenbach and Prichard. Owen based his theory of the unity of the human species on the observation that the osteological differences between the various human races "are much less in degree and very inferior in importance" compared to most of the distinctions between "the lowest varieties of Man" and "the highest of the Ape tribe"; also, the osteological and dental features which distinguish humans from the apes are the same for all races of humankind. "I have come to the conclusion that Man forms one species," Owen proclaimed. Moreover: "Man is the sole species of his Genus, the sole representative of his Order."[24] And at the conclusion of several of his "man-ape" lectures, Owen repeated his agnostic view of the origin of species: "Of the nature of the creative acts by which the successive races of animals were called into being we are ignorant."[25]

Delineating the Cerebral Divide

In the late 1850s, some ten years after he had begun the osteological contrasting of primates, Owen added another string to his bow in separating "man from monkey." This consisted of a study of the nervous system, begun as part of the Hunterian lectures for 1842, when he had outlined the main modifications of the mammalian brain and demonstrated that these can be used for taxonomic purposes because they go hand in hand with modifications of other organs and systems. To use cerebral characteristics for classificatory purposes was not a new thing; it had been done before on the Continent and been introduced to the Linnean Society by Owen's friend Charles Lucien Bonaparte.[26] The classification of mammals was a hot topic, and Owen had crossed swords with, for example, Geoffroy about the precise affinities of marsupials and monotremes. In a paper to the Linnean Society, read on February 17 and April 21, 1857, Owen proposed a cerebral system of classification of the mammals and an entirely new subdivision of the mammalian class into four major subclasses. He again presented this novel classificatory system in his Rede lecture and also in one of his Fullerian lectures at the Royal Institution.[27] The lowest subclass, the Lyencephala, or loose-brained mammals, are characterized by the fact that the two cerebral hemispheres, of small relative size and smooth exterior, are only loosely connected by an anterior band of nerve substance (or commissure) and by a fornix. Marsupials and monotremes make up this most primitive of the mammalian groups. In the next subclass, the Lissencephala, or smooth-brained mammals, a proper corpus callosum is present, but the brain is no further developed and exhibits a smooth surface with at most a few, simple convolutions, as, for example, in rodents. The third stage in the modification of the cerebrum shows an increase in relative size, extending more or less over the cerebellum, and "[s]ave in very few exceptional cases of the smallest and inferior forms of *Quadrumana*," the surface is folded into convolutions, or gyri, hence the name Gyrencephala. Examples include the hoofed animals, the carnivores, and all monkeys and apes.

This left only humans, and Owen proposed that the human species not only be placed in a genus, a family, and an order of its own but be assigned to a separate subclass, for which he coined the name Archencephala, the "ruler brains." The human brain—Owen stated—differs markedly from that of all other mammals: (1) it shows a sudden increase in the relative and absolute size of the cerebral hemispheres, corresponding, of course, to an increase in cranial capacity; (2) both front and back of the cerebrum are well developed; the posterior part of each hemisphere in particular extends backward beyond the cerebellum to form a posterior lobe; (3) the backward extension of the lateral ventricle in each hemisphere is curved so as to form a posterior horn; (4) in each of these a "hippocampus minor" is present; and (5) the surface area of the cerebrum is increased by many and deep convolutions. In his *Anatomy of Vertebrates* Owen later explained that each hemisphere can be envisaged as a bag of neural matter, opening out from the brain stem, the hollow of the bag representing the ventricle. In the embryo the ventricle is large and simple, and the bag's wall is very thin. It gradually becomes thicker, mostly at the upper and outer sides. The thickening wall protrudes at certain places into the cavity, or ventricle, which eventually takes on an elongate, hornlike curvature. The molding of its posterior part is completed by the development of an internally protruding wall to which the term "hippocampus minor" had been given.[28]

Owen believed that the degree to which the human cerebral hemispheres show the development of a posterior lobe, posterior horn, and hippocampus minor justifies using these features for taxonomic purposes. Owen's meaning and precise words soon became a matter of controversy, and we need to quote the relevant passage from his original Linnean presentation:

> In Man the brain presents an ascensive step in development, higher and more strongly marked than that by which the preceding subclass was distinguished from the one below it. Not only do the cerebral hemispheres overlap the olfactory lobes [at the front] and cerebellum [at the back], but they extend in advance of the one, and farther back than the other. Their posterior development is so marked, that anatomists have assigned to that part the character of a third lobe; it is peculiar to the genus *Homo*, and equally peculiar is the "posterior horn of the lateral ventricle," and the "hippocampus minor," which characterize the hind lobe of each hemisphere. The superficial grey matter of the cerebrum, through the number and depth of the convolutions, attains its maximum extent in Man.
>
> Peculiar mental powers are associated with this highest form of brain, and their consequences wonderfully illustrate the value of the cerebral character; according to my estimate of which, I am led to regard the genus *Homo*, as not merely a representative of a distinct order, but of a distinct subclass of the Mammalia, for which I propose the name of "*ARCHENCEPHALA*."[29]

For purposes of taxonomy, the cerebral characteristics did not take the place of the osteological ones but had to be taken in conjunction with them. Linnaeus had classified *Homo sapiens* as no higher than a separate genus of Primates; Cuvier had upgraded the taxonomic rank of humans to the ordinal status of Bimana; now Owen placed *Homo sapiens* even higher, in a subclass of its own, although he did not go as far as, for example, Isidore Geoffroy Saint-Hilaire, who in his *Histoire naturelle générale des règnes organiques* (1854–62) suggested that humans, because of their mental abilities, should be classified as a separate kingdom, taxonomically equivalent to that of plants and animals.[30]

The culmination of Owen's taxonomic exercise in separating humanity from the rest of the animal kingdom was the published version of his Rede lecture *On the Classification and Geographical Distribution of the Mammalia*, to which were added two appendices, "On the Gorilla" and "On the Extinction and Transmutation of Species." Part of what motivated Owen in placing humans on such an elevated taxonomic pedestal and thus keeping them out of the reach of the hypothesis of the gradual transmutation of species is revealed by a comparison of the Linnean paper, read to an inner circle of colleagues, with its modified version, presented as the Rede lecture and intended for the wider circle of friends from whom he derived his political support. One difference between the two versions simply reflected Victorian propriety; from his original paper Owen omitted the following characteristics of the human male: "The testes are scrotal; their serous sac does not communicate with the abdomen; they are associated with vesicular and prostatic glands. The penis is pendulous, and the prepuce has a fraenum."[31] This little catalogue of anatomical detail would not do in the Senate House at Cambridge.

More to the point, however, Owen also left out a footnote and added a peroration. If Owen's separation of humans from the apes was part of a strategy to keep at bay the consequences for human origins of his homological program and of his belief in the fact of organic evolution, then Cambridge was the place to make full use of it. Here his audience included Whewell (Sedgwick was ill at the time)[32] and other prominent figures deeply worried about the materialist tendencies of the age; and indeed, Owen used the occasion of his Rede lecture and of the relatively neutral-sounding subject of mammalian classification to stage the taxonomic crowning of *Homo sapiens* as the only representative of the subclass Archencephala and dramatically to present humankind's elevated status as a legitimation of its claim to spiritual uniqueness. The footnote he left out did not fit this neuroanatomical investiture of humanity, because it provided a tantalizing glimpse of Owen's belief in the fundamental unity of humans with the apes when he had stated that "I cannot shut my eyes to the significance of that all-pervading similitude of structure—every tooth, every bone, strictly homologous,—which makes the determination of the difference between *Homo* and *Pithecus* [orangutan] the anatomist's difficulty."[33] The additional peroration put the ideological implications of Owen's work on the cerebral classification of mammals beyond any doubt when he forcefully reminded his audience of its spiritual meaning. The human frame and brain are indications of humanity's moral and religious status. Owen's long and in many places technical lecture ended with a crescendo of "the sentiments that have naturally flowed from the contemplation of the highest of the gradations of Mammalian structure":

> Oh! you who possess it in all the supple vigour of lusty youth, think well what it is that He has committed to your keeping. Waste not its energies; dull them not by sloth: spoil them not by pleasures! The supreme work of Creation has been accomplished that you might possess a body—the sole erect—of all animal bodies the most free—and for what? for the service of the soul.[34]

This was the most inspired religious peroration of any of Owen's scientific discourses. There was nothing disingenuous in his admonition to the Cantabrigians. It fairly expressed his true spiritual convictions. Yet at the same time it suited his strategy of ambiguity, and this in turn put him in a decidedly false position. The elevation of *Homo sapiens* to a separate subclass, Owen's emphasis on the inadequacy of the transmutation theory to account for the anatomical differences between humans and gorillas, and the moralizing conclusion to his Rede lecture spoken in traditional, Christian phrases left on many of his contemporaries the impression that he supported the old doctrine of the miraculous creation of humans. Of course, he never actually stated this.

The Hippocampus Controversy Begins: Oxford, 1860

Owen's lack of clarity about human origins, serving the pursuit of his museum plans, would have proved unproblematic had it not been for the appearance of Darwin's *Origin of Species*, which made Owen's ambiguous position untenable. Critics of the book reasserted human uniqueness and appealed to Owen's authority for support; followers, on the other hand, sought to establish humanity's animal ancestry and attacked Owen as the main obstacle. Wilberforce emphatically stated in his critique of Darwin's book for the *Quarterly Review* that an origin of humans by natural selection is "absolutely incompatible" with the Bible and with the "moral and spiritual condition of man": "Man's derived supremacy over the earth; man's power of articulate speech; man's gift of reason; man's free-will and responsibility; man's fall and man's redemption; the incarnation of the Eternal Son; the indwelling of the Eternal Spirit,—all are equally and utterly irreconcilable with the degrading notion of the brute origin of him who was created in the image of God, and redeemed by the Eternal Son assuming to himself his nature."[35]

The Wilberforce-Huxley encounter about humankind's relationship to monkeys during the Oxford meeting of the BAAS in 1860 has been told, retold, and turned into a legend of Darwinian heroism and, more recently, has become the subject of fine, demythologizing scholarship by John Lucas and others.[36] Huxley's famous retort to the bishop of Oxford, who asked him whether it was through his grandfather or his grandmother that he claimed descent from a monkey—"He was not ashamed to have a monkey for his ancestor; but he would be ashamed to be connected with a man who used great gifts to obscure the truth"—was probably never uttered, and it is even less likely that the audience could have heard whatever was actually said, given the general commotion. By all contemporary accounts it was J. D. Hooker, not Huxley, who most effectively presented the pro-Darwinian case to the Oxford meeting.

Moreover, Huxley's real *bête noire* was not Wilberforce but Owen, whose studies of the contrast between humans and apes were singled out by Huxley as a new and sharp focus of Darwinian apologetics. The resulting Owen-Huxley clash must rank as one of the fiercest, bitterest, and most publicly sensational battles between scientific rivals of the nineteenth century. Its acute phase stretched over a period of more than two years, and it was principally fought out at the annual gatherings of the BAAS. It began in 1860, at Oxford; it rumbled on in 1861, in Manchester; and it culminated during the 1862 meeting at Cambridge. Both men also used the Royal Institution to air their contrasting views. Huxley had his own magazine, the *Natural History Review*, in which he published his attacks, whereas Owen used the older, more conventional *Annals and Magazine of Natural History* to state his position. Professional and general magazines, daily papers, weeklies and quarterlies, all reported on the clash of the titans or took sides. The *Athenaeum* in particular functioned as a mouthpiece for the two parties. On more than one occasion *Punch* made fun of the Owen-Huxley fight, and other magazines, too, were inspired by the controversy to literary parody. Abroad, especially in Germany, France, the Netherlands, and America, colleagues followed the controversy with bemused fascination and occasionally went into print to take sides.

Even more than Owen's stance with respect to Darwin and *The Origin of Species*, his connection with Huxley and *Man's Place in Nature* has been consistently misrepresented. Our perception of it has been distorted by a prejudiced, pro-Darwinian account of the episode, and nearly the entire body of secondary literature on the Owen-Huxley relationship portrays Owen as the dark knight whose evil intent was thwarted by Huxley as the knight in shining armor.[37] This is both inaccurate and leaves out some truly interesting aspects of the rivalry. The story cannot be told by assigning simple black and white roles. Only a detailed anatomy of the Owen-Huxley controversy over humanity's relationship to apes will suffice to get away from the distorted simplicity to which many historians customarily reduce it.

The hippocampus controversy rapidly deteriorated into a stubborn repeating, by both parties, of their previously stated positions. At the risk of making the narrative tedious, I shall trace the debate through its successive circles of repetition. The reason for doing this is that the very repetitiveness becomes informative when we see it as an indication of a "confusion of tongues," demonstrating that each party spoke its own language. Owen used the language of Christian theism, Huxley that of positivism; Owen's accent reflected authoritarianism, Huxley's a provocative rebelliousness.[38]

It has been alleged that Owen's work on humans and apes was motivated by his anti-Darwinian stance.[39] This cannot be true in view of the fact that he had begun his comparative primate studies a dozen years before *The Origin of Species* was published. It also has been alleged that the attack by Huxley on Owen's Archencephala did not derive from Huxley's Darwinian sympathy but purely from his sense of scientific integrity.[40] This, too, is incorrect. Although apparently he had expressed disagreement with Owen's novel classification of mammals to his students at the Royal School of Mines before

Darwin's book came out, Huxley formally initiated the argument with the superintendent of the Natural History collections, not after the latter's Linnean paper of 1857, but in a lecture on *The Origin of Species* early in 1860, entitled "On Species and Races, and Their Origin." People were afraid, he stated, of the logical consequences of Darwin's doctrine. "If all species have arisen in this way, say they—Man himself must have done so; and he and all the animal world must have had a common origin." To this Huxley answered: "Most assuredly. No question of it."[41]

Owen was not present at the Wilberforce-Huxley encounter in Oxford but two days before had attended a meeting of section D, during which Charles Daubeny had read a paper entitled "On the Final Causes of the Sexuality of Plants," in which the Oxford professor of botany and chemistry had shown himself sympathetic to, though not unreservedly supportive of, Darwin's new theory. As a sequel to this paper, a "sharp passage of arms" took place; Huxley gave a "direct and unqualified contradiction" to an anti-Darwinian assertion by Owen "that the brain of the gorilla 'presented more differences, as compared with the brain of man, than it did when compared with the brains of the very lowest and most problematical of the Quadrumana.'"[42] This at least is the account that Huxley and other Darwinians have given of Owen's utterances; and, to be fair to them, this is also how the *Athenaeum*, for example, reported Owen's words.[43]

It is surprising that Owen should have put the human-ape differences in those terms. He certainly had not done so in any of his gorilla papers before. He had gone as far as maintaining that the jump in cerebral organization from ape to human, that is, from Gyrencephala to Archencephala, was larger than the one from sloth to whale, that is, from Lissencephala to Gyrencephala. More commonly, however, Owen had demonstrated that the difference between human and gorilla is greater than that between gorilla and chimpanzee and also more than that between the various human races. This was of course true and not easy to argue with. The only person who had put the problem the way Owen purportedly did at the Oxford meeting—though reversed—was Huxley himself, who in his lecture "On Species and Races, and Their Origin" had stated that it is a demonstrable fact "that the anatomical difference between man and the highest of the *Quadrumana* is less than the difference between the extreme types of the Quadrumanous order."[44] This was an exceedingly clever reformulation of the question at issue, which, differently put, stated that such prosimians as lemurs and tarsiers differ more from the great apes than these differ from humans and that, if one puts prosimians and anthropoids in one and the same taxonomic category, it is illogical not to include humans. In this sense, St. George Mivart agreed in his later paper "On the Appendicular Skeleton of the Primates" with Huxley that man "evidently takes his place amongst the members of the suborder *Anthropoida*."[45] Huxley, however, intended to draw more than a merely taxonomic conclusion. If one is prepared to consider prosimians as the ancestors of apes, there is no reason for not accepting apes as the forebears of humans.[46] It is of course possible that Owen, in a moment of bravura and in the heat of the argument, simply contradicted Huxley's statement, even though subsequently he repeatedly argued that a comparison of "the extreme types of the Quadrumanous order" was meaningless and that only immediately adjacent steps should be compared, that is, not human-gorilla versus gorilla-lemur but human-gorilla versus gorilla-chimpanzee.[47]

Huxley made good his public contradiction of Owen, in a paper "On the Zoological Relations of Man with the Lower Animals." "Theologians and moralists, historians and poets," he asserted, have tended to set man apart from the animal world, whereas "the students of physical science" have gravitated toward the opposite opinion of "the closeness of the bond which unites man with his humbler fellows." In rephrasing the way he had formulated the issue, he asked, is man less distant from the gorilla than this ape is from the lemur? Not by Owen's cerebral criteria, Huxley maintained: "I hold it to be demonstrable that the Quadrumana differ less from man than they do from one another," in particular with respect to cerebral characteristics. Three of Owen's supposedly distinctive human brain features are not unique but present also in apes and monkeys; Huxley asserted:

1 That the third lobe is neither peculiar to, nor characteristic of man, seeing that it exists in all the higher Quadrumana.
2 That the posterior cornu of the lateral ventricle is neither peculiar to, nor characteristic of man, inasmuch as it also exists in the higher Quadrumana.
3 That the *Hippocampus minor* is neither peculiar to, nor characteristic of man, as it is found in certain of the higher Quadrumana.[48]

Huxley cited a variety of Continental studies which had shown that a third lobe covering much, if not all, of the cerebellum is present in several simians and that among the lowest, platyrrhine (flat-nosed) monkeys the lobe may be "relatively greater than in man":

> Thus, every original authority testifies that the presence of a third lobe in the cerebral hemisphere is not "peculiar to the genus Homo," but that the same structure is discoverable in all the true Simiae among the Quadrumana, and is even observable in some lower Mammalia; and any one who chooses to take the trouble to dissect a monkey's brain, or even to examine a vertically bisected skull of any of the true Simiae, may convince himself, on the still better

> authority of nature, not only that the third lobe exists, but that it extends to the posterior edge of, if not behind the cerebellum.[49]

From the same Continental authorities Huxley cited examples of simians that have both a posterior cornu and a hippocampus minor. It is precisely these structures that are most variable in the human brain and therefore unsuitable for taxonomic purposes. This being so, Owen's Archencephala would have to be discarded:

> Having now, as I trust, redeemed my pledge to prove that neither the third lobe of the cerebrum, nor the posterior cornu of the lateral ventricle, nor the hippocampus minor, are structures distinctive of and "peculiar to the genus *Homo*," I may leave it to the reader to decide the fate of the "sub-class *Archencephala*," founded upon the supposed existence of these three distinctive characters.[50]

Furthermore—Huxley insisted—the differences between the brains of the highest and the lowest of the human races are of the same order as those that separate the human from the simian brain. In other words, if a European brain be given the letter A, the Bosjesman's (Afrikaans for "Bushman's") brain a B, and an orangutan's brain a C, "the differences between A and B, so far as they have been ascertained, are of the same nature as the chief of those between B and C."[51] In this way, racist views became bound up in Huxley's evolutionary scheme, the various human races being seen as links in an evolutionary chain connecting Caucasians to apes or, more sophisticatedly, as the present-day end products of different lineages of ape ancestry.

The Hippocampus Controversy Broadens: Manchester, 1861

For his data Huxley depended entirely on early-nineteenth-century anatomists in France, Germany, and the Netherlands. The only British authority on simian anatomy until then was Owen. Acutely aware of his strategic disadvantage, Huxley encouraged a number of colleagues from his own generation of young Turks to dissect whatever simian brains they could lay their hands on, and in quick succession a remarkable series of "simian brain papers" appeared, all part of a feverish anti-Owen campaign. Huxley's own paper "On the Zoological Relations of Man with the Lower Animals" included the results of a dissection of the brain of a young chimpanzee by a friend, Allen Thomson.[52] This was followed by a particularly fine account of the dissected brain of a young chimpanzee by another Huxley friend, John Marshall, who worked as a surgeon at University College Hospital. He noticed that in the young specimen a posterior lobe covered the cerebellum and even extended beyond it. He also identified a posterior cornu with all the features of the human one, including the hippocampus minor. He concluded by stating that the sole purpose of his paper was

> to record the results of an anatomical investigation. I have no theory, zoological, or physiological, to support; I have no leaning towards any of the developmental hypotheses of the origin of species. But, on the question of facts, and the interpretation of those facts, my results, as to the existence of a posterior lobe, of a posterior cornu, and of a hippocampus minor, in the Chimpanzee, will be found to harmonize with the investigations and conclusions of Prof. Huxley and of Prof. Allen Thomson, already published in this Review.[53]

The paper was illustrated by the still-rare medium of photographic representation, showing the dissected brain and enhancing the claim by the Huxley camp to objective reportage.

The anti-Owen, simian-hippocampus bandwagon was now unstoppable. George Rolleston, the newly elected Linacre professor of anatomy at Oxford, had already jumped on it with a study of the brain of an orangutan. He, too, concluded that the internal anatomy of the simian brain does not present features which sharply differentiate it from the human brain and on which Owen had based his Archencephala. Moreover, the difference between "the soul of man" and "the life of the beast which perishes" "is not one which can be weighed or measured, be drawn or figured, be calculated in inches or ounces."[54] Soon Huxley himself, even though he preferred the writer's pen to the anatomist's scalpel, followed suit with a dissection of the brain of a New World spider monkey, *Ateles paniscus.* He demonstrated the presence of a posterior cornu and argued about the precise definition of a hippocampus minor, which—he insisted—should not be confused with the *eminentia collateralis*, another fold inside the lateral ventricle which in his monkey was indeed far less developed than in the human brain. Huxley went a step beyond his previous position and claimed that three of Owen's peculiarly human brain characteristics, more than just being shared by humans and Old and New World simians, actually set them apart from all other mammals and that these structures were not rudimentary but "often more largely developed, in proportion to other parts of the brain, in Apes than in Man."[55]

In the meantime, Owen returned to the topic of "the gorilla and the Negro" in a lecture at the Royal Institution, intended as a sequel to his lecture of 1859 in which he had discussed the osteological characteristics of the gorilla. He restated what he believed were the cerebral differences between humans and the anthropoid apes, clarifying that the posterior lobe can be defined as "that part of the hemisphere which covers the posterior third of the cerebellum, and passes beyond it." In the chimpanzee, this lobe and its internal features are less developed than the corresponding parts in *Homo sapiens* and should not be named by using the established nomenclature for the human brain. The difference is analogous to that between the Manx cat with its stunted tail and a regular cat; a fully developed tail can be said to be peculiar to the latter even though the former possesses a rudimentary tail. Although "there is a gradation of cerebral development from the lowest to the highest vertebrate species," there are also "interruptions in this gradation." In the mammalian class, for example, the development of the brain takes place in both small and large strides. Thus, the several steps leading from the lemur to the gorilla (via the marmoset, the New World monkeys, the Old World monkeys, the gibbon, the orang, and the chimp) are small compared with the gap which separates the gorilla from humankind. To say, with Huxley, that the difference lemur-gorilla is larger than gorilla-man is formulating the problem incorrectly; one might as well say that the difference opossum-gorilla or fish-gorilla is greater than the distance which separates the gorilla from the human species.[56]

The *Athenaeum* report of Owen's lecture was illustrated by accurate cross sections of a human skull and a gorilla's skull but also by two highly diagrammatic sections of a human brain and a gorilla's brain. In the latter, some of the differences from the human brain were exaggerated. Huxley immediately wrote a letter to point out the inaccuracy of the illustration of the ape's brain, in particular the fact that the cerebrum did not extend far enough over the underlying cerebellum.[57] Owen tersely admitted the inadequacy: "Your Reporter," he wrote to the editor, "gave faithfully the substance of my Lecture at the Royal Institution; but the Artist has been less successful in the copy of the cerebral diagrams. In their details the most careful draughtsman requires the revision of the anatomist."[58]

It seemed that, for the first time, Huxley had drawn blood and, as if encouraged by the smell of it, he pressed on with a second, vehement letter, "Man and the Apes," opening with: "Prof. Owen's admission of his responsibility for the very serious errors respecting matters of anatomical fact. . . . " Owen's definition of the posterior lobe—Huxley contended—"is new to science"; but even if accepted, it would not change the fact that a posterior lobe does exist in all Old and New World simians. The figure of a chimpanzee's brain in Owen's Rede lecture, intended to illustrate that the cerebrum does not fully cover the cerebellum, was copied from the Dutch anatomists Jacobus Schroeder van der Kolk and Willem Vrolik, who allegedly had depicted an imperfectly preserved, deformed specimen. Trespassing on the area of Owen's greatest expertise, Huxley asserted that a backward development of the cerebrum, better developed than in humans, is true "for the Gorilla itself." Because he did not possess a gorilla's brain but only a skull, he used the latter to argue from its inner configuration that the posterior cerebral lobes "projected beyond and behind the cerebellum."[59] Huxley supported his claim with a line drawing of the gorilla's skull cut vertically from front to back. The rhetoric of factualness was used once more and underscored by a *nota bene* that the illustration represented "a perfectly accurate *camera lucida* sketch of this skull, of half the size of nature; and of this the cut is, in all essential respects, *a fac-simile*."[60]

However, the area of cerebral overlap with the cerebellum shown in Huxley's figure was artificially produced. In order to establish the degree of overlap it is crucial, of course, to define the horizontal position of the brain and, in this case, of the skull. Huxley chose the plane of the membrane that separates the cerebellum from the overlying cerebrum as his horizontal axis. This gave the cross section of the skull a substantial and excessive backward tilt, and the objectivity of the tracing technique used in producing his figure did not alter the fact that it was as much a representational artifact as Owen's distorted diagram. Admittedly, craniometric methods were only just being standardized, and the "Frankfurt horizontal," for example, which defines the skull's horizontal position by means of a line connecting the base of the eye orbit with the top of the earhole, is of a later date (1877) than Huxley's letter. Yet Owen, in his very first description of a gorilla's skull in 1848, had depicted the skull in the position that the Frankfurt horizontal would later stipulate. Moreover, none of the other half-dozen definitions of a skull's level plane, one of which had been introduced by Spix as early as 1815, produced anywhere near the same backward tilt that Huxley engineered.[61] The conclusion is inevitable that Huxley, in his eagerness to give the gorilla a posterior lobe similar to the one in humans, committed the very distortion of fact of which he accused Owen. The latter could have hoisted Huxley with his own petard but he responded by ignoring his adversary.

Furthermore, Huxley accused Owen of an intentionally false translation from the Dutch account by Schroeder van der Kolk and Vrolik of their dissection of a chimpanzee's brain, Owen having turned the nominative "pes hippocampi minor" into the genitive "pes hippocampi minoris." Owen later responded to the accusation; but in any event, the point was irrelevant, because what the Dutch anatomists had said, namely that the chimpanzee's brain showed "an indication" of the hippocampus minor, was not disputed by either party. To Huxley "an indication" was enough to bolster his position that simians do have the feature, whereas to Owen it sufficed to prove that they do not have a proper hippocampus minor. Huxley ended his angry letter by stating "that I shall hereafter deem it unnecessary to take cognizance of assertions, opposed to my own knowledge, to the concurrent testimony of all other original observers, and already publicly and formally refuted. Life is too short to occupy oneself with the slaying of the slain more than once."[62]

Without mentioning any of his detractors by name or citing their simian brain papers, Owen answered in the form of a short letter to the *Annals and Magazine of Natural History*. He reiterated his consistently ignored point that, in the series from lemur to human, the last step, from anthropoid ape to "negro," is marked by a sudden rise in cerebral volume, constituting "the most important of the differential structural characters between the Human and Ape kinds." As far as the other three criteria were concerned, the difference in opinion between himself and "my fallible fellow-labourers in anatomical science" was a matter of definition, not of fact, although Owen asserted as fact that, in the orangutan, chimpanzee, and gorilla brains that he had dissected, the relationship of cerebrum to cerebellum was as represented by the Dutch anatomists. More cautious than before, Owen concluded: "There is no continuation of the lateral ventricle curving backwards, outwards, and inwards, nor any eminence accompanying it in those directions and extent, in any known Ape."[63] Schroeder van der Kolk and Vrolik reacted to this letter, confirming that their figures had been exact, but at the same time they affirmed the presence in both chimp and orang of the posterior lobe, the posterior cornu, and the hippocampus minor, even though these features are more fully developed in humans; moreover, they acknowledged that their specimen had indeed been somewhat deformed, leaving the cerebellum more uncovered than in the natural state.[64]

Huxley did not keep to his decision to remain silent. The controversy rumbled on during the 1861 meeting of the BAAS, which that year was held in Manchester; but it did not produce open conflict, because Huxley was absent. In the discussion that followed a paper entitled "On Some Objects of Natural History from the Collection of M. du Chaillu," Owen was asked about the features that separate the gorilla from *Homo sapiens*. He repeated the osteological and cerebral criteria, including the hippocampus minor, which—he added in qualification—exists only in an "undeveloped" condition in monkeys.[65] Rolleston, who more than any other member of Huxley's band of rebels involved himself in this particular matter, read extracts from a letter by Huxley "in reference to the brains of the Quadrumana" which characterized Owen's position as an "obstinate reiteration of erroneous assertions" which could "only be nullified by as persistent an appeal to facts,"[66] that is, to Huxley's own position in the matter.

The work that seemed to clinch the argument in Huxley's favor was the last in the 1861 string of brain dissections, by William Flower, then working at the Middlesex Hospital but later as Owen's successor, first at the Royal College of Surgeons and subsequently at the British Museum. Flower had acquired a series of brain specimens of all three quadrumanous families as then classified: the Catarrhina (Old World apes and monkeys), the Platyrrhina (New World monkeys), and the Strepsirrhina (lemuroids). He, too, concluded that posterior lobes are present, in certain instances larger than in humans, particularly in the smaller members of the Platyrrhina, and that such lobes could be defined by reference to a particular cerebral groove or furrow, the calcarine sulcus; also, that the hippocampus minor is one of the most striking characteristics of the simian brain, best developed in certain catarrhine genera such as vervet and rhesus monkeys, less in the anthropoid apes, and least of all, in proportion to the mass of the cerebrum, in humans. Like all members of Huxley's anti-Owen army, Flower used the rhetoric of scientific objectivity and factualness, declaring that his work was "of a purely anatomical character," "undertaken without reference to any theory as to the transmutation of species, or origin of the human race."[67]

The Hippocampus Controversy Culminates: Cambridge, 1862

The Owen-Huxley clash reached its climax in 1862 during the Cambridge meeting of the BAAS. Huxley was president of section D, on "zoology and botany, including physiology," to which Owen presented his anti-Darwinian paper "On the Characters of the Aye-aye" as well as "On the Zoological Significance of the Cerebral and Pedal Characters of Man." Owen confined his discourse to the external features of the brains of primates, reiterating his previous observation that the transition gorilla-Negro is accompanied by a jump in the size of the brain. A Negro's brain is as large as the average brain of a Caucasian, and Owen agreed with the antiracist conclusion of the Heidelberg anthropologist Friedrich Tiedemann "that there had been no province of intellectual activity in which individuals of the pure Negro race had not distinguished themselves."[68] The gorilla's cerebrum did not entirely cover the cerebellum, let alone extend beyond it. Its frontal lobes were relatively poorly developed and so were its convolutions. This was increasingly more so in the series from the gorilla's brain down to the smooth and unconvoluted brain of the marmoset, although there were exceptions such as the baboon, in which the cerebrum does advance beyond the cerebellum. Owen likened this exception to the long nose of *Semnopithecus nasicus*, which serves to prove the rule that simians lack a prominent nose. The *Medical Times and Gazette* reported Owen as having said:

> The sudden advance of so supremely important an organ as the brain, in the human race, and the marked hiatus between that highest grade of its structure and the next step below, attained by the orangs, chimpanzees, and gorillas, was one of the most extraordinary in the whole range of Comparative Anatomy. It was associated with the intellectual capacities, the power of framing general propositions, and of expressing thought in articulate speech.[69]

A parallel feature—Owen reminded his audience—was the modification of the human foot, in particular the great toe and associated tarsal bone. In the gorilla the disposition of the hallux as a hind thumb is as strong as in other apes and presents a contrast with the human foot, in which the innermost digit deserves the name hallux, or great toe.

Without mentioning his detractors by name—as so often—Owen answered some of their criticisms. Huxley's assertion that the difference gorilla-man is not as large as highest simian–lowest simian did not state the problem correctly, as Owen had argued before; this proposition is just as true and meaningless, he stated, as saying that the difference in the structure of the foot between human and gorilla is not as large as that between the foot of a gorilla and the ventral fin of a fish; only directly adjacent, successive steps must be compared, not steps that are remote from each other.[70]

As the foremost proponent of homological research, Owen concurred as a matter of course that counterparts to the features of the human brain and foot occur in simians. It is not that the bones of our toe are unique but that their disposition, if properly defined, may function as an "arbitrary" term of zoological classification. "So with regard to the posterior lobe of the brain and its contained structures in man . . . for brief and concise zoological differentiations."[71]

Upon the conclusion of Owen's paper, Huxley "came down like the wolf on the fold": some years ago, Owen had asserted that three features were "peculiar to and characteristic" of the human brain, namely the posterior lobe, the posterior cornu, and the hippocampus minor. Huxley had maintained that these structures not merely are present in apes but often are better developed in apes than in humans. "He [Prof. Huxley] now appealed to the anatomists present in the Section whether the universal voice of Continental and British anatomists had not entirely borne out his statements and refuted those of Prof. Owen." With respect to the foot Huxley stated "that the structural differences between man and the highest ape are of the same order and only slightly different in degree from those which separate the apes one from another."[72] He concluded by emphasizing that the man-animal differences are mental, not physical. Flower supported this counterattack by stating "that the distinction between the brain of man and monkeys did not lie in the posterior lobe or the hippocampus minor, which parts were proportionately more largely developed in many monkeys than in man."[73] Rolleston, in the most emotive attack, accused Owen of ignoring the work done by foreign anatomists such as the Frenchman Louis Pierre Gratiolet. Their work had established that the truly distinctive features of the human brain are its great weight, its great height, the multitude of frontal lobes, and the absence of a deep cleft or perpendicular fissure that is found in the posterior part of the cerebral hemispheres in apes. Rolleston apologized for the vehemence of his attack, yet there was a sting in his apology when he observed that "he felt there were things less excusable than vehemence; and that the law of ethics, and the love of truth, were things higher and better than the rules of etiquette or decorous reticence."[74]

These reactions by the young upstarts of section D indicate the extent to which Huxley had succeeded in defining the issue on his terms. After all, Owen had initially hinted and later explicitly indicated that when he referred to the posterior lobe and its internal features—the posterior cornu and the hippocampus minor—he had in mind their full and particular development in humans; he had never denied a partial presence of these structures in simians and had even been the first to point out exceptions such as the marmoset, in which the cerebrum fully covers the cerebellum. Huxley, not Owen, had focused the debate on only the three features of posterior lobe, posterior cornu, and hippocampus minor. Owen, in his original papers, had drawn attention to the diagnostic size of the human brain, its highly developed frontal lobes, and its convolutions. Rolleston was not attacking Owen but the straw man that Huxley had made of him. Yet the Cambridge fracas also indicated the extent to which Owen refused to enter into the argument's complicated details as revealed by the Continental literature and by new anatomical dissections. The two parties failed to find a common frame of reference, and Owen, in his rebuttal, merely restated his position, in part by reading from his old Rede lecture.

The clash attracted much press attention, and a report of the meeting in the *Medical Times and Gazette*, in the writing of which Owen had a hand and in which his position was more fully and coherently presented than that of his adversaries, led to further, angry letters. Rolleston was first with a long letter "On the Distinctive Characters of the Brain in Man and in the Anthropomorphous

Apes" which served as an elaboration on his impromptu speech at the Cambridge meeting. It revealed, among other things, that his mention of Gratiolet was not primarily motivated by Owen's alleged ignoring of foreign colleagues but by the fact that Owen had paid no attention to a Royal Institution talk by Rolleston himself, given early in 1862, "On the Affinities and Differences between the Brain of Man and the Brains of Certain Animals," in which he had paid much attention to Gratiolet's work on cerebral convolutions.[75] To the four distinctive features of the human brain that he had cited, Rolleston added yet a fifth: "the enormous preponderance of the human over the simious corpus callosum." He also cited extracts from the Continental literature stressing the variability of the posterior lobe in humans. The young Oxford professor ended his long epistle with a public affirmation of his Christian faith and by stating "that the true relationship of man's body to his soul, to the world in which he lives, and to the Governor of it, can never be fully elucidated either by physiological or psychological researches, nor yet by both combined."[76]

Huxley was next, with a letter that added no new arguments but presented a summary of the history of the controversy. It was remarkable mainly for its well-written simplicity and lucidity but also for its pugnacious language. He referred to "an unworthy paltering with truth," from which he wanted to distance himself; to the fact "that the untenability of his position ought to have been as apparent to Professor Owen as to every one else"; to Owen's "grave errors" and his refusal to make "the smallest of concessions to justice"; to "the utter baselessness of the assertions of Professor Owen"; and to "the two years through which this preposterous controversy has dragged its weary length." Huxley's supporters were, by contrast, "able and conscientious observers" who "have with one accord testified to the accuracy of my statements." The issue—Huxley asserted once more—was not one of interpretation or definition but of fact. He accused Owen of having "incautiously tried to press into his own service" Schroeder van der Kolk and Vrolik, who, he claimed, were on his, Huxley's, side, adding: "Even the venerable Rudolph Wagner, whom no man will accuse of progressionist proclivities, has raised his voice on the same side; while not a single anatomist, great or small, has supported Professor Owen."[77]

Outflanked by Huxley

The truth about pressing colleagues into one's own service, however, was quite the reverse. We have no evidence to suggest that Owen tried to enlist the help of any of his colleagues, either at home or abroad, although in places he cited their published work. Huxley, by contrast, not only ran a feverish campaign at home to enlist fellow workers in his band of anti-Owen rebels but also made use of Continental authorities on neuroanatomy and neurophysiology in maneuvers to cut Owen off from any relief supplies that might be shipped across the English Channel. Huxley's *Natural History Review* discussed "St. Hilaire on the systematic position of man," for example, while denigrating Owen as merely "an accomplished osteologist";[78] the magazine also carried "Note sur l'encephale de l'orang-outang" by Schroeder van der Kolk and Vrolik, in the introduction to which Owen was accused of having abused the authority of the Dutch anatomists, "their facts denied, their words misquoted, and their very figures misinterpreted";[79] and in a further instance of such criticism, an anti-Owen paragraph by the German botanist and cofounder of the cell theory, M. J. Schleiden, was quoted in full.[80]

Moreover, Huxley initiated a correspondence with the redoubtable Rudolph Wagner for the sole purpose of turning the Göttingen professor against Owen. This was a devious, well-planned, and also successful move, showing Huxley as the superior tactician. Wagner was one of Germany's leading comparative anatomists and physiologists, and in the course of the 1850s, he had become increasingly worried about the rising tide of scientific materialism. In 1857, the year in which Owen presented to the Linnean Society his taxonomic upgrading of humanity, Wagner published *Der Kampf um die Seele*, one of several treatises concerned with the question of whether the human brain can be regarded as an organ of the soul. Ever since the latter half of the 1830s, Owen and Wagner had occasionally been in friendly correspondence, and in 1860 Wagner sent a copy of the first volume of his *Vorstudien zu einer wissenschaftlichen Morphologie und Physiologie des menschlichen Gehirns als Seelenorgan* (". . . on the Human Brain as Organ of the Soul") to Owen, adding in two consecutive letters that he was "contra Darwin and Huxley" and sided with Owen in his views expressed at the Oxford meeting of the BAAS, which views Wagner interpreted as an anti-Darwinian insistence on a fundamental human-ape difference.[81]

In print, Wagner moderated his stance to a certain extent. He disagreed with Owen in that he regarded the overall organization, or "architecture," of the brain as a more reliable taxonomic criterion than any individual morphological feature. In his view, both Owen and Huxley "express certain truths"; both, also, "go too far."[82] No sooner had Wagner published his evenhanded assessment of the Owen-Huxley clash in a footnote to a paper in the *Nachrichten von der Gesellschaft der Wissenschaften zu Göttingen* than Huxley began a series of letters to Wagner in an effort to win him over to his side. Perceptively, Huxley grasped that Wagner's sympathy for Owen was part of a shared dislike of Darwinian ideas and that, if he were to win over the Göttingen physiologist, he needed to disown Darwin as much as possible. This is what he did, repeatedly assuring Wagner that he—Huxley—was not a Darwinian and that his work on apes had no direct bearing on Darwin's theory. He began by complimenting Wagner, by now just into his fifty-seventh year, expressing his "admiration of the manner in which a veteran in science like yourself, keeps his place in the first rank among the press of younger investigators." But with respect to Wagner's published position vis-à-vis the Owen-Huxley clash, how could Wagner say that Owen expressed certain truths "when every assertion which is peculiarly his own, is demonstrably false?" "I do not think you quite apprehend my position as regards Mr Owen. The question between us is not one of Darwinism or anti-Darwinism—not one of the importance or unimportance of certain characters—: but it is a question of anatomical fact and of personal veracity."[83]

This initial letter was followed by a much longer one giving a full exposé of the controversy as Huxley saw it and in which, once more, he dissociated himself as much as possible from Darwin. "In heaven's name let us be neither Darwinians nor anti-Darwinians but truthful men of science, as your poet says 'Im Ganzen, Guten, Wahren, resolut zu leben.'"[84] Huxley's letter was not intended merely to put his case to Wagner but to get the latter's vote in the campaign to defeat Owen. In a further letter, accompanying an offprint of his paper "On the Brain of *Ateles paniscus*," Huxley made this clear: "I should be very glad if you would give your opinion on the question of *anatomical fact*, à propos of my paper, in which you will find nothing Darwinian."[85]

In a narrow sense, Huxley's assertion that he was not a Darwinian may have been correct. As Michael Bartholomew and, in further detail, Mario di Gregorio have shown, Huxley entertained serious doubts about the efficacy of the mechanism of natural selection. For several years he kept it out of his own work and only in 1868, probably influenced by his twin Darwinian "bulldog" Haeckel, did he explicitly incorporate the concept of evolution in his research papers.[86] In the sense, however, in which Wagner was likely to understand Huxley, namely that Huxley entertained grave doubts about the theory of evolution and especially of human evolution, the latter's disavowal of Darwinism was apt to deceive. To put it differently, it was factual but untruthful. This is not to imply that Huxley's intent was one of calculated dishonesty. The tenor of his letter would seem to indicate that Huxley wrote it in a state of self-righteous indignation. Yet he himself had started the hippocampus attack on Owen as part of his defense of Darwin and, not long after, in *Man's Place in Nature* he explicitly linked the human-ape issue to the Darwinian controversy when he asserted that "the question of the relation of man to the lower animals resolves itself, in the end, into the larger question of the tenability or untenability of Mr. Darwin's views."[87]

The result of Huxley's correspondence was more than he could have hoped for. First, Wagner went public in support of Huxley in the latter's disagreement with Owen,[88] and Huxley responded by acknowledging that "your public support in this matter will be very useful," adding: "it is a great comfort to me to observe that one does not necessarily become prejudiced and obstructive as one grows old."[89] Second, at the end of 1862, on Wagner's recommendation, the thirty-seven-year-old Huxley was elected to the Königliche Gesellschaft der Wissenschaften zu Göttingen,[90] an honor that had come Owen's way only three years earlier. Poor Wagner received nothing in return. He held out great hopes that Huxley would discuss his *Vorstudien* in the *Natural History Review* but this never happened in spite of a reminder Wagner sent to Huxley of "our scientific harmony . . . against Owen on account of the Cerebrum of the Apes."[91]

Meanwhile, in a short answer to Huxley's letter in the *Medical Times and Gazette*, Owen, in the clearest manner yet, once again stated that the issue was one of definition, not of fact. For example, the bones in the foot of the gorilla are all well known and homologous to those in the human foot. The inner toe in humans, however, is more developed than, and differently directed from, the innermost digit in the simians, and the point in question is whether this difference justifies the use of different zoological names: "great toe" for humans and "hind thumb" for apes. Owen believed that such a segregating nomenclature was indeed warranted. Similarly, although the human posterior lobe with its cornu and hippocampus minor has its homologous counterpart in simian brain structures which "are and were known to me, and as freely recognised as by any other comparative anatomist," its "more developed and differently configurated" manifestation in humans allows for applying the names of posterior lobe, etc., exclusively to features of the human brain—as Tiedemann had done before Owen. Taking this procedure to mean that Owen had failed to observe and had denied the presence of homologous structures in simians was a misrepresentation.[92]

Although the logic of Owen's rejoinder was difficult to fault, it was to no avail, as neither party was willing to listen to the other and agree on the terms in which the matter should be discussed. Despite his rhetoric of being on a fact-finding truth mission, Huxley was hell-bent on beating Owen, on tarnishing his opponent's reputation, and on simultaneously bolstering the pro-Darwin drive. Even though in his correspondence to Wagner he had vouched that his stance in the matter had nothing to do with his attitude to Darwinism, the victorious culmination of his clash with Owen was its representation in the three most important, early Darwinian treatises on the age and origin of man, namely Lyell's *Antiquity of Man*, Huxley's *Man's Place in Nature*, and, several years later, Darwin's *Descent of Man.* Lyell emphatically sided with Huxley in focusing on the cerebrum-cerebellum relationship in primates. He accused Owen of having "overlooked" the fact that the chimpanzee picture by Schroeder van der Kolk and Vrolik represented a distorted specimen in which the cerebellum was displaced and appeared more uncovered than it is in real life. He quoted from Huxley that every marmoset, American monkey, Old World monkey, baboon, or manlike ape has its cerebellum entirely covered and possesses a large posterior cornu and also a well-developed hippocampus minor.[93] That this conclusion was only a difference in emphasis from Owen's modified position was ignored in the drive to appear to have defeated the despised "British Cuvier." Owen reacted angrily in a letter to the *Athenaeum* in which he gave his own version of the controversy and suggested that Lyell was acting more like a lawyer pleading a partisan case than a student of nature seeking truth.[94]

As W. F. Bynum has shown, the brain chapter in Lyell's book was not his own but Huxley's, more fully developed in the latter's *Man's Place in Nature.*[95] In this, the similarities of the human and chimpanzee brain were demonstrated by means of an illustration which might just as well have been used by Owen to demonstrate that in the human brain the posterior cornu and the hippocampus minor are more perfectly developed than in a chimp's brain. Huxley also made a case for giving up the old distinction between quadrumana and bimana. The hind-limb of the gorilla, for example, does not end in hands but in feet, although prehensile ones. They are constructed on the plan of a proper foot and differ in no fundamental character from human feet. The difference "is only skin deep," as Huxley catchingly phrased his point.[96]

Having defined the issue as one of fact, not of definition, and portrayed Owen as having ignored plain facts, Huxley finally drove his argument home with the implicit accusation, already made in one of his letters to Wagner, that Owen had acted without personal integrity and thus was undeserving of scientific credibility. "The question has thus become one of personal veracity. For myself, I will accept no other issue than this, grave as it is, to the present controversy."[97]

His second promise of future silence notwithstanding, Huxley returned to the subject of the brain in humans and apes with a note to Darwin's *Descent of Man,* using the opportunity to contradict, once more, his long-term adversary.[98]

7
Frames of Mind

For Huxley the hippocampus controversy came to a triumphant conclusion with the publication of *Man's Place in Nature.* In the secondary literature his victory has been romanticized as the slaying of "Goliath Owen" by "David Huxley." William Irvine concluded in his widely read *Apes, Angels and Victorians* (1956): "Huxley triumphed at every point and only gained further glory from the wild flailings of his adversary. . . . Owen had become almost an historical curiosity by the time Huxley's *Evidence of Man's Place in Nature* appeared in 1863."[1] This and later similar judgments about the Owen-Huxley controversy are merely a condensed repetition, however, of Huxley's side of the story. The jury of historians has reached a verdict by listening exclusively to the plaintiff, ignoring the testimony of the defendant and of any witnesses for the defense.

In Owen's Defense

Witnesses for the defense there were. To begin with, there was the written testimony of Continental authors from earlier in the nineteenth century. Owen was neither the only one nor the first to have attributed the disputed cerebral features uniquely to humans, or rather to have regarded their degree of development in humans as taxonomically distinctive. In 1820 the German naturalist Heinrich Kühl had characterized the part in a platyrrhine monkey that is homologous to the human posterior cornu as only a "beginning" of that organ.[2] In his classic *Icones Cerebri Simiarum* (1821), Friedrich Tiedemann had called the same part in the pigtailed monkey "a small furrow in lieu of the hinder horn"; he had also named it "a vestige of the hinder horn."

The hippocampi minores, he stated, do not occur in simian brains and are typical of human brains; his explicit conclusion was: "Homini ergo propii sunt" (They are therefore characteristic of man).[3] The Parisian anatomist Jean Cruveilhier, too, had stated in his *Anatomie descriptive* (1834–36) that by virtue of the large occipital development of the human brain, the posterior cornu of the lateral ventricle and the hippocampus minor reach their proper development only in *Homo sapiens.*[4]

Closer to home, the Edinburgh anatomist Robert Knox gave his assent to another of Owen's characterizations of human brains, stating that "in man the posterior lobe of the cerebrum overlaps the cerebellum, whilst in other animals (with scarcely an exception) *it does not.*"[5] In London, moreover, no less a journal than the *Lancet* fully supported Owen by condemning Huxley for changing the issue from one of definition into one of fact, and equally for countering Owen's restrained manner of argumentation with misleading and unseemly rhetoric. It commented, in reaction to the Owen-Huxley correspondence of October 1862 that had appeared in the *Medical Times and Gazette*:

> We observe with regret that Mr. Huxley has repeated his attack on Professor Owen, relative to the structures at the back part of the brain in man and apes. Anatomists and the profession became well aware, from Owen's reply to the first of these attacks, that the question was one of terms and definitions, not of facts. . . . We recommend Professor Huxley to try to imitate in these discussions the calm and philosophical tone of the man whom he assails. The fling and the sneer, however smart, will only recoil upon himself.[6]

By far the most detailed and competent defense of Owen came in the form of a review of Huxley's *Man's Place in Nature*, carried by the *Edinburgh Review.* Curiously, no review of Huxley's book appeared in any of the other leading periodicals. Even the *Westminster Review* announced the book's appearance merely in the form of a fairly minor notice.[7] The author of "Professor Huxley on Man's Place in Nature" was Charles Carter Blake, a former student of Owen's and a lecturer in comparative anatomy and zoology at Westminster Hospital School of Medicine. Nearly the entire review of twenty-nine pages was a systematic refutation of Huxley's attack on Owen. First, Blake entered the cycles of repetition through which the debate had been wound and addressed Huxley's "entirely illogical fallacy" of stating that man is much nearer to the gorilla than the gorilla is to the lowest of the simians, the lemur. "Huxley plunges from the gorilla down to the lemur," whereas he should be comparing "successive links in the quadrumanous series," in which we can see the brain change "by slow and gradual steps"; only the step from gorilla to man is a sudden one, without intervening links:

> When we analyse particularly the exact significance of these successive changes, we see that in the order *Quadrumana* there is a certain range of progressive increment in the ratio of development of the various brains. We may roughly say that the brains of the Quadrumana increase in development and complication through the series indicated in the arithmetical terms 1, 2, 3, 4, 5, 6, 7, 8, 9, 10, and that the human brain may be represented by the term 15. Of course a controversialist may assert that the number 10 (gorilla) is nearer the term 15 (man) than is the number 1 (lemur), the respective differences being 5 and 9. But the fact remains that between the gorilla and the lemur there is a series of well-defined but short steps; between the gorilla and man no intervening link has as yet been discovered. This, we apprehend, is the true aspect of the question; and, until zoology or geology shall have demonstrated to us the existence of intervening links, we are justified in placing man, as he is at present, in a separate sub-class.[8]

Blake then addressed Huxley's assertion that the human foot is too similar to that of an ape to be of taxonomic value. He granted that "the hinder extremity of the gorilla is formed by bones homologous with those of the human foot" but argued that the difference between gorilla and human is shown by the muscles and tendons rather than by the bones. For example, the great toe in humans is connected to a tendon which passes along the sole of the foot and is connected to a muscle, a long flexor, in the lower part of the leg. This arrangement makes it possible for a dancer to pirouette on tiptoe. In the orang, however, the homologous muscle terminates in three tendons, connected to the second, third, and fourth toe respectively, whereas in the gorilla the same three tendons are connected to the great toe and to the third and fourth toe—a very different arrangement that enhances the prehensile grasping ability of the simian foot.[9]

Blake now did what Owen had stubbornly refused to do, namely, discuss in detail the results of the brain dissections that Huxley and his collaborators had carried out. He praised John Marshall for having contributed "the most valuable monograph on the brain of the chimpanzee yet before us"; and William Flower, too, was complimented "for the ability and lucidity with which he has put before us the facts at his disposal." Blake admitted that in some quadrumana the cerebrum covers the cerebellum or even projects "slightly beyond it." Yet he concluded: "in no ape is the portion which projects beyond the cerebellum in any degree equal in bulk or substance to the far larger structure which is termed the posterior lobe of man." He further concluded that the posterior cornu is more elaborately developed in humans than in any apes or monkeys, and that the hippocampus minor, "taking it in the sense in which the term is used by human anatomists, is, strictly speaking, absent in all

Quadrumana, in none of which is there that characteristic inversion of the grey cortical brain matter, *coincident in its direction with the floor of the posterior cornu*, which forms the hippocampus minor of anthropotomy."[10]

Blake explained that Owen had never denied that the disputed features of the human brain can be traced in animals; "the whole scope of his public teaching," he reminisced, "so far as we are acquainted with it, has involved the recognition in the highest *Gyrencephala* [anthropoid apes] of organs admittedly homologous with those of man, but differing from the structure of the highest mammal [*Homo sapiens*] by their minor degree of development and their less amount of complexity."[11] Homologically identical organs in the various vertebrate classes, such as the limb extremities, do require zoologically different names, for example, hand, hoof, paw, flipper, or paddle. For Huxley to claim that every ape possesses "a well-developed hippocampus minor" is an exaggeration; the description used by Jacobus Schroeder van der Kolk and Willem Vrolik for the chimpanzee—namely, "indication of the hippocampus minor"—"far more accurately expresses its real signification." Huxley's illustration comparing the dissected brain of a human being with that of a chimpanzee was not entirely correct either, because it likened the brain of an adult human to that of a young chimp; adult should be compared with adult, because in immature brains the ventricles are proportionately larger. Like the *Lancet*, Blake also objected to "the injudicious and offensive manner" in which Huxley conducted himself.[12]

Owen's own most elaborate defense of the position he had taken appeared in his *Memoir on the Gorilla* (1865). In this he discussed at some length the taxonomic value of the characters of the brain and limbs of *Troglodytes Gorilla.* He particularly emphasized that, along with a number of his Continental colleagues, he had shown that all parts of the human brain have homologous counterparts in various apes and monkeys. More comprehensively this time, he concluded: "In the Gorilla, as in other latisternal [having a broad breastbone] Apes, the homologue of every organ and of almost every named part in Human anatomy is present."[13] Yet these homologues do not necessarily reach the same level of development. The cerebrum, for example, does not entirely cover the cerebellum, let alone extend beyond it, being coextensive therewith in length but not quite in breadth. The gorilla shows the same degree of limited development of the posterior cornu and the hippocampus minor as observed in the chimpanzee. "It is the beginning of those structures which are backwardly extended, with the parts of the hemispheres containing them, and fall short, in the same degree, of their extended and differently curved homologues, in Man."[14] Owen reiterated that such differences can legitimately be used for classificatory purposes, and that the difference between "the highest Ape and lowest Man" is much greater than that between any two genera of apes and monkeys.[15] In his trilogy *On the Anatomy of Vertebrates* Owen also made it clear, for the umpteenth time, that he had always been aware of homologous beginnings in simian brains of structures which reach their fullest development in humans.[16]

Although Owen, as a man of his time, referred to "higher" and "lower" races of mankind, he objected to the extreme racism of evolutionary anthropology by which the difference "Homer-Hottentot" was believed to be larger than the gap "Hottentot-gorilla." Owen's perception of the major gulf separating all humans from the apes and, by contrast, the fundamental similarity of the brains of all human races went hand in hand with his belief that blacks, for example, are capable of distinction in all areas of intellectual activity. He warmly supported the award by the University of Oxford, in 1864, of the degree of doctor of divinity to a former African slave who had become a bishop, adding: "I record with pleasure the instruction I have received in conversation with this sagacious and accomplished gentleman."[17]

Glowing Rivalries

In summary, and weighing the evidence for the defense, it is clear that Owen's case was not without good arguments and Huxley's victory less complete than we are led to believe in the secondary literature. It is a myth to maintain that Owen was either unable to see plain facts or unwilling to admit them. He did not deny that the gorilla had a hippocampus minor, as Mario di Gregorio alleges, but he asserted that it is not as well developed as its human counterpart and therefore does not deserve the name hippocampus minor. Owen's logic was flawless when he pointed out that, in comparing humans and beasts, homologically identical parts, if unequally developed, may be given different names.

Yet there was a certain sterile, scholastic subtlety to his logic. By giving the names "posterior lobe," "posterior cornu," and "hippocampus minor" exclusively to the relevant modifications of the human brain, and by then stating that these features are characteristic of the human brain, the false impression was created that nothing comparable is present in the brains of apes and monkeys and that the separation of humans from the rest of the animal world can be based on absolute, rather than relative, criteria. In his original Linnean Society presentation of 1857, Owen had not made it clear that "indications" or "traces" of a posterior cornu and hippocampus minor are present in simian brains. In the course of the controversy he retreated to a more moderate position when he qualified his original assertion that the hippocampus minor is peculiar to *Homo sapiens*; he explained that this meant, not that its occurrence as such was peculiar, but only that its particular, full development was. For example, in the 1866 edition of Brande and Cox's *Dictionary of Science*, Owen added a qualification to his definition of the hippocampus minor: "No such structure, corresponding in position and shape to the anthropotomist's definition, has yet been discovered in any known ape."[18]

Furthermore, the considerable amount of new information about the anatomy of the brains of apes and monkeys that became known after 1857 showed that Owen's disputed cerebral criteria did not systematically vary as smoothly as he had envisaged and were thus less useful for taxonomic purposes. Flower's study in particular demonstrated that the extent of the posterior development of the cerebrum over the underlying cerebellum does not change consistently through the primate order and is of little help in distinguishing human from simian. Owen's retort that the very bulk of the posterior lobe in humans is a characteristic feature, and his insistence on its taxonomic value, were not factually false, as Huxley fumed, but they were contrived all the same. For the practice of classification, the three disputed cerebral features proved inadequate, and in the long run even Owen himself quietly let go of them. In the preface to a handbook, *Zoology for Students* (1875), written by his loyal pupil Blake, he wrote:

> The most abrupt and marked rise in brain-development is presented by Man, and again more especially in the cerebrum, outswelling vertically, laterally, antero-posteriorly, extending forward above and beyond the olfactory lobes, overarching and extending backward beyond the cerebellum, with concomitant growth of the great commissure or corpus callosum within, and multiplication and deepening of the convolutions without.[19]

By emphasizing the exceptional size and complexity of the human brain, Owen was on safe ground, and although many of his colleagues might have felt that these features do not justify placing humans in a separate subclass of Archencephala, they did agree that a major gap exists between the human and the simian brain. To illustrate this, Owen cited the fact that the weight of the brain of a full-grown male gorilla is less than one-third of the average weight of the brain of the adult human male, even though the bulk of the body is greater in the gorilla than in *Homo sapiens*. Rudolph Wagner's work on brain size confirmed this; Louis Pierre Gratiolet's work on brain convolutions supported it; and George Rolleston's emphasis on the size of the human corpus callosum added to it. The *Westminster Review* also pointed it out.[20] Later biologists, such as Adolf Portmann, reconfirmed it and added the differentiating criterion of the growth curve.[21] In this respect, Huxley was out on a limb, and the major gap from ape to human remained empty until the much later discovery of various hominid fossils filled it with "missing links."

In view of this, the question arises: why did Owen not fall back much sooner on this unassailable position? After all, it was already present, and prominently so, in his lecture of 1857;[22] as we saw, Huxley was the person who narrowed the issue to the three specific cerebral features. Why not let go of these or at least de-emphasize their significance and thus not give Huxley a stick with which to be beaten? Was it Owen's vanity, as has been alleged? It is true that Owen found it difficult to admit to having been wrong, but it is a myth, one of many about Owen, that he never owned up to a mistake. In several of his publications he corrected previously expressed opinions if new data so required.[23] Yet Owen had grown increasingly autocratic and resented challenges to his authority. It was because Huxley represented such a challenge—I believe—that Owen dug in his heels and failed to show the flexibility to outclass or outwit his opponent. It should be borne in mind that the hippocampus controversy was the culmination of a decade of attacks by Huxley on Owen, which, from Owen's point of view, were ungrateful, disrespectful, and mostly spurious.

When in November 1850 Huxley returned to London from his four-year voyage aboard HMS *Rattlesnake*, he brought with him letters of recommendation to Owen by Australian naturalists. One of them wrote of Huxley: "He is talented and a gentleman."[24] Another, William MacLeay, suggested that Owen help Huxley with the publication of his observations.[25] In response, Owen wrote to Francis Baring, First Lord of the Admiralty, asking that Huxley be given the opportunity to work up his

> notes and drawings of numerous minute observations which he has made on rare and remarkable animals of the Australian seas during the surveying voyage from which he has just returned. . . . In order to enable Mr Huxley to prepare for the press his materials, which are really of great value to natural history, without incurring a loss of time, and pay, an appointment as assistant surgeon on "particular service," to a grand ship, would be requisite, and if he could be found such an appointment for twelve months the end would be gained.[26]

Huxley obtained from the Admiralty the appointment of assistant surgeon with leave of absence (subsequently extended to 1854), and at about the same time he was elected a fellow of the Royal Society. Both of these career advances he owed to a large extent to Owen, and Huxley reported back to MacLeay that Owen had been very civil to him: "Personally I am greatly indebted to him. . . . During my absence he superintended the publication of my paper, and from the moment of my arrival until now he has given me all the help one man can give another. Why he should have done so I do not know, as when I left England I had only spoken to him once."[27]

In spite of Owen's generosity, Huxley almost immediately began chipping away at his patron's work and reputation. Already in 1851 he attacked Owen's notion of the alternation of generations.[28] From there on it became something of an annual ritual for Huxley to involve himself with one of the subjects to which Owen's name was prominently connected, focusing on any real or imagined mistakes in Owen's work. These hostile forays into Owen territory included, apart from further criticism of Owen's work on parthenogenesis, attacks on Owen's anti-Lyellian advocacy of the theory of progressive development, his vertebral theory of the skull, and his study of the intestinal tract in brachiopods, of the vertebral column in certain plesiosaurs, of the lining of the cavity in the pearly nautilus, of the teeth of diprotodon, and so on.[29] When this first decade of hostility culminated in the early 1860s with an attack on Owen's cerebral classification of mammals, Huxley also tried to wreck Owen's plans for a national museum of natural history. After this, Huxley's hostile obsession with Owen and his work continued for at least another decade with a variety of new disagreements about the *Archaeopteryx*, the dinosaurs, and other mainly paleontological topics.[30] Huxley, by focusing on imperfections or mistakes, obscured the fact that Owen's oeuvre represented a major advance on previous work and that Owen's many publications had opened up new areas of investigation, which formed, to a certain extent, the framework of Huxley's correctives.

Initially, Owen continued to support Huxley but, as Adrian Desmond shows in his well-balanced assessment of the Owen-Huxley relationship, it was Huxley's lack of civility that finally turned Owen against his persistent detractor.[31] The turning point was Huxley's review of the tenth edition of *Vestiges* with its theory—to which Owen had publicly subscribed—that the creation of species occurred by nonmiraculous, natural means. This edition included new material in support of the doctrine of progressive development, taken from Owen's anti-Lyellian review in the *Quarterly*.[32] Thus, Huxley's scathing attack on *Vestiges* doubled as a swipe at Owen. Huxley's language was ferocious. He referred to the book as "a mass of pretentious nonsense," "charlatanerie," full of "blunders" and "whining assertions of sincerity," "the product of coarse feeling operating in a crude intellect," "a lumber-room of second-hand scientific furniture," and so on. Huxley denounced the notion of creation by natural law as "unworthy of serious attention" and singled out for detailed criticism the theory of fossil progression, which, he asserted, had no foundation in paleontological fact. He ended his review by praising the thus far most anti-Vestigian reviewer, namely Adam Sedgwick, "a genial man, who has made truth the search of his life," and returned to Owen, whom he accused of an "underhanded attack" in the *Quarterly* article upon the microscopist J. T. Quekett, the assistant conservator at the Hunterian Museum.[33]

The appearance of Owen's *Lectures on the Comparative Anatomy and Physiology of the Invertebrate Animals* in 1855 provided Huxley with a further and readily grasped opportunity to criticize the Hunterian professor.[34] Two years later, in 1857, Huxley, by his own account, broke off all personal contact with Owen because the latter had himself entered in the *Medical Directory* as professor of paleontology at the Royal School of Mines, where in 1854 Huxley had become a lecturer. Owen's post-Hunterian eagerness to continue in a professorial capacity made him present his position at the Royal School of Mines, which was effectively that of a visiting professor, as a formal professorship. What this incident shows is not only Owen's presumptuousness but also Huxley's ready antagonism, in this case brought about by the fact that Owen had managed to get a foothold in Huxley's institutional power base.

Thus, in the course of the 1850s, it became gradually apparent that Huxley was choosing a systematic attack on Owen as an avenue for self-advancement. Given the fact that Owen, more than any other naturalist, represented the establishment, Huxley's rise to eminence and to institutional power needed to be gained—he seemed to think—by bringing the ruling potentate down. His willingness to take on Owen brought him increasingly into the limelight and enhanced his prospects of career advancement.[35] Owen fully understood the ambitious motivation of his young adversary and resented it deeply. In a bitter postscript to a letter to Henry Acland on the hippocampus controversy he added: "Do you remember the story of the clever young Athenian who had the itch of notoriety. He sought the Oracle, and asked 'What shall I do to become a great man?' Answer: 'Slay one'!"[36]

To Owen, Huxley's picking of the hippocampus fight was just another personal assault, deviously carried out under the banner of disinterested truth. This attack, coming on top of a decade-long series of previous ones, must have been experienced by Owen as extreme provocation. Most of all, he felt indignant at the fact that Huxley turned a scientific debate into a campaign of malicious misrepresentation and defamation of character. Owen never forgave Huxley for this. Several years later, in the summer of 1871, the two men happened to be given adjacent seats at an official dinner. John Tyndall, who also attended the dinner, noticed that the two bitter enemies exchanged friendly words, and the next day he wrote to Owen: "I urgently trust that you will allow the spirit of peace and friendliness which operated last night to have further scope."[37] But Owen answered in what C. Davies Sherborn believed to be the only letter in which an opponent was severely criticized:[38] "Prof. Huxley disgraced the discussions by which scientific differences of opinion are rectified by imputing falsehood on a matter in which he differed from me. Until he retracts this imputation as publicly as he made it I must continue to believe that, in making it, he was merely imputing his own (base and mendacious) nature."[39]

Owen largely ignored the young upstart Huxley and rarely went beyond stiffly repeating his views, considering it undignified to do anything more. Nothing could have served Huxley's purpose better. Owen's reaction was a reminder of his autocratic character, which many colleagues had run up against and which the younger men feared and resented. Huxley himself had been "on my guard" toward Owen right from the start, in the early 1850s, because of the hatred that Gideon Mantell and others felt toward the Hunterian professor.[40] It was because of Owen's autocratic rule that Rolleston and several others sided with Huxley. Their participation was more anti-Owen than pro-Huxley. The controversy served as a rallying point for resentment of the "imperator's" aloofness, as shown by a set of documents recording Acland's involvement in "the great hippocampus question."[41]

Acland, by now the head of the medical faculty at Oxford, had kept quiet during both the Oxford and the Cambridge meetings of the BAAS. As a friend of both Owen and Rolleston, however, the turn of events at the Cambridge meeting greatly upset him, and immediately afterward he wrote a letter to Owen which took the form of a small pamphlet, printed for private distribution: it was to Owen that he turned for clarification, because Owen had not given a proper answer to the questions raised by the younger men; and had Owen not changed his original position, at least to a certain extent? At the Oxford meeting of the BAAS, scientific hypotheses such as the Darwinian theory which Huxley had defended should have been refuted "and not forcibly put down." Yet such suppression had taken place as the result of the "[a]mazing eloquence of the Bishop of Oxford, aided by your pre-eminent scientific support." Acland reminded his former teacher "that even you cannot afford to be indifferent to the goodwill and regard of the generation that follows after you." A few brief sentences from Owen could have settled the question at the Cambridge meeting. Huxley's charges had not been answered clearly, and Rolleston's objections had not been addressed fairly. Acland himself believed that the controversy arose in part because Owen set too much store by characters that were too limited for the purpose of distinguishing humans from the higher quadrumana. He had never been able to satisfy himself that any anatomical definition of *Homo sapiens* was safe and complete. A human being's higher nature does not hang on physical features but on moral ones, among which must be counted a willingness to accept facts as we know them.[42]

It was up to Owen—Acland pleaded—"our chiefest anatomist, on whom the mantle of John Hunter fell," to remove doubt from the public mind and restore confidence in the fairness of our scientific leaders. Failing this, the scientists would be looked on "as partisans of opinion, not as beacons of truth." The prestige of the scientific profession was at stake, and Owen, because of his very eminence, could attract new and young practitioners by a fair appraisal of their work. Acland ended his letter with an appeal to Owen not to damage his profession by settling disputes by means of the authority of his powerful position: "With you it rests to gather round you a rising generation of willing disciples, or to repel them by the fear that their honest labours are repressed or doubted. You may either restore confidence to those who cannot judge for themselves, or make them mistrust the candour of scientific men, and doubt the soundness of science itself."[43]

In a further letter, Acland continued to imply that Owen's overbearing manner was alienating the younger generation of scientific practitioners. "I have no doubt," he wrote, "that by frank and kindly words you may get a host of younger men to follow you in calm discussion on facts that have any real or supposed bearing on Darwin's Hypothesis and especially in the still infant subject of Human Ethnology."[44] In his answers, Owen restated that the issue was one "of *terms* or *interpretations* of facts," not of facts as such, and that he had spoken in a friendly, disinterested spirit, whereas his adversaries had behaved like hired applauders at a political rally, disrespectful of "the anatomist who had chiefly advanced the knowledge of the mature characters of Apes." "Scorning the arts of private misstatements and canvassing of judgment, I gather no group of young 'claqueurs' such as rendered the support of their inarticulate noises to each sophism of my adversary, in the *Section* at Cambridge."[45]

Acland also wrote to the archbishop of Canterbury, Charles Longley, who had been his Harrow schoolmaster, explaining that Owen's position would lead nonexperts to suppose a greater physical separation of humans from apes than is actually the case; and he urged him to use his influence with the clergy not to take sides in scientific disputes, for which its members lacked the proper training.[46] Acland intended to publish his correspondence with Owen, but in the end the two men failed to agree upon the proposed "Eirenicon," and after Acland's Oxford colleague, the geologist John Phillips, had advised him "that the less you have to do with the affair the better,"[47] he reluctantly dropped his peacemaking plan. Owen had antagonized too many of his colleagues, for too long, with his autocratic behavior, and this time they did not regret seeing him thrown off his high horse by a young pretender.

Public Perception of the Owen-Huxley Joust

In the wider, public perception of the hippocampus controversy there was no clear winner. The Owen-Huxley fight, which had become something of an annual ritual, drew comments from many sides. Press reactions ranged from severe censure to lighthearted caricature, and both men were pretty much evenly reprimanded. Huxley gained by this, no matter what. Being only in his midthirties, he benefited from having his name constantly bracketed with that of the most famous naturalist in the land, who, by contrast, was in his late fifties and at the zenith of his fame and power. Owen's refusal to let go of his quarrel with Huxley diminished his stature and appeared to reduce him to the populist level of his adversary, however much Owen tried to act with haughty formality. Not long after the Cambridge meeting, the *British Medical Journal* sternly called for an end to what it saw as a reprehensible personality contest: "Is it not high time that the annual passage of barbed words between Professor Owen and Professor Huxley, on the cerebral distinctions between men and monkeys, should cease? . . . Continued on its present footing, it becomes a hindrance and injury to science, a joke for the populace, and a scandal to the scientific world."[48]

A joke for the populace it certainly was. Who could resist comparing the scientists to the very monkeys over which they were fighting? Not the *London Quarterly Review*, which described the Cambridge discussions between Owen, Huxley, Rolleston, and Flower as follows:

> Animation increased, "decorous reticence" was at an end; and all parties enjoyed the scene except the disputants. Surely apes were never before so honoured, as to be the theme of the warmest discussion in one of the two principal University towns in England. Strange sight was this, that three or four most accomplished anatomists were contending against each other like so many gorillas; and either reducing man to a monkey, or elevating the monkey to the man![49]

Punch, too, drew on the man-monkey comparison in the form of two poems written by Philip Egerton. Although he was a friend and supporter of Owen, his "Monkeyana," published in 1861 shortly after the first Owen-Huxley exchanges in the *Athenaeum*, took no sides and had a gorilla ask:

> Am I satyr or man?
> Pray tell me who can,

And settle my place in the scale.
A man in Ape's shape,
An anthropoid ape,
Or monkey deprived of his tail?

. . . .

Then Huxley and Owen,
With rivalry glowing,
With pen and ink rush to the scratch;
Tis Brain versus Brain,
Till one of them's slain;
By Jove! it will be a good match![50]

The following year, however, after the Cambridge altercations, "the gorilla's dilemma" is no longer whether he is a man or an ape but, in addressing Owen and Huxley, the gorilla wonders whether in some respects monkeys are not superior to man, not only "in power of jaw," "in gymnastics," "in gagging, grimacing and chaff," but also in being able to keep quiet.[51]

Just as Lewis Carroll's *Alice's Adventures in Wonderland* made the dodo known to a wide readership, so Charles Kingsley's *Water Babies* brought the hippocampus to the general public. The Cambridge meeting was the first BAAS jamboree attended by Kingsley, who took an interest in the proceedings of section D. In reaction to the Huxley-Owen "tournament" he wrote "a little squib for circulation among his friends," entitled "Speech of Lord Dundreary . . . on the Great Hippocampus Question," in which the noble lord, who had been to Eton, where he had been swished for getting his Latin wrong, "accurately" expresses the general sense of the meeting by thanking the gentlemen for quarreling, "and so eloquently, too," by confusing a hippocampus with a hippopotamus, and by getting the contending parties mixed up with one another. Owen, Huxley, and all the others know best—the bumbling aristocrat states—because they are "monstrous clever fellows"; but can't they settle their differences by tossing a coin? He is glad to hear that there is a gulf between ape and man, especially if the ape bites, but if the gulf was bridged over by a structure, Huxley might have told a little about it. Was it wood? Was it iron? But if Huxley can see the structure, why can't Owen? It can't be invisible and this is another reason for them to toss a coin. For his aunt's sake, would the scientific gentlemen please settle the question by accepting apes as men, because then they will be entitled to hippocampi;

> for she says that her clergyman says, that if anybody ever finds a hippopotamus in a monkey's head, nothing will save her great, great, great—I can't say how great, you see—it's awful to think of—quite enormous grandfather from having been a monkey too; and then what is to become of her precious soul?[52]

At about this time, Kingsley's fairy tale *The Water Babies* was published in installments in *Macmillan's Magazine.* It contained some of the same material that went into Lord Dundreary's speech and included a variety of references to Owen, Huxley, the BAAS, and other names from the world of Victorian science. Those who do not believe in the existence of water babies, Kingsley wrote, may argue that if such creatures did exist, somebody would have caught one and have put it in spirits "or perhaps cut it into two halves, poor dear little thing, and sent one to Professor Owen, and one to Professor Huxley, to see what they would each say about it."[53] Kingsley continued his evenhanded treatment of the two biologists when he introduced an amalgam of Owen and Huxley in the character of "Professor Ptthmllnsprts" (Put-them-all-in-spirits). In an apparent reference to Owen's territoriality and lack of collegial spirit, Professor Ptthmllnsprts was described as "very good to all the world as long as it was good to him":

> Only one fault he had, which cock-robins have likewise, as you may see if you look out of the nursery window—that, when any one else found a curious worm, he would hop round them, and peck them, and set up his tail, and bristle up his feathers, just as a cock-robin would; and declare that he found the worm first; and that it was his worm; and, if not, that then it was not a worm at all.[54]

Huxley, too, was mocked, in particular for his scientific materialism, when Kingsley attributed to Professor Ptthmllnsprts the belief that nothing was true "but what he could see, hear, taste, or handle":

> He [Professor Ptthmllnsprts] held very strange theories about a good many things. He had even got up once at the British Association, and declared that apes had hippopotamus majors in their brains just as men have. Which was a shocking thing to say; for, if it were so, what would become of the faith, hope, and charity of immortal millions?[55]

The passage then switches to Owen's hippocampus stance:

> No, my dear little man; always remember that the one true, certain, final, and all-important difference between you and an ape is, that you have a hippopotamus major in your brain, and it has none; and that, therefore, to discover one in its brain will be a very wrong and dangerous thing, at which every one will be very much shocked.[56]

The most biting satire of the hippocampus controversy came in the form of a short play, published anonymously by George Pycroft, entitled *A Sad Case, Recently Tried before the Lord Mayor, Owen versus Huxley* (1863). The author denigrated the two controversialists for the vulgarity of their hippocampus brawl by casting them in the role of common costermongers. Dick Owen and Tom Huxley, both in the same line of work, selling "old bones, bird skins, offal," are arrested for causing a public disturbance. The arresting policeman testifies before the lord mayor that a crowd had gathered in the streets, around the two accused men, who were shouting at each other in the most awful language, calling each other names, such as "lying . . . Bimanous Pithecus" and "thorough Archencephalic Primate." Huxley even pushed a monkey toward Owen while shouting: "Look at 'em, a'nt they like as peas?" Owen tried but failed to cope with this "violence":

> He behaved uncommonly plucky, though his heart seemed broke. He tried to give Huxley as good as he gave, but he could not, and some people cried "shame," and "he's had enough," and so on. Never saw a man so mauled before. 'Twas the monkey that worrited him, and Huxley's crying out, "There they are—bone for bone, tooth for tooth, foot for foot, and their brains one as good as t'other."[57]

The constable's testimony is followed by an angry altercation between the accused parties. The mayor reprimands them and declares that such behavior is scarcely human, "at which Huxley laughed, and Owen looked grave." Owen then testifies that Huxley waylays him in public, throws dirt at him, and lately, aided by "his low set," attacked him in Cambridge. Huxley, when called upon, retorts that "we was in the same line, and comfortable as long as Dick Owen was top sawyer, and could keep over my head, and throw his dust down in my eyes. There was only two or three in our trade, and it was not very profitable; but that was no reason why I should be called a liar by an improved gorilla, like that fellow." A witness elaborates that the whole neighborhood has been unsettled by the disputes of the two prisoners and their associates, although he has pleasure in stating that Darwin has been the quietest of the set. "They were always picking bones with each other and fighting over their gains."[58]

In his summing up the mayor decides that no punishment is called for but that "the public had a right to claim protection from annoyance." He puts it to Owen, first, that he could prove his dissimilarity from an ape "by the practice of kindness, gentleness, forbearance, and humility" rather than by reference to the ape's anatomical structure. "And he put it to Huxley, whether it really was truth, pure and simple, he was fighting for; whether it was not more, or at least partly, for the purpose of exposing a weak point in his rival." He advises them to be friends and "to follow the example of good Edward Forbes." As soon as the two accused have left the court, "an altercation ensued"; and that very evening Huxley addresses a large assembly of working men in Jermyn Street, vilifying Owen and holding up a placard on which is painted a series of skeletons, beginning with the gibbon and ending with man, and on which are written the words: "I'll let him know his place in nature."[59]

The memory of the eventful early 1860s meetings of section D lingered for some time. In 1865, at the occasion of that year's BAAS meeting, *Punch* depicted Owen and Huxley in a dancing embrace, each nearly throttling the other.[60] As late as 1872, *Lord Bantam*, a fine satire on various Victorian sensibilities, refers to a "Professor Foxley" who delivers an address to the Grand Eclectic Symposium and Aesthetic Soiree on the subject of "The Hippocampus Minor and its relation to the Mosaic Cosmogeny."[61]

Men, Monkeys, and Mind

The Owen-Huxley clash lent itself well to lighthearted banter. Yet those who engaged in it were aware that the hippocampus fracas had deeper causes than a shallow personal rivalry. The question of whether or not human uniqueness can be captured by physical attributes, such as the anatomical features of the brain selected by Owen, was a profound one. Several critics believed that Owen, by defining *Homo sapiens* merely in terms of particulars of size and shape of the cerebrum, trivialized what it is to be human. As we have just seen, Kingsley mocked the idea that a "hippopotamus major" could be a criterion of humanity. The real, major difference was of a moral and intellectual kind—it was felt—and this should not be reduced to a little more or a little less of some cerebral convolution. In two consecutive years, 1861 and 1862, the *Times* expressed this view and asked the scientists whether it was not "far more orthodox to discard such trifling considerations and seek the true characteristics of our race in those mental attributes by virtue of which man towers so high above his inferior companions on earth?"[62] The *London Quarterly Review* stated the same opinion; the impassable gulf between humans and apes existed with respect to moral and intellectual features, not material ones. The imponderables of mind and soul are what sets us apart from the gorilla: "Let material distinctions be dismissed:—the mind, the soul, the grand mystery of thought, the airy magic of fancy, the boundless range of imagination,—these are the true and noble distinctions of the human being."[63]

Less volubly, but with the same intent, the *British Medical Journal* added to its above-cited rebuke of Owen and Huxley an appreciative summary of the views of Jean-Louis-Armand de Quatrefages, who taught anthropology at the Muséum national d'histoire naturelle in Paris. In his recently published *L'espèce humaine* (1861), he argued that humans are distinguished from animals by having a notion of right and wrong and by a belief in the existence of a God, that is, by the possession of moral and religious awareness. He believed that neither anatomical nor physiological nor even such mental features as articulate speech separate humanity from the animal world; these merely present differences of degree, not of kind. The truly distinctive characteristics of humans are their moral awareness and religious nature, by virtue of which they occupy a special position in the natural world. In fact, these features set humans apart from animals just as much as the latter differ from plants; and Quatrefages joined his fellow naturalist and countryman Isidore Geoffroy Saint-Hilaire in placing *Homo sapiens* not just in a subclass of its own, as Owen had done, but far beyond, in a separate kingdom.[64]

This antianatomical criticism—mainly applicable to and directed at Owen—was of course naive in that it begged the very question of the nature of the human "mind" and "soul," and especially whether or not these are separate from our bodies. The critics appeared to advocate that humans have unique mental attributes without necessarily having corresponding, unique physical features. In this view, the soul is an independent, immaterial, and immortal entity that inhabits the body and uses the bodily organs in a way analogous to how a musician plays an instrument. That the soul's instrument was the central nervous system, in particular the two cerebral hemispheres, few really doubted.

In the course of the nineteenth century, however, doubt about the concept of a transcendental soul burgeoned, and the materialist view that "mind" and "soul" are not only confined to the nervous system but in fact are no more than its products, originating and dying with the brain, became more and more prevalent. Moreover, the notion of "mind" was increasingly separated from that of "soul," and the latter became relegated to the sphere of theological speculation. This soul-body problem (or, in more modern parlance, the mind-body problem) became the battleground on which vitalist dualism clashed with materialist monism—a battleground that was located at the confluence of a number of different currents of scientific research, in particular anthropology, neuroanatomy, neurophysiology, and, more popularly, phrenology. Several approaches were followed, such as the comparative one (comparing human races with each other and comparing humans with simians), the pathological one (studying arrested brain development in microcephalic cretins), and increasingly also the experimental one (vivisecting to localize cerebral functions).

Owen's work on anthropoid apes was very much part of these developments. Because the hippocampus controversy has always been looked at from Huxley's side, it has been made part of, even subsumed under, the events surrounding the publication of Darwin's *Origin of Species.* This Darwinian connection may elucidate Huxley's involvement in the controversy but not Owen's, who had worked on the human-ape distinction ever since he began publishing papers on comparative anatomy in 1830, nearly three decades before Darwin published his theory of evolution. In order to appreciate Owen's position in this controversy, we must not view him from the angle of his disagreement with Darwin and Huxley but compare him with his European colleagues who, like himself, had been discussing the human-ape differences and the soul-body problem in the pre-*Origin* decades of the nineteenth century. Such a comparison puts Owen in a very different light from that in which Darwinian historians have placed him.

Two prominent participants in the comparative anatomical approach to the soul-body debate, one from France, the other from Germany, may serve to illustrate the issue. Both men believed that distinctively human brain features do exist but these were to be found in characteristics of the brain as a whole, not in restricted anatomical parts. Gratiolet who, along with Owen, was one of the first to describe the brain of a gorilla,[65] established an international reputation with his classic *Mémoire sur les plis cérébraux* (1854), which was a meticulous study of cerebral morphology in humans and in a wide range of simians. In the more comprehensive *Anatomie comparée du système nerveux* (1839–57), the second volume of which was written by Gratiolet, he, along with Quatrefages and several others among his French colleagues, maintained that *Homo sapiens* is sufficiently different from the animal kingdom to constitute a separate human kingdom. To Gratiolet, this was a "règne de verbe" (kingdom of the word), founded on humanity's possession of a rational soul as manifested by the ability to speak.[66] He opposed the phrenological ideas of Gall and Spurzheim and sided with Flourens in agreeing that the brain is unitary and acts as an "organ d'ensemble."[67] Gratiolet believed in an absolute physical and mental separation of humans from animals. Although the cerebral convolutions in the adult state of both humans and monkeys are arranged on one and the same plan, the two groups are separated by a fundamentally different development of their brains. For instance, in humans the convolutions of the anterior lobes appear before those of the posterior lobes, which is the reverse of what happens in apes and monkeys. The validity of this criterion was borne out—Gratiolet maintained—by his study of microcephalic idiots, whose brains, in their small state of arrested development, may look less advanced than those of the anthropoid apes; yet microcephalic brains show early, characteristically human features. Most microcephalics also "have an intelligible language, not very rich, to be true, but articulate and abstract; their brain, which looks inferior to that of an orang or a gorilla, is still that of a speaking soul.[68]

Among the men who tried to stem the rising tide of materialism was also Rudolph Wagner. Already in his very early publications, such as *Lehrbuch der vergleichenden Anatomie* (1834–35), he had stated that humans share with animals the possession of body and soul but differ from them because humans have in addition a mind, "ein unmittelbares Geschenk Gottes" (a direct gift from God).[69] Wagner further had enumerated a variety of cerebral and skeletal features that characterize humans. Among the former he had listed, as Owen later did, the large development of the human brain, especially of the two hemispheres, which, more completely than in apes, cover the cerebellum; he had also cited the presence of a posterior cornu, a hippocampus minor, and various other cerebral features, such as the many and pronounced cortical convolutions. More explicitly than Owen was to do in his 1857 presentation to the Linnean Society, Wagner had qualified this list of human peculiarities by adding that in part these are present in the anthropoid apes.[70] In the early 1850s, Wagner turned his attention specifically to the question of the human soul. In a number of tracts he rounded on the materialists while carrying the banner of the traditional, biblical belief in the origin of humankind from a single, specially created pair and in a life after death.

In 1854 Wagner presented a paper entitled "Menschenschöpfung und Seelensubstanz" to the Versammlung deutscher Naturforscher und Ärzte, which met in Göttingen that year. He asserted that there are no physiological grounds for denying the existence of an independent immaterial soul and that the moral order of society requires us to assume the soul's existence. This address was published, and in just a few weeks' time its three thousand copies were sold; a new edition was printed, to which Wagner added further thoughts on the future of human souls. He stated that, although in scientific matters he sided with the skeptics, in matters of faith he preferred the simple "Köhlerglauben" (the faith of a person whose job it is to make charcoal, i.e., a simple faith).[71] This expression provided Carl Vogt, known for his scientific materialism, with the title of a scathing counter-booklet, *Köhlerglaube und Wissenschaft* (1856), in which he denounced Wagner's piety.[72] The mutual abuse in the Wagner-Vogt clash was such that the Owen-Huxley conflict of a few years hence, which in a number of ways paralleled this German precursor, paled by comparison. Both men continued to promulgate their opposing ideas. Vogt elaborated on his materialistic views in *Vorlesungen über den Menschen, seine Stellung in der Schöpfung und in der Geschichte der Erde* (published in the same year as Huxley's *Man's Place in Nature*), in which Vogt firmly sided with the Darwinians against Owen.[73]

Three years after the Göttingen meeting, in 1857, when Owen presented his cerebral classification of mammals to the Linnean Society, Wagner published his clearest exposition of the question in *Der Kampf um die Seele*. Although humans share both body and soul with the animals, humankind is unique in that we are capable of abstract thought and have a notion of morality

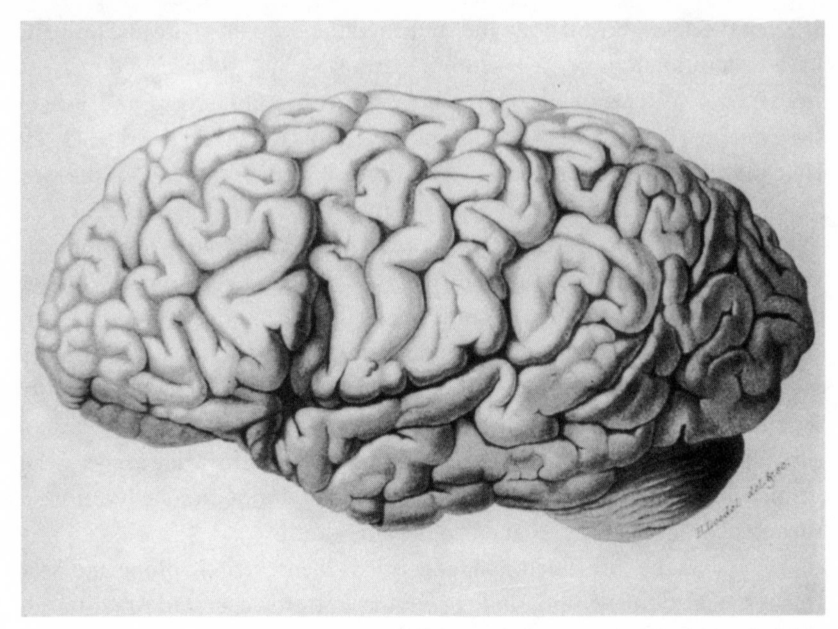

FIGURE 18. The brain of the great mathematician Carl Friedrich Gauss, which featured in the debate over whether human uniqueness can be founded on cerebral morphology. (Wagner, *Vorstudien*, vol. 1, pl. 4.)

and divinity, the "symbol" of which is speech. No serious objections can be brought against the assumption of a soul—he reiterated—nor against that of its immortality.[74] In his *Vorstudien zu einer wissenschaftlichen Morphologie und Physiologie des menschlichen Gehirns als Seelenorgan* (1860–62) Wagner continued the work by Gratiolet and others, concluding that the cerebral convolutions in humans and simians are constructed according to basic types that, although similar, cannot be reduced one another.

Wagner made capital out of the fact that he had in his collection of human brains that of his Göttingen colleague Carl Friedrich Gauss, one of the greatest mathematicians of all time, who had died in 1855 and with whom he had conducted pious deathbed conversations.[75] Gauss's cerebral organ was relatively large and characterized by pronounced cortical convolutions (fig. 18). The same was not true, however, of the brain of another highly gifted colleague, the mineralogist J. F. L. Hausmann, whose death in 1859 had provided Wagner with another prize specimen in his collection of brains, and Wagner left undecided the question of whether or not intellectual excellence is indicated by Gaussian brain features. He doubted in general that mental attributes could be correlated with specific cerebral parts. The various forms that brains take do not correspond to functions in the way that vertebrate limbs do. In mammals and birds, for example, widely varying "architectures" of brains can accompany very similar psychological abilities.[76]

Many other naturalists, in Germany and elsewhere, contributed to the debate ever since it had taken a new Cartesian turn when Samuel Sömmerring published his notorious booklet *Über das Organ der Seele* (1796), in which he localized the soul in the intraventricular cerebrospinal fluid.[77] Carl Carus, by contrast to Sömmerring, believed that body and soul were two expressions of one and the same unity, that the nervous system was the spatial representation of the soul, and that the soul was the manifestation in time of the living nervous system.[78] Carus objected both to the idea that the soul is located in a specific organ and to the notion that the soul can be divided into components, each of these to be assigned to specific cerebral organs, as Franz Gall did. To Carus, the soul reaches its fullness ("Vollendung") in the harmonic development of the entire organism.[79] Humans' exalted position in nature is reflected in the fact that in them the gradual development of the cephalic and thus the cerebral end of the body has reached its completion.[80] Emil Huschke, also influenced by *Naturphilosophie*, conceived of the soul as an "Einheitsprinzip" that expresses the unity of body with mental activity. It permeates the entire organism and stands in relation to the body the way God relates to the universe; that is, it is present throughout the body. Yet Huschke was more sympathetic than Carus to Gall's cerebral localization theory.[81] Other primate experts who held immaterialist and vitalist positions included Schroeder van der Kolk, who in cooperation with Vrolik described various prosimians and simians obtained from the Dutch East Indies.[82] Across the Atlantic, the Yale naturalist and coeditor of the *American Journal of Science* James D. Dana followed in Carus's steps by placing the crucial human-animal distinction in the degree of dominance of the head. Our posture is erect, and our anterior limbs are no longer for locomotion but serve the purposes of the head: "The *cephalization* of the body,—that is, the subordination of its members and structure to head-uses—so variously exemplified in the animal kingdom, here reaches its extreme limit. Man, in this, stands *alone* among Mammals."[83]

Owen on the Mind-Body Problem

How did Owen compare with his Continental colleagues on the issue of "soul" and "mind"? Did he stand on the side of Gratiolet, Wagner, and the other biologists of his generation who attempted to uphold the notion of an immaterial, indestructible soul against the reductionist materialism of Vogt and his allies? One might be inclined to assume so, if only because the protagonists of an independent soul were unanimously antitransmutation and anti-Darwin, whereas the materialist thinkers such as Georg Büchner and Vogt became champions of the Darwinian cause. Yet, surprising as it may seem, Owen explicitly rejected the concept that "soul" and "mind" are entities independent from "body."

The two issues of a "vital principle" and the "soul" were closely related, as the line of reasoning for or against was similar in both cases. Before examining why Owen rejected the concept of a soul as something separate from the body, let us glance at his opinion about the *vis vitalis*. In the concluding section of his *Anatomy of Vertebrates* Owen took a nonvitalist stance. The way in which an amoeba, for example, attracts extraneous matter for absorption is analogous to a magnet's attracting objects around it. To call the amoeba "living" is a term correlative with "magnetic." "Devitalise the sarcode, unmagnetise the steel, and both cease to manifest their respective vital or magnetic phenomena."[84] Just as magnetic, electrical, and other modes of force originate from natural conditions, so the vital mode has come into existence without the interposition of a miraculous, creative act. Owen did not maintain that the amoeba's vital properties can be reduced to physicochemical processes, but rather that life is a fundamental property of matter, just as magnetism is, each produced under different but natural conditions, and that life is not based on an independent, separately created vital principle. For this reason, it may be more accurate to describe Owen's position as nonvitalist or physicalist rather than antivitalist or materialist.

To move from the phenomenon of "life" to that of "mind," a nervous system is needed that allows for sentience and volition. Higher mental states are dependent—Owen stressed—on an increase in the size and complexity of the central nervous system. Add to this that nerve energy is convertible into electricity and vice versa, and the conclusion may be drawn that thought relates to the brain of humans in the way that electricity does to the electric apparatus of the stingray. Psychological phenomena in both animals and humans are not produced by an independent, indivisible, immaterial mental principle or soul but are the products of brain functions. Owen argued that his theory was supported by the phenomena of altered states of mind produced in sleep and by drugs and diseases. Owen objected to calling this view "materialistic." Our sensation of outside objects is the result of forces, and so is inner volition. Nothing is gained by calling the one "material" and the other "immaterial." "Our ideas of things without as within the 'ego' are the action and reaction of forces, as 'material' or 'immaterial' as the ideas themselves."[85]

This denial of the existence of an independent soul did not mean that Owen rejected faith in a future life and the resurrection of the dead, but his belief in these rested on their being parts of a divine revelation. Owen believed that a day of judgment would come and that God would restore the just. The notion of an independent soul and of a purgatory, however, was an invention of the priesthood, which derived benefit from it, as did the even "baser brood" of table-rappers. Owen twice cited Locke, intimating that he saw himself in the tradition of British empiricism and "sensationalism." In view of the stubborn mistake of classifying Owen with the advocates of a *vis vitalis* and citing him as a representative of the party of orthodoxy and traditionalism,[86] it may be worth quoting a letter in which Owen concisely and unambiguously repeated his disbelief in "the 'vital principle' and the 'soul.'"

> The present phase of physiology substitutes "brain" for "mind." The latter signifies the sum of the acts and powers of its organ, as "Life" is the sum of those of the organism. When I was an Edinburgh student, in 1822 [*sic*; 1824], our best anatomist Dr Barclay, put forth his work in defence of a "vital principle." What "soul" or "mind" was to "brain," "vital principle" was to "frame." That "principle" and "soul" were held to be entities, not mere "abstract ideas," and to be alike "imperishable." In the other view "vital principle" and "soul" are terms for sums of functions of organs and organic wholes, which end with death. But only as regards this small globe, one of millions of such "worlds."
>
> . My faith is that *The Power* which created, by whatever secondary actions, an "honest man," will not throw away "His noblest work." He who made us can, and, I believe, will remake—such individuals at least, as have fulfilled the Supreme intention.
>
> Now, compare the two corresponding modifications of "faith in a resurrection."
>
> One—the common one—holding the brain actions to be those of an independent imperishable entity, have to dispose of their ever living "soul" somehow; until the date of what they term the "final judgment." During the intervening period "soul" must be somewhere—a profitable faith to the priest who, for a consideration, has power to get a poor soul out of purgatory.
>
> Legislators are at one that a merited punishment is deterrent in the ratio of its proximity to the committal of the crime. But, whether there be "purgatory" or not, a day of judgment will come, God alone knows when, but certainly not so soon as some trusting Christians once expected.
>
> On the other hand, the re-creation of the individual with a body—whether "glorified," according to St Paul or not—will be to him or her, directly after the individual has taken a last leave of his sorrowing surrounders. Of any length of what we term "past time," he, or she, will be unconscious. The transit from the presence of weeping relatives to that of the "Glorious One" will be immediate.[87]

Owen expressed these views fairly late in his career, after *The Origin of Species* and *Man's Place in Nature* had appeared. Were his earlier beliefs more traditional and did he take this nontraditional position in order to keep up with the Darwinians, as has been alleged about his explicit expressions of evolutionary belief in the concluding chapter to the *Anatomy of Vertebrates*? I have shown above that Owen was an evolutionist long before 1859, and there are reasons to believe that his unorthodox views about the soul also predate the emergence of Darwinism.

First, Owen's belief in a natural, nonmiraculous origin of species, including *Homo sapiens*, meant a rejection of the creationist doctrine of continual divine intervention in the workings of nature. Given that imparting souls to all new human beings would imply just such heavenly intervention, it is likely that Owen did not subscribe to it. Second, he never expressed the orthodox belief in souls, even though this would have stood him in good stead at the College of Surgeons, where powerful figures such as Benjamin Brodie proclaimed the existence in humans of an indivisible "mental principle."[88] Third, if Owen did indeed not believe in the existence of independent, immortal human souls, he had good reason to keep quiet about this in view of the brouhaha at the college over the Lawrence affair, which did not die down until years after Owen had taken up his Hunterian employment. From 1816 to 1818 William Lawrence delivered one of the two courses of Hunterian lectures at the college that were unified into a single series when Owen was appointed Hunterian professor in 1837. In the second introductory lecture of 1816 Lawrence explicitly rejected the concept of an independent living principle or vital force. Physiology ought to be based on "observation and experience." Invoking a *vis vitalis* to explain the phenomena of life is like appealing to the pagan god Jupiter to explain thunderbolts: "It seems to me that this hypothesis or fiction of a subtle invisible matter, animating the visible textures of animal bodies, and directing their motions, is only an example of that propensity in the human mind, which has led men at all times to account for those phenomena, of which the causes are not obvious, by the mysterious aid of higher and imaginary beings."[89]

To Lawrence, the vital property was as fundamental to organic bodies and without need of "extrinsic aid" as chemical and physical properties are to matter. John Abernethy, at the time the other of the two Hunterian professors, attacked his former pupil, accusing him of working in concert with materialist French physiologists and "of propagating opinions detrimental to society, and of endeavouring to enforce them for the purpose of loosening those restraints, on which the welfare of mankind depends."[90] In the introduction to his *Lectures on Physiology, Zoology, and the Natural History of Man*, delivered in 1817 and 1818, Lawrence answered Abernethy's charges and made matters worse by extending his antivitalist analysis to the human mind. Just as animal functions are inseparable from animal organs, so mental functions are inseparable from the nervous system. In other words, mental activity is a function of the brain, not of an immaterial, immortal soul:

> Where then shall we find proof of the mind's independence of the bodily structure?—of that mind, which, like the corporeal frame, is infantile in the child, manly in the adult, sick and debilitated in disease, frenzied or melancholy in the madman, enfeebled in the decline of life, doting in decrepitude, and annihilated by death?
>
> Take away from the mind of man, or from that of any other animal, the operation of the five external senses, and the functions of the brain, and what will be left behind?[91]

These lectures, published in 1819, generated a storm of criticism, especially from among Lawrence's senior colleagues at the college.[92] Within a month of its release, Lawrence was compelled to withdraw the book. Yet it went through new printings, and extracts from it dealing with the mind-body problem were republished in 1832 and 1840, together with a fiercely combative introduction in which Lawrence was praised for "boldly displaying truth" in spite of the fact that "he may be borne down by a powerful faction."[93]

The similarity of Owen's position in the *Anatomy of Vertebrates* with Lawrence's iconoclastic views is striking, and if Owen, during his Hunterian tenure, had indeed already been in agreement with Lawrence, it is not surprising that the punitive reaction within the college made him decide to keep quiet. There is yet a fourth reason, however, why Owen is likely to have held Lawrence's view of the human mind in pre–*Origin of Species* days. This is that Owen's method of distinguishing humans from simians on the basis of specific anatomical features can be seen as a direct continuation of Lawrence's programmatic *Lectures on Physiology, Zoology, and the Natural History of Man*. As W. F. Bynum highlights, Lawrence emphasized that, just as life should not be thought of as independent of an organic body, so it is absurd to speak of a vital function without an appropriate organ.[94] Accordingly, the number and kind of intellectual phenomena in different animals correspond closely to the level of development of their brains. Mental manifestations in animals and humans alike are proportional to the degree of perfection of the cerebral organ. Thus, humankind's superiority over animals can be defined by means of specifics of organization, especially of the brain.[95]

This came close to the theory of Franz Gall and Johann Spurzheim, and although Lawrence expressed himself unable to pronounce on its accuracy, he praised the two authors for "rescuing us from the trammels of doctrines and authorities, and directing our attention to nature."[96] So in order to distinguish humans from animals, in mental terms, it suffices to focus on specific cerebral features; and this is exactly what Owen did. Implicit in his attempt to distinguish humans from apes by means of specific features of the brain was therefore the assumption that mental phenomena—the mind, the soul—are dependent on specifics of cerebral organization. This undercut the belief in a unitary mind and thus in an independent immaterial soul and was the reason why Wagner, for example, objected to Owen's use of the posterior lobe, posterior cornu, and hippocampus minor, citing instead the overall, unitary cerebral "architecture" as distinctive of brains.[97]

While Owen's clash with the Darwinians may appear to place him in the orthodox camp, a comparison with his Continental colleagues shows how very wrong this is. With respect to the questions of the origin of species and of the mind-body problem, Owen did not belong in this camp. While working in the empiricist tradition of Locke and, closer to the Hunterian Museum, of Lawrence, he advanced an opinion similar to that of some Continental materialist philosophers. The fact that Owen held physicalist views about the mind-body problem agrees well with his initial definition of the vertebrate archetype, not as a Platonic idea, but as a reflection of a material force.

It is true that for a long time Owen displayed a cautious reluctance to state his position lucidly and unambiguously, and he never warmed to the idea of human evolution, always ready to emphasize the major gap that separates humans from animals. As late as 1883, in what was probably his last public statement on the topic of "our origin as a species," he scolded the "manifest desire in some quarters to anticipate the looked-for and, by some, hoped-for, proofs of our descent—or rather ascent—from the Ape."[98]

This ambiguity may reflect the Janus-faced nature of Owen's museum politics. His scientific position was close to that of the materialists, and especially close to the pantheistic materialism of John Tyndall and other protagonists of the cause of metropolitan science.[99] Yet Owen's political friendships were not with them. Many of the materialists were left-wing radicals who had been active in the political turbulence of 1848–49. Vogt's involvement in the Revolution, for example, forced him to flee from Giessen to Switzerland. At the other end of the political spectrum, Gratiolet led troops of the National Guard against the republican insurgents of 1848. Owen, through his sociopolitical connections, belonged in the Christian royalist camp, and he actually drilled with the Honourable Artillery Company. Whereas his insistence on a wide separation of humans from apes may in part have been fueled by concerns of a religious and social nature, it was certainly also prudent in view of his institutional ambitions. In fact, the hippocampus controversy was inextricably intertwined with the museum question.

The Du Chaillu Affair

The debate about humanity's place in nature reflected to a significant extent a clash about the place in society of the parties who conducted the debate.[100] Each party's stance represented a different sociopolitical location and a distinct strategy of self-advancement. As Pycroft's play *A Sad Case* perceptively satirized, Huxley's advocacy of man's proximity to monkeys doubled as an effort to drag down his enemy Owen, whereas Owen's emphasis on humankind's elevated position in nature was inseparable from his effort to secure the support of the Anglican establishment for the realization of his museum plans. We should recall that the early 1860s, when the hippocampus controversy raged, were also the most crucial period in the development of Owen's museum project: the Gregory Committee reported on the state of the British Museum, Owen lectured on the aims of a "National Museum of Natural History," and the Commons was the scene of vehement exchanges about Owen's proposals. The fact that the hippocampus controversy coincided with this critical phase of Owen's institutional ambitions increases the likelihood that the two issues were connected, directly as well as indirectly. A direct link was the gorilla. Its close resemblance to humans made it central to the debate about humanity's place in nature, and this in turn enhanced its significance as a museum specimen, adding a strong argument to the case being made by Owen for adequate space to display the natural history collections. This was dramatically highlighted by the furor caused by the explorer and anthropologist Paul B. Du Chaillu, the main purveyor of gorillas and gorilla parts for London's museum world.

Du Chaillu was a French American who spent his boyhood on the west coast of Africa, where his father was an agent for a Parisian firm. Here, at a station in Gabon, Paul received some education in a Jesuit mission school and acquired an early interest in the country, its natural history, and its natives, with whose languages and customs he grew acquainted. In 1852, he went to the United States, where he secured the support of the Academy of Natural Sciences in Philadelphia for an expedition into Central Africa. The expedition began in 1856, when he was only twenty years old, and lasted nearly four years, during which time he reportedly traveled some eight thousand miles through tropical Africa. Du Chaillu established several facts about the course of rivers, made innumerable ethnographical observations, discovered the Fan tribe of cannibals, collected many exotic birds and mammals, most sensationally the western lowland gorilla, and brought back the most vivid and detailed observations of this anthropoid giant's behavior in the wild to date. He wrote a narrative of his expedition, *Explorations and Adventures in Equatorial Africa*. Apparently, in America he had difficulty selling his specimen collection and his manuscript, and in December 1860 he wrote to Owen that he intended to go to England "to place my specimens of the Gorilla and other African apes at your service."[101]

Some two months later, in February 1861, Du Chaillu met Owen in whom he found an effective patron. With Owen's support *Explorations and Adventures* were published by John Murray. Through Owen's connection with Roderick Murchison, Du Chaillu was invited to lecture at the Royal Geographical Society and also at the Royal Institution and the Ethnological Society; Owen himself lectured on Du Chaillu's collection at the Manchester meeting of the BAAS. At Owen's insistence, the British Museum bought, for a considerable sum of money, part of Du Chaillu's collection of gorilla skins. During this time "Du Chaillu was a frequent visitor at Sheen Lodge [Owen's home]."[102] It would be naive to regard Owen's motivation in extending his patronage to the young explorer as purely altruistic. Owen's help served the dual purpose of strengthening his position in the fight against Huxley and strengthening, too, his case for a separate natural history museum. Owen firmly connected the two issues. After Du Chaillu's lecture "Travels in the (Gorilla) Region of Western Equatorial Africa," presented to the Royal Geographical Society, Owen rose to emphasize the importance of the gorilla:

> Professor Owen said that natural history had never received a more remarkable acquisition than had been imparted that evening. Hitherto we had only obtained a few raw materials of this great Gorilla; but now, for the first time, the naturalists had heard from one who had seen the Gorilla in its native country some authentic account of its power and habits. In natural history, as we went on comparing form with form we soon became impressed with the idea of a connected scale, and the interest increased as we ascended; but when we came so near to ourselves as we did in the comparison of this tailless anthropoid ape, the interest became perfectly exciting.[103]

Owen then argued that the gorilla came closer to humans than do the other anthropoid apes but that the cerebral differences were striking. He concluded with an outline of the progress in the study of natural history, pointing to the numerous accessions which had been made, and he expressed the hope that the government would provide a suitable building for the arrangement and display of these accessions. Several of the trustees of the British Museum were present, for example, Murchison, who as chairman of the meeting warmly thanked Du Chaillu, and also Owen's staunchest ally in museum matters, the chancellor of the exchequer, Gladstone, who was effusive in his praise of both Du Chaillu and Owen.[104]

The symbolism of this platform, on which Owen was flanked by Du Chaillu and by British Museum trustees, was unmistakable. As A. E. Gunther has commented, Owen was seen to be "at one" with both the gorillas and with the orthodox party.[105] The Royal Geographical Society provided a room in its apartments for the display of Du Chaillu's collection of natural history. "The fashionable and scientific world flocked to his museum. Du Chaillu became the lion of the day; and his work, ushered into the world by one of the most distinguished of our publishers, was read with avidity and circulated in thousands."[106] In his lecture "On a National Museum of Natural History," Owen again used Du Chaillu's gorillas to support his appeal for museum space.[107]

To the naturalist and the general reader alike, one of the most gripping passages in the *Explorations and Adventures* was the one in which the young adventurer describes his first confrontation with a large male gorilla. This description has gained in significance by the echo it produced in popular literature.[108] If one wanted to pinpoint the main fountainhead of the Victorian lore that gorillas are ferocious creatures rather than the humanized "gentle giants" into which David Attenborough, Dian Fossey, and others have transformed them, it would have to be the following paragraph of purple prose:

> The underbrush swayed rapidly just ahead, and presently before us stood an immense male gorilla. He had gone through the jungle on his all-fours; but when he saw our party he erected himself and looked us boldly in the face. He stood about a dozen yards from us, and was a sight I think I shall never forget. Nearly six feet high (he proved four inches shorter), with immense body, huge chest, and great muscular arms, with fiercely-glaring large deep gray eyes, and a hellish expression of face, which seemed to me like some nightmare vision: thus stood before us the king of the African forest. He was not afraid of us. He stood there, and beat his breast with his huge fists till it resounded like an immense bass-drum, which is their mode of offering defiance; meantime giving vent to roar after roar. . . . He advanced a few steps—then stopped to utter that hideous roar again—advanced again, and finally stopped when at a distance of about six yards from us. And here, just as he began another of his roars, beating his breast in rage, we fired, and killed him. With a groan which had something terribly human in it, and yet was full of brutishness, he fell forward on his face. The body shook convulsively for a few minutes, the limbs moved about in a struggling way, and then all was quiet—death had done its work, and I had leisure to examine the huge body.[109]

Yet Du Chaillu's discoveries by no means met with universal applause. In fact, when the narrative of his African *Explorations and Adventures* was published, a storm of disagreement broke over the question of the truthfulness of his account. In fact, "the great gorilla controversy" generated more public interest than its Siamese twin, "the great hippocampus controversy," and as Wilfrid Blunt states, it was nearly as vehement as the storm that followed Darwin's *Origin of Species.*[110] Du Chaillu's main critic was John Edward Gray, keeper of the Zoological Department at the British Museum. When the *Athenaeum* published a moderately appreciative review of Du Chaillu's "very amusing book, one likely to direct renewed attention to the wide and interesting region peopled by tribes descending from the sons of Ham,"[111] Gray reacted with a censorious letter on "the new traveller's tales." Du Chaillu's qualifications as a traveler—he charged—were "of the slightest description"; the map attached to his book was "one of the most primitive that I have seen for years"; his competence as a naturalist was "of the lowest order"; with respect to Du Chaillu's collection of mammals, "there is not a specimen among them that indicates that the collector had traversed any new region"; on the contrary, all "have been received long ago from the different trading stations on the west coast of Africa," in particular the gorilla, specimens of which had been arriving at "almost every museum in Europe" "for the past fifteen years"; the book contained "improbable stories," and "there is the same exaggeration in the illustrations," some of which were copied from other people's work and not from sketches made on the spot.[112]

In further letters and articles, Gray expanded upon these charges and added a new one: that the skin of a large male gorilla acquired by the British Museum showed no holes indicative of the gorilla having been shot in the chest while facing its hunter, as stated in *Explorations and Adventures.*[113] The innuendo of this remark was that Du Chaillu had not only lied but proved to be a coward to boot who had failed to act according to the code of ethics of big-game hunters, shooting his quarry in the back rather than head-on. Others joined Gray in his assault on the little Frenchman, among them Charles Waterton, the explorer of British Guiana, and the German geographer Heinrich Barth, famous for his explorations of Chad and other parts of Africa. Additional accusations included that the timetable of Du Chaillu's itinerary was inaccurate; that the gorilla hunter was a humbug because he was a poor shot; that gorillas did not thump their breasts; that a harp made of organic fibers as described by Du Chaillu was a physical impossibility; and that another of his stories about natives of equatorial Africa lighting fires was stretching the truth because the jungle was too damp for burning. Collectively, the accusations added up to the following: that Du Chaillu's narrative was fictitious; that he had never been to the regions he claimed to have explored; and that he had not shot the animals he had brought with him but purchased them in trading posts along the coast from native hunters who also were his source of the wild tales about savages and ferocious animals. "I am quite convinced, in my own mind," Waterton wrote, "that du Chaillu's adventures in the land of the gorilla are nothing but impudent fables."[114]

Du Chaillu responded with a two-pronged self-defense. On the one hand, he admitted that he was not the scientific traveler that some people had taken him for and acknowledged that in the course of compiling his book from rough field notes inaccuracies and mistakes might have crept into his narrative. On the other hand, he insisted that his book was a truthful narrative of real encounters.[115] Owen joined the fray and defended his protégé by confirming that the wounds of the large gorilla at the British Museum did indicate a frontal charge. For additional confirmation he cited his friend and British Museum trustee Philip Egerton, "a gentleman who combines an acuteness of observation which has placed him high in science, with a well-known reputation as a skilful marksman and deerstalker."[116] To this again Gray delivered a riposte,[117] which in turn produced a response that invoked the august names of Richard Burton and David Livingstone to underpin Du Chaillu's credibility. A variety of dailies, weeklies, and other periodical magazines pronounced on the issue of Du Chaillu's veracity. The *Literary Gazette*, for example, published attacks, and the *Critic* parried these.[118] After several months during which attack and retaliation followed one another like tidal movements, the controversy reached something of a climax when the immensely popular nonconformist preacher Charles Haddon Spurgeon expounded to a packed congregation in his six thousand–seat Metropolitan Tabernacle at Newington Causeway in support of Du Chaillu, who appeared with him on the platform along with the MP Austen Henry Layard, a supporter of Owen in museum matters.[119]

Spurgeon's Baptist sanction of Du Chaillu–Owen notwithstanding, the atheist W. Winwood Reade, nephew of the novelist Charles Reade, set sail for Africa to explore gorilla country, reporting in his *Savage Africa* (1864) that Du Chaillu's account of the gorilla was inaccurate, that the animal never beats its breast, for example, and that Du Chaillu had never killed a gorilla but was merely "an industrious collector of skins."[120] A dazzling statistic is that Du Chaillu, Spurgeon, and Reade were at the time youngsters in their mid-twenties.

More jocularly, Kingsley introduced Du Chaillu in his *Water Babies* in a section that shows apes to have developed from "the great and jolly nation of the Doasyoulikes" by a process of degeneration—Kingsley was serious about his "degradation" theory[121]—ultimately leading to the demise of all "Doasyoulikes,"

> all except one tremendous old fellow with jaws like a jack, who stood full seven feet high; and M. Du Chaillu came up to him, and shot him, as he stood roaring and thumping his breast. And he remembered that his ancestors had once been men, and tried to say, "Am I not a man and a brother?" but had forgotten how to use his tongue; and then he had tried to call for a doctor, but he had forgotten the word for one. So all he said was "Ubboboo!" and died.[122]

It is perfectly true that from a scientific point of view Du Chaillu's account was ramshackle. Few denied this. But how could one reasonably expect anything more than he had actually accomplished? The preparations for his trip were made when Du Chaillu was still a teenager. He had been brought up as a trader, without a grounding in the Humboldtian instrumental techniques and rigorous record keeping of contemporary exploration. David Brewster, who reviewed *Explorations and Adventures* for the *North British Review*, concurred with the critics that the timetable of Du Chaillu's travels was inaccurate and that some tales were apparently exaggerations. At the same time, he acknowledged that the "intelligent and enterprising traveller" had extended geographical knowledge and made important contributions to natural history.[123] The historical writer George William Cox, who discussed Du Chaillu's book for the *Edinburgh Review*, made similar points, even though he ended on a slightly more critical note.[124] Both reviewers presented Du Chaillu's tales about the horrors of indigenous slavery as contributions to the antislavery movement.

Why should Gray have wanted to launch such a hypercritical attack? He was no explorer himself and it cannot have been the *jalousie de mêtier* which may have added acid to the ink of Barth and Waterton. The answer is that Du Chaillu was the pawn by which Gray attempted to checkmate Owen. The latter was Gray's superior at the British Museum, and Gray resented the authority of the superintendent whose plans for museum reform he wanted to obstruct. A direct attack on Du Chaillu doubled as an indirect attack on Owen. After all, Du Chaillu helped advance Owen's work in several ways. He had procured a collection of virtually unique gorilla specimens, which expanded the basis on which Owen could found his claim for increased museum space; the possession of various gorilla specimens enhanced Owen's credibility as the leading expert on anthropoid apes; and Du Chaillu's observations of the savage behavior of gorillas in the wild added to the features that separate humans from apes (fig. 19). The description of the inhuman ferocity of the gorilla, baring its fearsome canines and thumping its breast with "vulcanic"

FIGURE 19. Adult male gorilla depicted as a fearsome monster. Owen's pioneering studies of the gorilla helped him in his fight with Huxley over humanity's place in nature and at the same time provided him with a popular museum exhibit, justifying his demands for a major new natural history building. (Owen, "Contributions to the Natural History of the Anthropoid Apes, no. VIII," pl. 43.)

fists, dramatically added to the belief that a major chasm separates us from the nearest anthropoids.[125]

What position did Huxley take in the Du Chaillu affair? He and Gray were on the same side of the museum question, against Owen, but Gray was not a Darwinian, and the gorilla was as useful to Huxley as it was to Owen in the debate about humanity's place in nature. This was made clear by Huxley's friend John Chapman, who reviewed Du Chaillu's book for the *Westminster Review*. He condemned the book's "chronological jumble" but judged the narrative to be "substantially true" and then went on to use the occasion for a recapitulation of the hippocampus controversy, following Huxley's line.[126] Initially, therefore, Huxley took a moderately appreciative stance. The *Natural History Review*, which in simian matters spoke with Huxley's tongue, commented that, although Du Chaillu's narrative contained unintentional errors due to "imperfectly kept notes," "a rather vivid imagination," and "a not very perfect memory," it was essentially trustworthy. He had "entered a region never before discovered by civilized man" and had "seen and hunted the Gorilla in his native wilds, and brought back a mass of information concerning this interesting 'anthropoid,' and his kith and kin among the apes."[127]

The temptation to undercut Owen's credibility by undermining that of his protégé Du Chaillu proved too strong, however, for Huxley to resist, and not long after the review of *Explorations and Adventures* had appeared, he changed his mind, and in a letter to Wagner on the hippocampus controversy inserted a paragraph ridiculing Spurgeon's support for Du Chaillu and accusing Owen of having attempted "to crush truth by popular authority."[128] Huxley's condemnation of Du Chaillu grew increasingly harsh. In *Man's Place in Nature* he did not quote Du Chaillu's work "because, in my opinion, so long as his narrative remains in its present state of unexplained and apparently inexplicable confusion, it has no claim to original authority respecting any subject whatsoever."[129] The following year, the *Natural History Review* took the side of Winwood Reade, commenting that he had "wiped out M. Du Chaillu and all the statements put forward by that too famous traveller as to the habits of the Gorilla," placing Du Chaillu under the suspicion of "*directly* intentional deceit."[130]

Throughout 1862, while the controversy abated, confirmation of Du Chaillu's disputed discoveries came trickling in. Yet this did not heal the young explorer's wounded pride, and in 1863 he returned to Africa for a second, two-year expedition, to collect new proofs of some of his earlier observations. This time he went better prepared, taking with him not just his compass and notebook but also a sextant, photographic equipment, and other tools of the trade. The result was another book, *Journey to Ashango-Land* (1867), in which he confirmed his account of the behavior of the gorilla[131] and reported the exciting discovery of a pygmy people inhabiting the equatorial forest. Owen added to the respectability of this book by contributing an appendix on skulls of three different tribes from western equatorial Africa, the measurements of which he cited in support of his belief that the various human races have one and the same unitary origin as a species.[132] The accusation that Du Chaillu's exploration story was fraudulent proved unfounded. "Never were we more in the right," Murchison wrote to Owen in the autumn of 1864, "than when we stood up for this fine little fellow."[133]

A. E. Gunther believes that Gray defeated Owen in the matter of the gunshot wounds to the gorilla.[134] Be this as it may, in order to appreciate what the outcome of "the great gorilla controversy" was, we need to consider it at a more general level than that of bullet holes in a gorilla's hide. On balance, Owen came out of it well. He was vindicated in his support for Du Chaillu, whom he turned into a deeply grateful supporter of the cause of "your New Museum,"[135] and the gorilla became a rallying emblem for Owen's museum supporters. Both the stuffed skins of gorillas and their skeletons became new exhibition icons. Especially sensational was the comparative display of the skeletons of a human being and a gorilla; and the photograph of these, taken by Roger Fenton for the trustees of the British Museum, created an international sensation.[136]

In an article on Du Chaillu's *Explorations and Adventures*, the *London Quarterly Review* encouraged its readers to visit both the Hunterian Museum and the British Museum "to compare the bony skeleton of the gorilla with that of man."[137] The overall outcome of the Du Chaillu affair can be illustrated by two contrasting pieces written by the journalist George Augustus Sala. As the editor of *Temple Bar* he was responsible for an anonymous satire in its November 1861 issue, "With Mr. Gorilla's Compliments," in the epilogue to which Owen is rebuked for giving shelter to "this little Frenchman."[138] Some two decades later, however, in 1880, having visited "the splendid palace dedicated to the Natural History Department at South Kensington," Sala wrote a bombastic piece for the London *Daily Telegraph* in which he praised Owen as "the Apostle" of a "new temple to nature."[139]

Owen's "Ministry of Truth"

To Owen, however, his new museum was more a cathedral of science than a temple to nature. He was a churchgoing Anglican who regarded scientific work as a "ministry of truth" and saw his museum as an architectural expression of the scientists' growing importance as "God's ministers," through whom natural truth was revealed. Owen was a genuinely religious man and a committed Christian theist, as St. George Mivart and other close followers of Owen affirmed.[140] When he received his knighthood, Owen chose for his armorial bearings the words *Scientia et Pietate*.[141] His private letters to his young son, William, which are beyond any suspicion of studied ambiguity, in places oozed pure religious sentiment, for example, when he wrote to "Will" that, whereas humans have but a single pair of arms and hands, the echinus has "a hundred": "And both we and it are the works of a great Creator who never loses sight of the working of his machines. Let nothing disturb your feeling of reverence for Him when in His house and engaged in His worship. God bless you, my dear Boy, prays daily your affectionate father."[142]

This perception of himself as one of God's ministers not only lay behind his disagreement with Darwin and the Darwinians but also put him on a collision course with the orthodox ecclesiastical establishment. What Owen strove for was a fundamental redistribution of cultural authority in society so as to acquire for himself and for his scientific profession a major share of it, especially that part which covers issues of science and religion.[143] He accordingly engaged in a series of skirmishes for intellectual leadership with representatives of the church but also of other institutions of Victorian society, which, in addition to the clergy, included the aristocracy and the judiciary.[144] Thus, his crossing of swords with theologians was part of a wider contest with members of those groups whose authority was traditionally recognized as extending over questions of science. Once again, there was much ambiguity in all this, given that Owen's rise to power depended on the support of various members of the three traditional cultural estates[145] and yet in the end had to take place at their expense. These skirmishes were fought more in the public press than in the restricted journals of scientific societies, and we must systematically study Owen's popular writings to get a clear picture of his territorial fights with representatives of the nonscientific institutions of cultural authority.[146]

The denial by Owen that the human soul is an immortal entity was not an isolated instance of clergy-challenging heterodoxy. On a variety of other issues he appropriated the role of arbiter of truth and took a nontraditional, freethinking stance. One of these issues concerned the longevity of humans, which, on the one hand, is a biological question and, on the other, a theological one, since the genealogies in chapter 5 of the book of Genesis attribute to the antediluvian patriarchs life spans of hundreds of years and in the case of the proverbial Methuselah as many as 969 years. The stage for Owen's attack on the clergy's ascendancy over this issue was set during the early part of the 1860s by two famous instances of the introduction into the Anglican Church of higher, or historical, criticism. The first was the publication of seven *Essays and Reviews* (1860), authored by members of the established church, among whom were the Oxford luminaries Benjamin Jowett, Mark Pattison, and Baden Powell. The second was the appearance of *The Pentateuch and Book of Joshua Critically Examined* (1862), written by the Welshman John William Colenso, bishop of Natal. This was the first book of a seven-part series, published over the period 1862–79, that cast doubt on the historical accuracy and the Mosaic authorship of the Pentateuch and introduced such recent products of Continental Old Testament scholarship as Abraham Kuenen's *Historisch-kritisch Onderzoek naar het Ontstaan en de Verzameling van de Boeken des Ouden Verbonds* (1851–65).[147]

Orthodox concern, if not panic, followed these invasions of Old Testament criticism into the established church, and countermoves were organized. Both *Essay and Reviews* and *Pentateuch* were each followed within a short period by well over a hundred perturbed pamphlets. Official synodical sentences against the heterodox authors were pronounced, although quashed on appeal. Owen was widely expected to lend his support to the orthodox cause, but he failed to do so. It is true that the *Replies to "Essays and Reviews"* (1862), composed by seven Anglican clergymen and with a preface by Samuel Wilberforce, sported "a note by Professor Owen"; but this note was no more than a letter by Owen in which he, at the time deeply embroiled in the hippocampus controversy, reiterated his stance vis-à-vis the difference human-ape, and reiterated, too, in his peculiarly convoluted way, that he believed in divine purpose in nature as well as in an origin of species by natural means.[148] In 1864, as part of the uproar caused by the essayists, nearly half of the clergy of England and Ireland (10,906) united in issuing the so-called Oxford declaration, which reconfirmed the orthodox position that the Bible was divinely inspired and not only contained but actually was the Word of God.[149]

In the wake of this ecclesiastical protest, a group of London chemists composed a further memorial which deplored "that researches into scientific truth are perverted by some in our own times into occasion for casting doubt upon the Truth and Authenticity of the Holy Scriptures."[150] The chemists wrote to fellows and members of all British scientific societies with a request for their supporting signature to the "Scientists' Declaration." When the following year the declaration was published, no fewer than 717 signatures had been obtained. On the list were such well-known names and Owen loyalists as Brewster, the Earl of Enniskillen, Sedgwick, and lesser figures such as Thomas Rymer Jones.[151] One of the declaration's instigators, the Anglican chemist Herbert McLeod, eagerly canvassed Owen but, significantly, Owen refused to sign. So did several others. The Duke of Argyll concurred with the message of the memorial but saw no purpose in putting his signature to it. Murchison simply begged to decline.[152] Owen's refusal was not couched in such irenic terms: he emphatically criticized the contents of the declaration. To him, it had little to do with standing up for religion. In his earlier letter, appended to the *Replies*, he had referred to scientists as "God's ministers," and this time as well, he described science as a truth mission, answering McLeod's letter as follows:

> The memorial of which you have sent me a printed copy commences with a charge against some of our contemporary searchers after scientific truth as perverters of such researches into occasion for casting doubt upon the Truth and Authenticity of Holy Scripture. I have the conviction that so grave a charge ought to be less vaguely made, and should be brought home by evidence against the accused, before its publication. It is, indeed, a matter of deep concern to me that so many estimable fellow Christians should, according to your letter, be regardless of the risk they run of bearing false witness against

> fellow-labourers by spreading abroad, under their signatures, so damaging and, as I trust, unfounded an accusation.[153]

Later entreaties by McLeod and another of the organizers, Capel Henry Berger, who together visited Owen at the British Museum, failed to change Owen's mind. "His interpretation of Scripture is rather free. He was very kind and gave us a copy of his lecture at Exeter Hall," McLeod noted in his diary.[154] Rather than join the orthodox party, Owen supported the heterodox movement and, in the lecture referred to in McLeod's diary, undermined the belief in the inspired authenticity and historical accuracy of the book of Genesis. In 1863 he was asked to give the opening lecture to a course of lectures for the winter season at the YMCA at Exeter Hall. Owen addressed his earnest audience on "instances of the power of God as manifested in His animal creation." He presented examples both of design in individual species such as the aye-aye, on which animal he had just published his classic monograph, and of abstract design as represented by the vertebrate archetype.

So far so good. It soon became clear, however, that by "Power of God" Owen meant nothing more nor less than the power of scientific inquiry. To the consternation of the organizers, Owen undercut various orthodox beliefs, such as that the serpent is a degraded reptile, cursed for its part in the "Fall of man" in Paradise and condemned to crawl on its belly and eat dust. The presence of vestigial hind-limbs in, for example, boas and pythons was seen by some as confirmation that the serpent is indeed the degenerated creature of the Genesis story. Hugh Miller, for one, made much of the supposedly degraded nature of serpents.[155] Owen, whose vertebrate archetype could account for rudimentary organs, held a different view. He insisted that serpents were beautifully adapted: "It is true that the serpent has no limbs; yet it can outclimb the monkey, outswim the fish, outleap the jerboa, and, suddenly loosing the close coils of its crouching spiral, it can spring so high into the air as to seize the bird upon the wing: thus, all those creatures fall its prey."[156] More importantly—Owen went on—paleontology showed that serpents with all their characteristic features, including a crawling mode of locomotion, "existed long ages before the creation of man" and thus long before any "Fall of man." Theologians had been wrong, misinterpreting the Bible and not knowing "the Power of God," that is, the results of paleontology.

The YMCA publications committee initially considered the lecture unacceptable but later relented. Independently, Owen published his own version, *Instances of the Power of God*, in which he included some fifteen pages that had been "omitted in the delivery on account of time" and proved to contain several other heterodox views.[157] In the added pages Owen concentrated on the geographical and paleontological distribution of birds and mammals to show that the belief in a universal deluge is untenable. A corollary of a worldwide flood in which all terrestrial animals drowned, apart from those taken aboard Noah's Ark, was that the spot where the Ark had landed must have become the center of distribution of all air-breathing species. Such a center did not exist, a fact borne out by the research of "single-minded honest men, gifted with the faculty of rightly observing God's works, and impelled to the exercise of such entrusted gifts,"[158] that is, by himself and his fellow naturalists.

In short, Owen used the results of his biological work to attack traditional interpretations of the Bible and through these the theological profession, gravely warning his audience: "Beware, therefore, of logically precise and definite theologies, accounting, from their point of view, for all things and cases natural and preternatural, claiming to be final and all-sufficient."[159] He admonished the young Christian men to free their minds from theological preconceptions and emancipate themselves from interpretations of the Bible imposed upon them by the church. They were to put more faith in the scientists, who are God's "predestined instruments" to illuminate his power.

Colenso expressed his approval and the *Guardian* called Owen a "Colensified Christian."[160] More negatively, Gray, unhappy with his superior at the British Museum, in a letter to a lady friend accused Owen of being a Johnny-come-lately, who did not want to be left behind in the race toward liberal thought.[161] This was not accurate, however. Owen's participation in the historical criticism of the Pentateuch should not be misunderstood as a tagging along with the atheism or agnosticism of Darwin and Huxley. It is true that in the post-*Origin* period Owen spoke his views with less ambiguity than he had done before; but his heterodox stance was nothing new. For example, Owen had expressed his critical view of the biblical story of the serpent as early as the third part of *History of British Fossil Reptiles*, which had been sufficiently sensational for the *Edinburgh New Philosophical Journal* (1850) to have reprinted the relevant extract.[162] Nor did Owen stop at his infamous 1863 lecture "Instances of the Power of God." When the first volume of *The Holy Bible, with an Explanatory and Critical Commentary* appeared in 1871, in the commentary to which E. Harold Browne, the bishop of Ely, reaffirmed traditional belief against Colenso *cum suis*, Owen pounced on the bishop's writings with seemingly gratuitous aggression. The offending passage was merely the following: "As to the extreme longevity of the Patriarchs it is observable that some eminent physiologists have thought this not impossible."[163] As an example of such physiologists, the comte de Buffon was mentioned and, implicitly, also Albrecht von Haller. Owen fumed: "Truth demands the contradictory statement, that no physiologist, of whatever degree of 'eminence,' at the present day, admits the possibility of an animal with the characteristics of the human genus and species living to any of the ages specified in the fifth chapter of Genesis."[164]

The bishop should have sought the opinion of contemporary physiologists—Owen insisted—and he proceeded to make himself the spokesman for modern physiology. His main argument against the possibility of multicentenarians was that their longevity requires physiological features that, if present, would make the creatures something other than *Homo sapiens*. Given his early and lasting expertise in the study of teeth, exemplified by *Odontography*, Owen made much of the effects of age on grinding teeth. In any one species, the particulars of dental growth are linked to the length of life. This is true in horses, elephants, and also in humans. Humans have two sets of teeth, and the mature ones "may do their work, under favourable circumstances, for thirty or forty years before being worn out and shed." A life of eight or nine centuries would require many successive sets of true working molars for masticating food.

> To meet the mechanical wear of mastication, the teeth of Methuselah must either have been renewed and changed many times, as in the elephant; or the tooth-matrix must have been modified after the plan of that of the Megatherium, whereby new tooth-material became added to one end of the molar in the ratio of its abrasion from the other end—in other words the teeth must have grown like the nails.[165]

Such dental features would characterize "a long-lived genus of the Bipedal order, zoologically distinct from the actual species of *Homo*."[166] Yet antediluvian humans had not differed in dental characteristics from their present-day relatives. Owen referred to his cave researches from the winter of 1863–64 when, engaging in rare field forays, he twice visited the Cave of Bruniquel in the south of France. Fragments of a human skull, including the upper and lower jawbones, were discovered under a stalagmite, together with Stone Age implements and fossils of extinct mammals. These afforded Owen "the opportunity of testing the zoological characters of palaeontological man." "They are the same as now," he affirmed.[167] How did he know that prehistoric people might not have had three or more sets of teeth? One only needed to cross-section the jaw bones, and the germs of successor teeth, had there been any, would have been detected. The dentition in skulls from other caves, all older than the period of the "multicentenarian bipeds of the Hebrew cosmogony," and the teeth in skulls from Egyptian tombs were the same as in today's human species.

To reiterate: the thrust of Owen's lengthy and detailed outburst against the bishop of Ely was directed less at the issue of human longevity as such than at the authority of the clergy. His explicit target was an ecclesiastical establishment that propped up "a crumbling edifice" of dogmas and thus prevented "intellectual progress."[168] They should take account of the science of the time, the practitioners of which, whether Buffon in his day, or later Cuvier, and now Owen himself, were ministers of truth. The clergy's insistence on the historicity of the Bible was just a concealed means of retaining intellectual ascendancy. At the conclusion of his article Owen defined what he believed formed the limited sphere of ecclesiastical authority with respect to human longevity, namely the spiritual meaning of our normal life span, stating that "the gifts that flow from a true knowledge and teaching of the power of God," in relation to longevity, are the following moral lessons:

> To accept death as the price of living, to see in the limitation of individual existence the necessity of succession, to recognise in that ordinance the condition of the highest pleasures of life, the source of our purest emotions, the basis of all social happiness, to balance against our own departure the dear responsibilities and yearnings towards offspring, the reverential affection towards parents, the closer and holier love of helpmate, to change the curse into the blessing, the deprecation into the thanksgiving.[169]

We have a further indication of what Owen's motives were in turning the topic of longevity into an issue. This comes in the form of his long written response to a letter by Josiah Crampton, son of Owen's old friend Philip Crampton, a noted Dublin surgeon, in which Josiah, himself a clergyman, expressed deep concern about the fact that Owen appeared to overthrow Genesis. Owen widened his previous Pentateuchal criticism to include the stories of creation and deluge. As "a Minister and Interpreter of Truth" Owen felt compelled to reject the miraculous creation of species and see them as the product of natural law. The deluge as told in Genesis, too, "is a Fable characteristic of an ignorant and semi-barbarous age" and was inspired by a local inundation. The clergy culpably resisted scientific truth: "As the number of those [the scientists] to whom such knowledge may be imparted and who are capable of receiving it, increases, the condition of the ordained teachers [the clergy] shutting out and resisting such knowledge will become the more lamentable."[170]

In a pamphlet by a clergyman named W. F. Hobson, *Longevity; or, Professor Owen and the Speaker's Commentary* (1872), the author launched a counterattack in defense of the bishop of Ely, putting the issue unambiguously in terms of a struggle for intellectual hegemony between theologians and scientists. The pamphleteer, engaged as chaplain to the armed forces, accused Owen of "real hostility to theology," and while he acknowledged that Owen's "own subject and speciality" was "his unquestioned dominion," he rebuked the professor for having trespassed on history and legend, which are not the sphere of the naturalist. In the face of historical documentation, physiology should "retire into its own province." Scientists are no less fallible than theologians.[171]

Owen's stance confused and disappointed many of his friends, who looked to him for support in defending traditional beliefs threatened by the rising tide of scientific naturalism and Old Testament criticism. His accommodating utterances on the subject of the creation of humans by a personal God never were intended, however, to support the Genesis story but expressed a belief in a spiritual destiny of humankind. Nor did Owen join those scientists who in reaction to scientific naturalism turned to spiritualism.[172] As we have seen, he did not believe in a soul separate from the body and regarded séances and table-rappings as fraudulent tricks. None of this in Owen was occasioned by the advent of Darwinism. As early as the beginning of the 1850s, he ridiculed séances and characteristically leveled his criticism less at a belief in spirits as such than at the mediums, who arrogate to themselves the authority of speaking a special truth. If spirits can tell what goes on among us—he reasoned—they surely know that mediums often exploit their clients, and spirits would be unlikely to lend themselves to such abuse:

> If a disembodied essence can know anything of the material sayings and doings on this planet it must know them most thoroughly, intimately and truly. The value of such knowledge to us must depend upon the power of communicating to us such knowledge, clearly, intelligibly and fully. If the selected agents are mercenary and use their privilege to tax their fellow-creatures, the spirit or essence must know *that* also; and truly it seems but a poor amusement or employment for spiritual essences to lend themselves to enrich the dubious characters who at present assume to themselves the chief privilege of summoning and questioning the spirits, and of revealing their replies.[173]

With age, Owen did not grow more orthodox, unlike his great political patron Gladstone, who toward the end of his life increasingly devoted himself to biblical apologetics; witness his *Impregnable Rock of Holy Scripture* (1890). In the mid-1880s Gladstone engaged in a famous exchange with Huxley in the pages of the *Nineteenth Century*, and for this he asked Owen's assistance. Owen responded fulsomely, but the content of his letters fell far short of what Gladstone must have hoped for, as the latter sought to reconcile Bible and science and in particular tried to harmonize the sequence of the Genesis days of creation with the succession of the modern stratigraphic column—even though Gladstone did not interpret the days of creation literally as periods of twenty-four hours.[174]

Owen reiterated his earlier critical positions on creation, the deluge, and the longevity of the antediluvian patriarchs. To him, the imagery of the Bible is that of the uneducated people of all times. The first chapter of Genesis teaches us nothing more than that an almighty creative cause directs the course of the history of life on this planet: "The divine Chapter teaching the fundamental Truth of the guidance to a Higher Life, illustrates the same by such references as were intelligible to the age it addressed, as they will be to the non-scientific- and the wage-classes of all time."[175]

To many of his contemporaries Owen's religious position was probably as puzzling as Thomas Carlyle's. Both men were opposed to the positivist tendencies of their day, yet they rejected many of the accepted, traditional creeds, shocking even liberal theologians. The fullest expression of this two-sided stance was given by Owen in his essay "On the Argument of 'Infirmity' in Mr. Lewes' Review of the *Reign of Law*." It combined a scornful attack on Comtean positivism with an equally emotive one on traditional Bible belief. If the clergyman from his pulpit preaches something different from what the scientist teaches from his professorial chair, the solution to the conflict can be brought about only when theologians let themselves be illuminated by the "light of truth" of the scientists.

Owen appropriated the religious language about the Bible as the Word of God and about its authors as divinely inspired writers, transferring it to the enterprise of science. The history of scientific discovery had been a process of gradual self-revelation by God, not accidental but guided by means of the illumination of "His faithful servants and instruments," the scientists. Their discoveries by definition did not undermine the Christian faith. What they had done was expose fallible belief systems. "No scientific discovery collides against any sentence of the divine Sermon on the Mount.[176]

The essence of Owen's view was that God's self-revelation has been a continuous, progressive process, with new "parcels of truth" being added "in God's good time." The clearest and most concrete truths are therefore the most recent and are not to be found in the Bible but in modern scientific literature. To the extent that the Bible contains God's revelation, it is to be found at a general level of moral lessons and spiritual meaning. For theologians to decree that the Bible is God's truth in a historical and physical sense is to stifle God's later, more specific revelations. Such biblical literalism is a matter, not of belief in, and service to, God, but of self-service by an ecclesiastical establishment that, by restricting God's self-revelations to its primitive, biblical phase, limits truth to the area of their own theological expertise.

Owen's religious stance is exemplified by an incident that took place in 1876, at the end of a Sunday church service he attended. The preacher apparently accused some naturalists of dethroning God and hoisting science into his seat. After the sermon, in the vestry, Owen sternly rebuked the preacher in

FIGURE 20. Richard Owen in 1881, portrait in oils by William Holman-Hunt painted shortly before the official opening of the Natural History Museum (© The Natural History Museum, London; used by permission of The Natural History Museum).

the presence of other assembled clergymen and churchwardens, exclaiming to their blank amazement: "My Christian Brethren! I trust with God's help, that Science will continue to do for you what she has always done, return you good for evil!"[177] Owen believed in God; Owen belonged to the established church; but the clergy had to remove some of its intellectual furniture to make room for that of the scientists.

As a lecturer and professor, Owen taught the public about the truth of the natural world, turning his social space into one contiguous with that of the clergy. The professorial chair and the pulpit became symbols of two professions competing for ascendancy.[178] In his Hunterian lectures Owen habitually congratulated his medical students on "the inestimable privilege" they enjoyed by entering their professional studies through the portal of anatomy, contrasting this "with the dry and unattractive preliminary exercises of the Lawyer and the Divine."[179]

The claim to ecclesiastical authority was also made by the very architecture of the Natural History Museum. In 1855, when Owen heard that a Gothic style had been selected for the Oxford University Museum, he wrote disapprovingly to Acland: "The sciences were not born nor nursed where that style originated."[180] Yet he went along with the ecclesiastical design of his own museum as a cathedral of science. The central part of the Natural History Museum where Owen envisaged his lecture theater was inspired by a Romanesque cathedral, complete with main arcade, triforium, and side chapels. The journalist Sala, who visited the new museum shortly before its official opening, enthused about "the cathedral-like proportions."[181]

Owen's investiture as bishop of his own cathedral took place at long last when in 1881 the Natural History Museum in South Kensington opened to the public and he himself was installed as the de facto director. Richard Owen, CB (soon to be created KCB), DCL, FRS, Foreign Associate of the Institute of France, etc. etc., had come into his own. That same year William Holman-Hunt produced what was probably the finest of the many Owen portraits (fig. 20). It was exhibited in a London gallery and compared in the *Times* to Holman-Hunt's "religious subjects." It indeed gave expression, with the purple of Owen's outfit, the sanctity of his facial expression, and the pious pose of his hands, to his priestly, if not episcopal, aspirations.[182]

Not long after, Owen retired from his museum employment. The nine remaining years of his life were spent in relative isolation at his grace-and-favour house (i.e., occupied by permission of the queen) in Richmond Park. He occasionally still went to town, continuing as president of the Palaeontographical Society. When in 1885 he resigned from this, his last office, "the close of all public work and responsibility" had come to Owen.[183]

In the summer of 1892, suffering from stomatitis (an inflammation of the mucous membrane of the mouth), his health went into a steep decline. Anticipating the end of the great man's life, the *British Medical Journal* began issuing weekly reports on his medical condition.[184] Upon Owen's death, on December 18, "literally of old age,"[185] the journal sentimentalized that Owen had possessed an intense love of animals and birds and that his garden contained shrubs and trees from all parts of the world that had been planted with his own hands and resounded with the song of birds.[186]

Appendix

Anatomy of Owen's Scientific Oeuvre

Owen's many publications can be subdivided according to a number of thematic interests that stretched over different lengths of time, although most ran concurrently. Each was the vehicle of different though connected and in places overlapping concerns and philosophies. The longest-running theme was that of Australian mammals, both recent and fossil, studies of which appeared for the most part in *Philosophical Transactions.* This theme ran virtually from the very beginning till the very end of Owen's career as a publishing naturalist. First there were the studies of living marsupials and monotremes, begun in 1832, which were compiled in 1847 in the form of two major contributions to Robert B. Todd's *Cyclopaedia of Anatomy and Physiology.* In 1835 Owen began his successful studies of the Pleistocene marsupials of Australia, which culminated in the 1870s with a series of papers presented to the Royal Society and collected as *Researches on the Fossil Remains of the Extinct Mammals of Australia* (1877). As late as 1888, not long before his death, a final paper on the subject flowed from his trembling pen. This theme primarily reflected a preoccupation with colonial natural history.

A related colonial theme was that of flightless birds, in particular the extinct moas of New Zealand. His publications on this theme began in 1839 and led to the longest series of single-genus papers of Owen's career, on the dinornis, published in *Transactions of the Zoological Society.* The bulk of these were collected in *Memoirs on the Extinct Wingless Birds of New Zealand* (1879), covering four decades of publishing. In addition to the colonial connection, there was the link with Cuvierian functionalism, for which the subject of flightless birds provided a spectacular medium of presentation.

A further, long-lasting theme concerned primates. It ran from 1831 and reached its peak with *Memoir on the Gorilla* (1865), occupying much space in

On the Anatomy of Vertebrates (1866–68). Most of these papers were published in *Transactions of the Zoological Society*. The material formed the main plank of his antitransmutation—though not antievolution—platform and at the same time was the vehicle for some of his most unorthodox, physicalist convictions about the human mind and soul. Owen's work on apes and monkeys also led to a major clash with Huxley over humanity's place in nature.

Fossil mammals—other than the Australian nonplacentals—formed another substantive topic, in particular the unusual fossil sloths from North and South America. Owen began to work on these in 1836, when he was invited to describe the fossil material brought back by Darwin from his *Beagle* voyage. The relevant papers appeared in *Transactions of the Royal Society*, and the ripest fruits of this research were two monographs: *Description of the Skeleton of an Extinct Gigantic Sloth* (1842) and *Memoir on the Megatherium* (1861). Both were in the finest tradition of Cuvierian and Bucklandian functional paleontology that culminated in 1853 with the emplacement of a megatherium model in the geological gardens of the Crystal Palace grounds.

Last, but not least, there were the fossil reptiles. Owen contributed little to the subject before his two-part BAAS paper "Report on British Fossil Reptiles," begun in 1839. From this time on, however, he steadily produced papers on a taxonomically wide-ranging collection of reptile material that included not only ichthyosaurs, plesiosaurs, dinosaurs, and pterosaurs but also the relatively primitive labyrinthodonts and the dicynodonts from South Africa. Nearly all of these papers appeared in the publications of the Geological Society and as memoirs of the Palaeontographical Society; the most substantial of them came out in the late 1850s. Owen used reptilian fossils to substantiate both transcendentalist and, more sensationally, functionalist views. Illustrating the latter epistemology were the dinosaur models that occupied center stage among the Crystal Palace geological "monsters."

An overarching theme was that of comparative osteology, both recent and fossil. This work was rooted in Owen's cataloguing activities of the 1840s, and it was in the course of these labors that he perfected the transcendentalist approach. The main publication outlets for this were a BAAS "Report on the Archetype and Homologies of the Vertebrate Skeleton" (1846) and his published lectures on the comparative anatomy of the vertebrates (1846; 1866–68).

Notably absent as major themes were invertebrates (even though Owen did some brilliant work on cephalopods and published a sensation-arousing booklet on parthenogenesis in aphids, flukeworms, etc.) and physical anthropology. Dispersed through Owen's many publications are various gold nuggets of originality to which present-day experts from time to time draw the historian's attention.[1]

Notes

Preface

1. Browne, "Natural Causes."

2. For a concise and engaging overview of the geography-of-knowledge approach, see Livingstone, *Putting Science in Its Place*; see also Finnegan, "The Spatial Turn." A particularly detailed case study of the locations of pre–*Origin of Species* evolutionary thought in Britain is by Secord, *Victorian Sensation*. Browne, too, stresses the importance of place in her *Charles Darwin: Voyaging* and *Charles Darwin: The Power of Place*. For further references, see Livingstone, *Putting Science in Its Place*.

3. Burrow ("In the Iguanodon Diner") disapproved of my "sociologising explanation"; Lenoir (review symposium "Imposing Owen"), by contrast, wished for more of the "institutional perspective"; while Levere approved of "the way in which the institutional framework is convincingly integrated with the style and content of Owen's science" (review symposium "Imposing Owen," 57).

4. Patterson, "Archetypes and Ancestors," 375.

5. Limoges, "Owen as Strategist."

6. The metaphor is borrowed from Porter, "Bones, Stones and Buckland," who used it to describe a similar narrative I constructed in writing a scientific biography of William Buckland (*Great Chain of History*).

7. Levere, review symposium "Imposing Owen," 56.

8. Such an Owen representation would accord with recent interest in the role of family and women in science. It is, perhaps, no coincidence that a majority of those critics who have asked for more foregrounding of Owen-the-man are women: Browne, "Natural Causes," 3; Neumark, "Man for the Job"; Ritvo, review of *Richard Owen*, 299; Camerini, "Power of Biography." Admittedly, Jacob Gruber, in his review of *Richard Owen*, also asked for more of "a personal context" (331).

9. Rupke, "Neither Creation nor Evolution"; Rupke, "The Origin of Species from Linnaeus to Darwin"; Rupke, "Darwin's Choice."

10. The expression is taken from Gregorio, "Wolf in Sheep's Clothing."

11. Amundson, "Typology Reconsidered"; Amundson, "Owen and Animal Form"; Camardi, "Richard Owen," 510–11; B. Hall, introduction to *Homology*; B. Hall, preface to Owen, *On the*

Nature of Limbs; Padian, "Form versus Function"; Padian, "Rehabilitation of Owen"; Padian, "Richard Owen's Quadrophenia"; Panchen, "Richard Owen"; see also Elwick, "Styles of Reasoning," 61–62.

Chapter One

1. A clipping of the article occurs in Owen's autobiographical scrapbook, BM(NH), L, OC 24.

2. Rev. Richard Owen, *Life of Richard Owen by His Grandson*, vol. 2, 383–86 (on the spine, the title *Life of Professor Owen* is given; abbreviated below as Rev. Owen, *Life*).

3. Ibid., 333–82.

4. "Owen Memorial," BM(NH), L, OC 38.

5. Gillispie, *Edge of Objectivity*, 313.

6. Rev. Owen, *Life*, vol. 1, 119–21, 208–9, 407–8; vol. 2, 38, 95–96. Jacob W. Gruber suspects that manuscript material related to the controversy with Darwin was systematically removed from the Owen papers; see Gruber and Thackray, *Owen Commemoration*, 16.

7. Rev. Owen, *Life*, vol. 2, 273–332.

8. Dowager Duchess of Argyll, *George Douglas*, vol. 1, 410.

9. "Men of the Day, No. 57, Professor Owen," *Vanity Fair*, Mar. 1, 1873, 20.

10. Hollander, *Scientific Phrenology*, 285.

11. Dowager Duchess of Argyll, *George Douglas*, vol. 1, 410.

12. J. Clark, *Old Friends*, 377. The *Oberon* story is discussed by Austin and Jones, "The Clifts," 69; see also Rev. Owen, *Life*, vol. 1, 178.

13. Dickens, *Our Mutual Friend*, 29. See also Rev. Owen, *Life*, vol. 1, 318.

14. Kingsley to Owen, July 16, 1867, CUL, OC (letter 123).

15. Rev. Owen, *Life*, vol. 1, [8]; see also vol. 2, 272.

16. Ibid., passim.

17. Owen to the Duke of Teck, Mar. 17, 1886, in Dobson, "Account of the Life and Achievements," 143; BM(NH), L, OC 86. See also Gruber and Thackray, *Owen Commemoration*, 21.

18. L. Huxley, *Life of Huxley*, vol. 1, 101.

19. Curwen, *Journal of Gideon Mantell*, 200, 225, 250, 280–81.

20. R. Fox, "Observations on Subterranean Temperature."

21. Pym, *Memories of Old Friends*, vol. 1, 256, 260.

22. Huxley to John Tyndall, May 13, 1887, in L. Huxley, *Life of Huxley*, vol. 2, 167; F. Darwin and Seward, *More Letters of Darwin*, vol. 1, 309.

23. Carlyle to Jane Carlyle, Aug. 26, 1842, in Ryals and Fielding, *Letters of Thomas and Jane Carlyle*, vol. 15, 51–52.

24. Beer, *Darwin and Huxley*, 61. A summary of Darwin's critical remarks about Owen is given by Hull, *Darwin and His Critics*, 171–75.

25. Dowager Duchess of Argyll, *George Douglas*, vol. 1, 410.

26. For example, John Goodsir to Owen, Jan. 23, 1846, BM(NH), L, OC 13 (fols. 184–85); Owen to Goodsir, Jan. 29, 1846, in Goodsir, *Testimonials*, 19. See also Gruber and Thackray, *Owen Commemoration*, 59–61.

27. *Dictionary of National Bibliography*, s.v. "Gould, John"; see further Tree, *Ruling Passion of John Gould*.

28. For example, Owen to the Secretary of the Royal Society, Jan. 18, 1859, RS, RR. 4 (155), on the paper "On the Ova and Pseudova of Insects" by the young amateur naturalist John Lubbock.

29. Cooper to Owen, Dec. 18, 1839, RCS, OC. Cooper's devoted admiration is still noticeable in Cooper to Owen, Sept. 19, 1859, RCS, MS Add. 192 (no. 70).

30. [Acland], "Obituary. Sir Richard Owen, K.C.B.," 11 (pamphlet reprinted from *British Medical Journal*, no. 2 [1892]: 1411–15). A less warmhearted obituary notice was written by Flower, "Obituary Notices," i–xiv. Flower also wrote Owen's original *Dictionary of National Bibliography* entry.

31. Davidson to Owen, Sept. 26, 1878, BM(NH), L, OC 9 (fols. 238–39).

32. It appeared in the form of a twinned review of Gordon, *Life and Correspondence of William Buckland*, and Rev. Owen, *Life*, both published in 1894: Mivart, "Century of Science."

33. Haupt, "Homologieprinzip bei Richard Owen." During the "decades of drought," E. S. Russell also contributed to keeping the trickle of Owen scholarship flowing by devoting a chapter of his *Form and Function* (102–12), to Owen.

34. R. Richards, *Romantic Conception of Life*, 514–54. See also his *Meaning of Evolution*, passim.

35. A. Desmond, *Archetypes and Ancestors*; A. Desmond, *Politics of Evolution*; E. Richards, "Question of Property Rights"; also E. Richards, "Political Anatomy of Monsters"; Sloan, *Richard Owen* (the quotation is on p. i); also Sloan, "Whewell's Philosophy of Discovery." See further A. Desmond and Moore, *Darwin*.

36. "Calendar of Owen Letters" (unpublished MS, 1980), BM(NH), L, OC 87. I gratefully acknowledge the use of this typescript. See also Gruber and Thackray, *Owen Commemoration*; Gruber, "Owen, Sir Richard (1804–1892)."

37. Conway Morris, *Crucible of Creation*; Conway Morris, *Life's Solution*. For an earlier wide-ranging discussion of the issue, see Hull, *Science as Process*, 200–276.

38. Amundson, "Owen and Animal Form," xliii–xlviii.

Chapter Two

1. Rev. Owen, *Life*, vol. 1, chaps. 1–4.

2. As early as the beginning of the nineteenth century a two-volume history was written of the Muséum d'histoire naturelle: Fischer, *Nationalmuseum*. See also Limoges, "Muséum d'histoire naturelle of Paris."

3. Rev. Owen, *Life*, vol. 2, 240–41.

4. Dobson, *Conservators*, 10–17.

5. Owen to the trustees of the British Museum, Nov. 27, 1883, BM(NH), L, OC 24.

6. On British museums outside London, see Ball et al., "Provincial Museums." The BAAS list of provincial museums was reprinted, together with an additional twenty-nine London museums (for only three of which the date of foundation was provided), by Murray, *Museums*, vol. 1, 291–312. On colonial museums of natural history, see Sheets-Pyenson, *Cathedrals of Science*. See also Kohlstedt, "Australian Museums of Natural History." On the Harvard Museum of Comparative Zoology, see Winsor, *Shape of Nature*.

7. From among the many secondary sources, see Alter, *Wissenschaft, Staat, Mäzene*, in particular 283–86; Cardwell, *Organisation of Science*; Morrell and Thackray, *Gentlemen of Science*, in particular 12–16; C. Russell, *Science and Social Change*; Sanderson, *Universities*.

8. Flower, "Obituary Notices," v. This view has been repeated ever since.

9. Rupke, *Richard Owen*, 16–22.

10. Rev. Owen, *Life*, vol. 1, 32–33.

11. Cope, *Royal College of Surgeons*, chaps. 2, 8.

12. Negus, *Hunterian Collection*, 107ff.

13. Owen to Blizard, Apr. 5, 1831, RCS, OC, 275/h.7 (U-2).

14. Owen to W. E. Gladstone, Oct. 25, 1885, BM, Add. MS 44,492 (fol. 236).

15. Owen, "Report to the Board of Curators of the Museum of the Royal College of Surgeons, on the *Muséum d'Anatomie Comparée* in the Garden of Plants, Paris," Sept. 1831, RCS, 275.h.7 (3).

16. Ibid.

17. Ibid. See also Outram, *Georges Cuvier*, 161–63.

18. Clift and Owen to J. G. Andrews, May 5, 1836, RCS, 275.h.7 (5).

19. Dobson, "Hunter's Museum"; Owen to Robert Peel, Apr. 20, 1846, BM, Add. MS 40,590 (fol. 113).

20. Owen's Report to the Museum Committee, Dec. 15, 1845, RCS, 275.h.7 (8).

21. "Copy or Extracts 'from Any Minutes of the Trustees of the Hunterian Museum, relative to the Means of Enlarging the Space for Receiving and Exhibiting the Collections Thereto Belonging, Gradually Increased and Increasing,'" *Parliamentary Papers*, 1851 (75), vol. 43, 395; see also RCS, 275.11.7 (10).

22. Owen's report on the scope and function of the Hunterian Museum, n.d., BM(NH), L, OC 90, vol. 1.

23. Anon., *London Interiors*, vol. 1, opposite p. 129.

24. C. Carus, *England und Schottland*, 117. See also Hambury, "Visit of Professor Carus."

25. Silliman, *Visit to Europe*, vol. 2, 437–41.

26. Keith, "Minutes of the Museum Committee," 86.

27. Museum Committee on Owen's Report, Jan. 6, 1848, RCS, 275.11.7 (9).

28. Dobson, "Hunter's Museum," 286, states that the parliamentary grant amounted to £12,000.

29. "Report of the Commissioners," *Parliamentary Papers*, 1850 (1170), vol. 24, 3(b); see also BM(NH), L, OC 90, vol. 1.

30. "British Museum," *Literary Gazette*, Dec. 6, 1851, 847.

31. L[ewes], "Professor Owen," 81.

32. Rev. Owen, *Life*, vol. 2, 14.

33. Trevelyan, *Life of Lord Macaulay*, 659. Also quoted in Rev. Owen, *Life*, vol. 2, 14–15.

34. Rev. Owen, *Life*, vol. 2, 19.

35. From among the various histories of the British Museum (Natural History), see Gunther, *Century of Zoology*; Stearn, *Natural History Museum*. Much of this section is taken from Rupke, "Road to Albertopolis."

36. Owen, "On the British Museum of Natural History," BM(NH), L, OC 59 (8), 4–5. An edited version of this manuscript was presented by Owen as his presidential address to section D of the BAAS: *BAAS, Report 1881*, Transactions of the Sections, 651–61.

37. Owen, "On the British Museum of Natural History," BM(NH), L, OC 59 (8), 4–5.

38. *The Times*, Mar. 27, 1868; a clipping of the article occurs in BM(NH), L, OC 90, vol. 4, 32.

39. Owen, *On the Extent and Aims*, 22–23.

40. Ibid., 29–30.

41. "Report from the Superintendent of the Departments of Natural History, 10 February 1859," *Parliamentary Papers*, 1859 (126), vol. 14, 73. Sixty copies of this "Report" were privately issued, BM(NH), L, Hist. Coll.

42. "Report from the Select Committee on the British Museum," *Parliamentary Papers*, 1860 (540), vol. 16, 262.

43. Panizzi to the Chancellor of the Exchequer, Nov. 24, 1862, BM(NH), L, OC 90, vol. 1, 53–55. The sequence of events was summarized by Panizzi, "Report from the Principal Librarian, 22 January 1864," *Parliamentary Papers*, 1864 (117), vol. 32, 46–49.

44. Owen to Pleydell-Bouverie, June 13, 1863, BM, Add. MS 42,181 (fol. 69).

45. Lewes, "Professor Owen," 79.

46. Owen's Hunterian lecture of Apr. 5, 1842, BM(NH), L, OC 38.

47. "Report of the Commissioners Appointed to Inquire into the Constitution and Government of the British Museum," *Parliamentary Papers*, 1850 (1170), vol. 24, 173.

48. Owen, *On the Extent and Aims*, 112.

49. Rev. Owen, *Life*, vol. 2, 33–34. See also Alfred Waterhouse to Owen, Aug. 14, 1873, BM(NH), L, OC 26 (fols. 162–63).

50. See Negus, *Hunterian Collection*, 3.

51. Rupke, "Richard Owen's Hunterian Lectures." See also Padian, "Missing Hunterian Lecture."

52. J. Clark, *Old Friends*, 369.

53. Ibid., 368.

54. "Professor Owen's Lectures," *Lancet*, no. 2 (1840–41): 111.

55. Minute on use of lecture hall, 1856, British Geological Survey Library Archives, GSM I/7P309.

56. Rev. Owen, *Life*, vol. 2, 59.

57. Ibid., 60.

58. Ibid., 61.

59. Murchison to anon., Feb. 27, 1857, in ibid.

60. For example, "Professor Owen's Lectures on Reptiles," *Medical Times and Gazette*, Nov. 22, 1862, 552–53; Nov. 29, 1862, 579–80; Dec. 6, 1862, 606–7; Dec. 13, 1862, 639–40.

61. Rupke, "Road to Albertopolis"; Rupke, *Richard Owen*, 36–40.

62. Owen, presidential address (1881), in *BAAS, Report 1881*, Transactions of the Sections, 656.

63. Günther, "Address," 593.

64. See, e.g., Palmerston to Gladstone, July 18, 1861, in Guedalla, *Gladstone and Palmerston*, 172.

65. See, e.g., Briggs, *Age of Improvement*, 454–62.

66. Most of the Gladstone-Owen correspondence about the British Museum took place during 1861–62. The letters about the Exhibition Building, Gladstone to Owen, Aug. 14, 1862, and Owen to Gladstone, Aug. 18, 1862, were transcribed by Owen in the 1879 autograph manuscript "On the British Museum of Natural History," BM(NH), L, OC 59, 210–14.

67. *Hansard's Parliamentary Debates*, vol. 171 (June 15, 1863), cols. 922–23.

68. (*a*) "Reports from the Select Committee Appointed to Inquire into the Conditions, Management and Affairs of the British Museum," *Parliamentary Papers*, 1835 (479), vol. 7; 1836 (440), vol. 10. (*b*) "Report of the Commissioners Appointed to Inquire into the Constitution and Government of the British Museum," *Parliamentary Papers*, 1850 (1170), vol. 24. (*c*) "Report from the Select Committee on the British Museum," *Parliamentary Papers*, 1860 (540), vol. 16. (*d*) "Fourth Report of the Royal Commission Appointed to Make Inquiry with Regard to Scientific Instruction and the Advancement of Science," *Parliamentary Papers*, 1874 (c. 884), vol. 22.

69. (*a*) "Memorial to the First Lord of the Treasury, Presented on the 10th Day of March, by Members of the British Association for the Advancement of Science, and of Other Scientific Societies, Respecting the Management of the British Museum, with the Names Affixed," *Parliamentary Papers*, 1847 (268), vol. 34, 253–56. (*b*) "Memorial Addressed to Her Majesty's Government by the Promoters and Cultivators of Science on the Subject of the Proposed Severance from the British Museum of Its Natural History Collections, Together with the Signatures Attached Thereto" (presented July 6, 1858), *Parliamentary Papers*, 1857–58 (456), vol. 33, 499–504. (*c*) "Memorial Addressed to the Right Hon. the Chancellor of the Exchequer" (presented Nov. 19, 1858), *Parliamentary Papers*, 1859 (126), vol. 14, 64–67. (*d*) "Memorial Presented to the Right Hon. the Chancellor of the Exchequer" (dated May 14, 1866), *BAAS, Report 1879*, lxi. (*e*) "To the Right Hon. the First Lord of the Treasury" (dated Mar. 25, 1879), *BAAS, Report 1879*, lx–lxi.

70. See n. 68 (*b*) above, 7.

71. *Quarterly Review* 88 (1850): 151.

72. See n. 69 (*a*) above, 253.

73. See n. 68 (*b*) above, 173; also Rev. Owen, *Life*, vol. 1, 294.

74. [Ford], "British Museum," 148.

75. Owen, "Address" (1858), xcvii.

76. See n. 69 (*c*) above, 66.

77. See n. 69 (*d*) above, lxi.

78. Sclater, "On Certain Principles," 124.

79. See n. 68 (*d*) above, 9–10, 29.

80. See n. 69 (*e*) above.

81. "To the President and General Secretaries of the British Association for the Advancement of Science" (dated July 22, 1879), *BAAS, Report 1879*, lxii.

82. MacLeod, "Whigs and Savants."

83. [C. Lyell], "Scientific Institutions," 156; Vigors made his view known before the select committee "appointed to inquire into the condition, management and affairs of the British Museum"; *Parliamentary Papers*, 1836 (440), vol. 10, 115–17. See also Gunther, *Founders of Science*, 79–80.

84. Gunther, *Founders of Science*, 79–80.

85. See n. 69 (*a*) above, 254.

86. See n. 69 (*b*) above, 500–501.

87. [Ford], "British Museum," 155–56.

88. "Papers Relating to the Enlargement of the British Museum," *Parliamentary Papers*, 1859 (126), vol. 14, 55–56.

89. Darwin to Murchison, June 19, 1858, *Parliamentary Papers*, 1859 (126), vol. 14, 61.

90. Lyell to Murchison, June 21, 1858, *Parliamentary Papers*, 1859 (126), vol. 14, 60–61.

91. Murchison to Owen, Nov. 3, 1867, in Rev. Owen, *Life*, vol. 2, 182.

92. *Quarterly Review* 88 (1850): 502.

93. [J. Jones], "British Museum," 222–23.

94. Owen, "Address" (1858), xcv–xcvii. See also "Report from the Superintendent of the Department of Natural History, Feb. 10, 1859," *Parliamentary Papers*, 1859 (126), vol. 14, 72–75.

95. A. Desmond, *Archetypes and Ancestors*, passim.

96. See n. 69 (*c*), above.

97. See n. 68 (*c*) above, 301.

98. Ibid.

99. "The Natural History Collection of the British Museum," *Lancet*, Nov. 16, 1861, 484.

100. Owen, "On a National Museum," 119.

101. Owen, *Inaugural Address at Leeds*, 16.

102. The full title of the lecture was "On the Whale Lately Stranded on the Caithness Coast and on the Whale-Kind in General," RCS, Cabinet III (5) (B). A newspaper account of the lecture of Sept. 29, 1863, at the Lancaster Athenaeum occurs in BM(NH), L, OC 38; see also BM(NH), L, OC 90, vol. 4, 67. See further Owen, "Whales and Whaling," *Athenaeum*, Aug. 10, 1861, 195.

103. *Hansard's Parliamentary Debates*, vol. 166 (May 19, 1862), 1915.

104. See n. 68 (*c*) above, 303.

105. J. Gray, "Address," 75–80; Günther, "Address"; Flower, *Essays on Museums*, 1–53.

106. *British Medical Journal*, no. 1 (May 24, 1862): 550.

107. Beer, *Darwin and Huxley*, 61.

108. [Owen], "Darwin on the *Origin of Species*."

109. See Rupke, "Road to Albertopolis."

110. A. Desmond, *Archetypes and Ancestors*, 119.

111. See chap. 3.

112. Owen, Hunterian lecture of April 5, 1842, BM(NH), L, OC 38.

113. Owen, "On the *Archaeopteryx*."

114. Owen, *Memoir on the Dodo*. See further Rupke, *Richard Owen*, 71–74.

115. Owen, *On the Extent and Aims*, 6.

116. There was nothing fundamentally new about this. For instance, in 1696 John Woodward had issued *Brief Instructions for Making Observations in All Parts of the World* (London).

117. W. Buckland, "Instructions."

118. *Arcana of Science and Art* 4 (1831): 231–33.

119. Owen to John Herschel, Dec. 31, 1847, RS, HS. 13 (letter 189).

120. Owen to John Lubbock, Apr. 20, 1839, RS, LUB. O. 66.

121. Ibid.

122. Herschel, *Manual of Scientific Enquiry*.

123. Owen, "Instructions for the Zoologist of the Zambesi Expedition," RS, MM 14 (16); "Zoological Instructions for the Naturalist of the Expedition to Vancouver's Island," RS, MM 14 (18).

124. Owen drew attention to this in the opening paragraph of his "Remarks on the 'Observations sur l'Ornithorhynque' par M. Jules Verraux," 317. On vertebrate paleontology and imperialism, see Buffetaut, *History of Vertebrate Palaeontology*, 162–31.

125. Fitton to Owen, May 11, 1838, BM(NH), L, OC 12 (fols. 231–32).

126. Gijzen, *'s Rijks Museum*, 86ff.

127. Gruber and Thackray, *Owen Commemoration*, 34ff. See also Andrews, *Southern Ark*, 101–22; Rev. Owen, *Life*, vol. 2, 24–25, 238–39. An instance to which Owen himself gave much prominence was the assistance of the colonial secretary of Mauritius, H. Sandwith, in acquiring the aye-aye; Owen, *Monograph on the Aye-aye*, 7–9.

128. Rev. Owen, *Life*, vol. 2, 45.

129. On May 17, 1834, Owen produced a handlist of all the specimens which Bennett by then had sent to the College of Surgeons; RCS, OP, Cabinet VIII (1) (b).

130. Bennett himself produced several accounts of Australasian natural history; see, e.g., his *Gatherings of a Naturalist.* Among the secondary sources are Moyal, "Sir Richard Owen"; Moyal, *Scientists in Nineteenth Century Australia*; Moyal, *Bright and Savage Land.*

131. Blumenbach, "Ueber das Schnabelthier."

132. Home, "Some Observations" (1800); Home, "Anatomy of *Ornithorhynchus paradoxus*" (1802). The 1802 article was also separately published (London, 1802).

133. Appel, "Henri de Blainville," 312–13.

134. The *Archiv für Anatomie und Physiologie*, edited first by J. F. Meckel and later by Johannes Müller, published both the French contributions, e.g., by Geoffroy Saint-Hilaire, in translation, and the German ones by von Baer and Meckel; see, e.g., 1827, 14–27, 568–76; 1830, 119–29; 1834, 51–52.

135. Grant to Geoffroy Saint-Hilaire, Sept. 14, 1829, in É. Geoffroy Saint-Hilaire, "Eier des Ornithorhynchus," 123–26.

136. This paper was never published. See Gruber's exhaustive "Does the Platypus Lay Eggs?"

137. [Lewes], "Life of Geoffroy St. Hilaire," 175.

138. Owen, "Marsupialia"; Owen, "Monotremata."

139. The story was told in some detail in the contemporary press; a clipping from the *Naturalist* occurs in BM(NH), L, OC 90, 76. See also Blunt, *Ark in the Park*, 129–30.

140. Timbs, *Yearbook of Facts* (1852), 207. This particular issue had a frontispiece that showed Owen holding a moa femur.

141. Mitchell, "Limestone Caves." Mitchell was not the only one to supply British museums with Australian cave bones or to write about the cave fauna; see Lang, "Discovery of Bone Caves," in which Lang interpreted the Wellington Valley caves as Bucklandian hyena dens.

142. Owen to Mitchell, May 8, 1838, in Mitchell, *Three Expeditions*, 365–69. Later cooperative contact is evident from, e.g., Owen to Mitchell, Mar. 29, 1840, Mitchell Library, T. L. Mitchell Papers, vol. 4, 61–64. See also Foster, *Sir Thomas Livingston Mitchell*; Holland, "Thomas Mitchell."

143. Owen, "On the Discovery of the Remains of a Mastodontoid Pachyderm"; Owen, "Additional Evidence."

144. Quoted in Branagan, "Richard Owen (a Review)," 98.

145. Owen, "Description of a Fossil Molar Tooth." See also his *Researches on the Extinct Mammals of Australia*, viii.

146. In addition to Mitchell, Charles Gould, son of Owen's friend John, placed Owen on the Australasian map by naming a peak in Tasmania after Owen, and Julius von Haast did the same in the South Island of New Zealand. See Branagan, "Richard Owen (a Review)," 98–99.

147. Owen, *Researches on the Extinct Mammals of Australia*, vol. 1, viii.

148. [Owen], "Mr. Cumming's *Hunter's Life*," 1–2. Charles Dickens's *Household Words*, vol. 5 (1852), 157, too, made the distinction between "hunter" and "student" and cited Cumming as an example of the former and Owen of the latter.

149. See R. Desmond, *The India Museum.*

150. For more on the period's museum culture, see among others Forgan, "Building the Museum"; Lenoir and Ross, "Naturalized History Museum"; Spary, *Utopia's Garden*; Yanni, *Nature's Museums.*

151. Owen's speech at the Hunterian anniversary dinner, RCS, Misc./Hunt. D (i).

152. Ritvo, *Animal Estate*, 205.

153. A. Desmond, *Politics of Evolution*, 135.

154. Owen, *On the Extent and Aims*, 126.

Chapter Three

1. The expression is taken, *mutatis mutandis*, from A. Desmond, *Politics of Evolution*, 365.

2. "Report of the Commissioners Appointed to Inquire into the Constitution and Government of the British Museum," *Parliamentary Papers*, 1850 (1170), vol. 24, 176 (128).

3. J. Clark, *Old Friends*, 376 ("His bows were not easily forgotten. His enemies said, and his friends could not deny, that they varied with the rank of the person to whom he was presented.") A. Desmond, *Archetypes and Ancestors*, 40 ("the Reverend Owen ruinously edited his letters to leave *The Life of Richard Owen* an interminable succession of bishops and dukes, skillfully manipulating material to help grandfather jockey his way into the highest echelons. No doubt the stereotype contains a good deal of truth: Owen seemed happiest among royals and curiously insecure among colleagues.")

4. M. Brock and Brent, "Oxford of Peel and Gladstone." For a discussion of the "Noetic school," see also Corsi, *Science and Religion*, 83–140.

5. Atlay, *Sir Henry Wentworth Acland*, 98, 144.

6. Owen, autograph notes for an obituary of Peel, BM(NH), L, OC 90, vol. 1.

7. Murchison to Owen, Jan. 25, 1850, American Philosophical Society. See also Gruber and Thackray, *Owen Commemoration*, 78–80.

8. Egerton to Owen, Mar. 3, 1859, BM, Add. MS 49,978.

9. Owen to his sister Eliza, Feb. 28, 1859, RCS, OC MS Add. 262 (letter 496). On Broderip, see [J. Parker], "William John Broderip."

10. See David Thomas Ansted to Owen, Dec. 19, 1841, BM(NH), L, OC 1 (fols. 155–56); Owen to his wife, Caroline, Dec. 27, [1841], BM, Add. MS 45,927 (fol. 38).

11. Owen to Buckland, Jan. 11, 1842, BM, Add. MS 40,499 (fol. 252); Buckland to Peel, Jan. 12, 1842, BM, Add. MS 40,499 (fol. 250).

12. Buckland to Peel, Oct. 4, 1842, in C. Parker, *Sir Robert Peel*, vol. 3, 445.

13. Peel to Owen, Nov. 1, 1842, BM, Add. MS 40,518 (fol. 24).

14. Owen to Peel, Nov. 1, 1842, in C. Parker, *Sir Robert Peel*, vol. 3, 445–46.

15. Whewell to Owen, Nov. 3, 1842, BM(NH), L, OC 26 (fols. 283–84); also Rev. Owen, *Life*, vol. 1, 204–5.

16. Owen to his sister Maria, June 17, 1844, RCS, Owen Misc. Letters (118–93) (no. 152). Nearly four decades later, Owen wrote another account of this memorable dinner, for inclusion in a biography of Sydney Smith: Owen to [Stuart J. Reid], Oct. 15, 1883, in S. Reid, *Rev. Sydney Smith*, 368–70.

17. Owen to his sister Maria, June 17, 1844, RCS, Owen Misc. Letters (118–93) (no. 152).

18. Peel to Buckland, n.d., in Rev. Owen, *Life*, vol. 1, 247.

19. Owen to Peel, Dec. 26, 1844, BM, Add. MS 40,446 (fol. 294).

20. Owen to his sister Eliza, Oct. 9, 1846, RCS, Owen Misc. Letters (118–93) (no. 60).

21. "Men of the day. No. 57. Professor Owen," *Vanity Fair*, Mar. 1, 1873, 20.

22. A clipping of the newspaper story, with Owen's autograph annotations, occurs in his autobiographical scrapbook, BM(NH), L, OC 24.

23. Owen to Playfair, June 28, 1883, in T. Reid, *Lyon Playfair*, 327–29. Also Owen to Gladstone, n.d., BM, Add. MS 44,485 (fol. 156): "I feel, in this relation, towards you, as I did towards Sir Robert Peel, who, after honouring me with occasional attendances at my Hunterian Lectures, was pleased to supplement the salary allotted by the Royal College of Surgeons, by the Pension which you have done me the honour and favour to augment."

24. Passages in Timbs's *Club Life of London* are taken verbatim from [T. Taylor], "Clubs of London."

25. Cowell, *The Athenaeum*, 11.

26. Huxley's X Club has attracted more than its fair share of historians' interest: Barton, "'An Influential Set of Chaps'"; Barton, "Huxley, Lubbock"; Jensen, "X Club"; MacLeod, "X-Club."

27. Rev. Owen, *Life*, records several meetings of the Literary Society dining club, e.g., vol. 1, 225–26, 261–62, 302–5, 327–28.

28. Timbs, *Club Life of London*, vol. 1, 206–7.

29. Tennyson to the Duke of Argyll, Feb. 17, 1865, in Tennyson, *Alfred Lord Tennyson*, vol. 2, 20.

30. Owen to Taylor, Feb. 2, 1878 [1879], Bodleian Library, MS Eng. lett., d. 13 (fol. 252).

31. Henry Reeve to Owen, Apr. 16, 1886, in Rev. Owen, *Life*, vol. 2, 265.

32. See n. 30 above.

33. Rev. Owen, *Life*, vol. 2, 73. Among his dining club companions over the years, Owen recorded the Duke of Argyll; the bishop of London Charles J. Blomfield; Lord Clarendon; the politician David Dundas; the historian James Anthony Froude, editor of *Fraser's Magazine* (1860–74); Gladstone; the historian of Greece George Grote; Hallam; the headmaster and later provost of Eton B. C. Hawtrey; the colonial governor Edmund W. Head; the society physician Henry Holland; the statesman the Marquis of Lansdowne; the dean of Westminster and later dean of St. Paul's Henry H. Milman; the historian T. B. Macaulay; Peel's protégé and friend of the prince consort Thomas Pemberton Leigh; Henry Reeve, editor of the *Edinburgh Review* (1855–90); the master of the rolls Lord Romilly; Lord John Russell; the lexicographer William Smith, who was editor of the *Quarterly Review* (1867–93); the historian the Earl Stanhope; the fifteenth Earl of Derby Edward Henry Stanley; Mr. Stirling; Spencer Horatio Walpole, who was home secretary in several Derby ministries; and the master of Trinity College William Whewell. A few foreign luminaries, such as the duc d'Aumale, were also mentioned. See Rev. Owen, *Life*, vol. 1, 260–61, 396–97; vol. 2, 73, 169, 186–87, 189.

34. T. W. Reid, *Richard Monckton Milnes*, vol. 2, 277, 279, 406.

35. Morrell and Thackray, *Gentlemen of Science*, passim.

36. "Table Showing the Places and Times of Meeting of the British Association, with Presidents, Vice-Presidents, and Local Secretaries, from Its Commencement," *BAAS, Report 1858*, [xx–xxii].

37. Compare Rosse, "Address" (presentation of the Copley Medal to Owen), with [Broderip and Owen], "Progress of Comparative Anatomy"; [Broderip and Owen], "Generalizations of Comparative Anatomy."

38. Buckland to Peel, Jan. 12, 1842, BM, Add. MS 40,499 (fol. 250).

39. Rupke, *Great Chain of History*, 21–26; Rupke, "Oxford's Scientific Awakening." On the uses of natural theology, see further Brooke, "Natural Theology of the Geologists"; Blaisdell, "Natural Theology."

40. Buckland and Cuvier met, for example, on June 19, 1818, at the Royal College of Surgeons in London; RS, MS 251 (no. 4). See also Sarjeant and Delair, "Irish Naturalist."

41. See Rupke, *Great Chain of History*, 39. Connecting the deluge with the last of Cuvier's geological catastrophes had been done before by Robert Jameson in his introduction to Cuvier's *Essay on the Theory of the Earth*, v-ix.

42. See Morrell and Thackray, *Gentlemen of Science*, 226–29.

43. W. Buckland, "On the Fossil Remains of the Megatherium," 104–5; the complete manuscript version of this abbreviated report is in the Devon Record Office, Buckland Papers.

44. W. Buckland, *Geology and Mineralogy*, vol. 1, 142.

45. Sedgwick, *Discourse*, 23.

46. Whewell, *History*, vol. 3, 468.

47. [Brewster], "Life and Works."

48. Anon., "Fossil Reptiles of England," 519.

49. [Broderip and Owen], "Progress of Comparative Anatomy," 363.

50. Rosse, "Address," 103.

51. Bompas, *Life of Frank Buckland*, 157. These are not the only references to Owen as the "English Cuvier." From among the various other ones, see a review of Owen's *Lectures on the Comparative Anatomy and Physiology of the Vertebrate Animals* in the *Provincial Medical and Surgical Journal*, 1847, 73: "To our illustrious countryman, John Hunter, to the immortal Cuvier, and to Owen, upon whom the mantle of Cuvier has been gracefully described as having descended, are we indebted for the most valuable discoveries in this department of science."

52. Owen to Buckland, July 28, 1832, in Rev. Owen, *Life*, vol. 1, 64.

53. Buckland to Clift, n.d., in ibid., 65.

54. An assessment of the value of Owen's nautilus study is given by Ward, *In Search of Nautilus*, 26–29.

55. Curwen, *Journal of Gideon Mantell*, 212, 221.

56. Carlisle to Owen, n.d., in Rev. Owen, *Life*, vol. 1, 66.

57. W. Buckland, *Geology and Mineralogy*, vol. 1, 296.

58. Cited in H. Woodward, *Geological Society of London*, 117.

59. In Daubeny, *Fugitive Poems*, 88.

60. W. Buckland, *Geology and Mineralogy*, vol. 1, 297–98. See also Owen, *Lectures on the Comparative Anatomy of Invertebrate Animals* (1843 ed.), 329–30. For a recent functional interpretation of the nautilus, see Wells, "Living Fossil."

61. Rupke, *Great Chain of History*, 31–41.

62. In his Bridgewater Treatise, Buckland did publish a paleoecological reconstruction of pterodactyls; *Geology and Mineralogy*, vol. 2, pl. 26(P). This reconstruction may well have been an imitation of a reconstruction by G. A. Goldfuss; see Langer, "Frühe Bilder"; also Rupke, "Metonymies of Empire."

63. G. Cuvier, *Ossemens fossiles* (1812), vol. 3, 3–4; the English translation is taken from W. Buckland, *Geology and Mineralogy*, vol. 1, 81.

64. Owen, "Exhibition of a Bone of an Unknown Struthious Bird," 170–71; Owen, "Notice of a Fragment"; also Owen, "On the Bone of an Unknown Struthious Bird."

65. Owen, *Memoirs on the Extinct Wingless Birds*, vol. 1, v; also in Rev. Owen, *Life*, vol. 1, 149.

66. Owen, *Memoirs on the Extinct Wingless Birds*, vol. 1, v (he mistakenly mentions the year 1838).

67. Broderip to Buckland, Jan. 20, 1843, BM, Add. MS 38,091 (fol. 193).

68. Ibid.

69. [Broderip and Owen], "Progress of Comparative Anatomy," 402.

70. Gruber, "From Myth to Reality," 340.

71. Owen, "On an Extinct Genus"; Owen, "On *Dinornis Novae-Zealandiae*." See also *Annals and Magazine of Natural History* 12 (1843): 444–46. In Jan. 1843, Owen wrote to his sisters: "it [the moa] turns out, however, to have been much bigger" (Rev. Owen, *Life*, vol. 1, 208).

72. Gordon, *Life and Correspondence of William Buckland*, 181–82.

73. Curwen, *Journal of Gideon Mantell*, 225.

74. Pantin, *Science and Education*, 19–25.

75. John Rule to Owen, Oct., 18, 1839, Turnbull Library, MS Papers 3046; see also J. W. Harris to Rule, Feb. 28, 1837, ibid. Also Dobson, "Life and Achievements," 23–24.

76. Silliman, *Visit to Europe*, vol. 2, 439–40.

77. Owen, *Memoirs on the Extinct Wingless Birds*, vol. 1, passim.

78. Gruber, "From Myth to Reality," 342.

79. Owen, "On *Dinornis* (Part IV)."

80. Owen to De la Beche, Nov. 1, 1843, National Museum of Wales, De la Beche Archive.

81. Owen's *Memoir on the Dodo* carried a posthumously published "Historical Introduction" by Broderip.

82. [Broderip and Owen], "Progress of Comparative Anatomy," 403.

83. J. Clark, *Old Friends*, 373.

84. "Lectures by M. de Blainville on Comparative Anatomy," *Lancet*, no. 1 (1839–40): 221.

85. Ibid., 222.

86. Broderip to Buckland, Jan. 20, 1843, BM, Add. MS 38,091 (fol. 193).

87. Owen to Silliman, Mar. 16, 1843, in *American Journal of Science* 45 (1843): 185–87.

88. In a letter to his sister Maria, June 17, 1844, RCS, Owen Misc. Letters (118–93) (no. 152), Owen describes a dinner at Robert Peel's: "Sydney Smith and Buckland soon began to grow jocular, and opened on me about the *big-bird*. 'Ah!' said S.S., with reference to some remark on my joy at the safe arrival of the box from New Zealand, 'that was Owen's *magnum bonum*.'" See Rev. Owen, *Life*, vol. 1, 232.

89. Owen, "Fossil Mammalia." For a general history of the recovery of fossil mammals from South America, see Simpson, *Discoverers*.

90. Parish, "Account."

91. W. Buckland, *Geology and Mineralogy*, vol. 1, 60.

92. Owen and Parish, "Note on the Glyptodon." See also Owen, "Description of a Tooth and Part of the Skeleton of the *Glyptodon*."

93. Owen, "On the Megatherium." Earlier, Owen had presented to the BAAS "Notices of Some Fossil Mammalia of South America."

94. See Owen, *Memoir on the Megatherium*, 78–80.

95. W. Buckland, *Geology and Mineralogy*, vol. 1, 146.

96. Ibid., 160.

97. See Owen, *Memoir on the Megatherium*, 7–9.

98. Owen, *Description of the Skeleton of an Extinct Gigantic Sloth*, 146–47.

99. [Broderip and Owen], "Progress of Comparative Anatomy," 397.

100. See Morrell and Thackray, *Gentlemen of Science*, 551. Oxbridge encouragement of Agassiz is apparent from, among other things, [Broderip], "Agassiz on Fossil Fishes."

101. Owen, "Report on British Fossil Reptiles."

102. Owen, "Report on British Fossil Reptiles, Part II." A controversy has arisen over the precise date of Owen's introduction of the name "dinosaur"; was it 1841, when Owen delivered his second report on British fossil reptiles, or 1842, when the report was published? Or again, was it in 1841, in a published preprint of the report? See Torrens, "When Did the Dinosaur Get Its Name?" On Mantell's contributions, see Dean, *Gideon Mantell*.

103. Hume, *Learned Societies*, 39 of the supplement.

104. See Rudwick, *Scenes from Deep Time.*

105. Woodward, *Geological Society of London*, 117.

106. See Broderip, *Zoological Recreations*, 364.

107. "The Crystal Palace, at Sydenham," *Illustrated London News* 24 (Jan. 7, 1854): 22. See also Barber, *Heyday of Natural History*, 177–79.

108. "Gigantic Bird of New Zealand," *Illustrated London News* 24 (Jan. 7, 1854): 22.

109. "Men of the day. No. 57. Professor Owen," *Vanity Fair*, Mar. 1, 1873.

110. See, e.g., Casier, *Iguanodons de Bernissart.*

111. A. Desmond, "Designing the Dinosaur."

112. Dear, "Invention of the Dinosaur," 33.

113. [Owen], "Dinosauria," in Brande and Cox, *Dictionary of Science*, vol. 1, 684. The generic definition, however, was mainly founded on the teeth; see [Owen], "Iguanodon," ibid., vol. 2, 192, and "Megalosaurus," ibid., 488.

114. See Rupke, *Great Chain of History*, 175–76.

115. See Bowler, *Invention of Progress*, 46–47.

116. [Owen], "Ancient Animals in South America," 197.

117. Ibid., 200.

118. Rev. Owen, *Life*, vol. 1, 373–74.

119. BM(NH), L, OC 39.

120. W. Buckland, *Geology and Mineralogy*, vol. 1, 249–53.

121. See ibid., 252–53.

122. Owen, "On the Teeth of *Labyrinthodon*"; Owen, "Description of Part of the Skeleton and Teeth of *Labyrinthodon.*"

123. Hitchcock, "Footmarks of Birds (Ornithichnites)."

124. See Mantell to James Deane, Feb. 13, 1843, in "Ornithichnites of the Connecticut River Sandstones and the Dinornis of New Zealand," *American Journal of Science and Arts* 45 (1843): 184–85.

125. Ibid.

126. Broderip to Buckland, Jan. 20, 1843, BM, Add. MS 38,091 (fols. 193–94).

127. Owen to Silliman, Mar. 16, 1843, in "Ornithichnites of the Connecticut River Sandstones and the Dinornis of New Zealand," *American Journal of Science and Arts* 45 (1843): 187.

128. [Owen], Descriptive and Illustrated Catalogue of the Fossil Organic Remains of Mammalia and Aves, 376–81.

129. [Harkness], "Fossil Footprints."

130. Owen, *Palaeontology*, 293.

131. Anon., "Popular Course of Geology," 7.

132. Logan, "On the Occurrence"; Logan, "On the Footprints."

133. Lyell, "Anniversary Address, 1850."

134. Lyell, "Anniversary Address, 1851."

135. Owen to Lyell, n.d., in ibid., lxxvi.

136. Hopkins, "Anniversary Address," lxxi.

137. Ibid., lxxx. See Owen, "Description of Protichnites from the Potsdam Sandstone."

138. [Broderip and Owen], "Generalizations of Comparative Anatomy," 47.

139. [Owen], "Lyell—on Life," 417.

140. Ibid., 438. This is reminiscent of Sedgwick's remark about how Lyell "in the language of an advocate" had forgotten "the character of an historian"; Sedgwick, "Anniversary Address, 1831," 303.

141. *Proceedings of the Geological Society* 1 (1834): 439–43.

142. Lyell to Owen, n.d., in Rev. Owen, *Life*, vol. 1, 372–73.

143. Owen named the fossil in a letter of Dec. 15, published under the heading "Vertebrate Air-Breathing Life in the Old Red Sandstone," *Literary Gazette*, Dec. 20, 1851, 900. Mantell's resentment is apparent from Curwen, *Journal of Gideon Mantell*, 278.

144. Benton, "Progressionism in the 1850s."

145. *Hansard's Parliamentary Debates*, vol. 166 (May 19, 1862), col. 1915.

146. Ostrom, "Meaning of *Archaeopteryx*," 174.

147. [Owen], "Lyell—on Life," 450.

148. Agassiz, "On the Differences."

Chapter Four

1. Rev. Owen, *Life*, vol. 1, 327. Owen had met Guizot once before, informally, in 1840, at the zoo; ibid., 173.

2. See, e.g., Gillispie, *Genesis and Geology*, 217ff.

3. Rupke, *Great Chain of History*, 265.

4. W. Buckland, *Vindiciae Geologicae*, 5.

5. *Dictionary of National Bibliography*, s.v. "Acland, Sir Henry Wentworth."

6. [Mozley], "Oxford Commission," 168.

7. Playfair, "Study of Abstract Science," 48.

8. Rupke, *Great Chain of History*, 274. On Frank Buckland, see Burgess, *World of Frank Buckland*.

9. Rehbock, *Philosophical Naturalists*, 56.

10. See also Kaplan, *Thomas Carlyle*; Ashton, *German Idea*, chaps. 2–3.

11. Carlyle, *Oliver Cromwell's Letters*.

12. Pym, *Memories of Old Friends*, vol. 1, 264. On Carlyle's influence on the scientific naturalists, see J. Moore, "1859 and All That"; F. Turner, "Victorian Scientific Naturalism." See further C. Moore, "Carlyle and Goethe."

13. Rev. Owen, *Life*, vol. 1, 268–69, 283. Owen's interest in Milton was apparent from, among other things, an article, "Milton and Galileo," that he published in 1869.

14. Rev. Owen, *Life*, vol. 1, 261.

15. Owen to his wife, Caroline, Oct. 12, 1860, BM, Add. MS 45,927 (fol. 40).

16. Owen, "Speech at the Literary Fund Dinner, May 8, [1863?]," RCS, Misc. MS/Q (viii).

17. Carlyle, *On Heroes, Hero-Worship*, 80.

18. Pym, *Memories of Old Friends*, vol. 2, 59–60.

19. Carlyle to John Sterling, Aug. 29, 1842, in Ryals and Fielding, *Letters of Thomas and Jane Carlyle*, vol. 15, 55. For more on the Carlyle-Owen connection, see Ulrich, "Thomas Carlyle, Richard Owen."

20. Carlyle to Owen, Sept. 15, 1862, in Rev. Owen, *Life*, vol. 2, 55–56.

21. Howarth, *British Association*, 91. See also MacLeod, "X-Club," 306–7.

22. Brian G. Gardener, "Forbes, Owen and the Red Lions," abstracts of papers, Richard Owen Centenary Meeting, Natural History Museum (London, 1992), 10.

23. Cannon, *Science in Culture*, 49.

24. Whewell, "Comte and Positivism."

25. Morrell and Thackray, *Gentlemen of Science*, 21.

26. Ospovat, "Perfect Adaptation."

27. See the extensive list of references on comparative anatomy in Carus, *Lehrbuch der Zootomie*, vii–xx. For a general history of nineteenth-century comparative anatomy, see Bernard Balan's impressive *L'ordre et le temps.*

28. See Rupke, "Caves." On Camper, see Visser, *Zoological Work of Camper.*

29. On Cuvier, see Outram, *Georges Cuvier*; Coleman, *Georges Cuvier.*

30. See, e.g., Bourdier, "Geoffroy Saint-Hilaire versus Cuvier."

31. Appel, *Cuvier-Geoffroy Debate.*

32. The story is told by, among others, Lewes, *Life of Goethe*, 550–51. On Goethe's interest in the Academy debate, see Kuhn, *Empirische und ideelle Wirklichkeit.*

33. See, e.g., A. Desmond, *Politics of Evolution*, 246–48.

34. Owen to Buckland, Jan. 11, 1842, BM, Add. MS 40,499 (fol. 252).

35. Owen, *Odontography*, vol. 1, lxvii.

36. Owen, "On the Structure of Fossil Teeth from Old Red Sandstone"; Owen, "On the Structure of the Teeth of *Dendronus strigatus.*" On the origin of the Microscopical Society, see [Cooper], "Brief Sketch of Microscopic Science." See also *Athenaeum*, Mar. 7, 1840, 194.

37. Owen, *Odontography*, vol. 1, xli.

38. Ibid., xlix–lvi. See also the lengthy "Mr. Owen and Mr. Nasmyth," *Lancet*, no. 2 (1839–40): 486–93; and [Broderip and Owen], "Progress of Comparative Anatomy," 387–92.

39. Private communication, Wolf Reif, Institut für Geologie und Paläontologie, Universität Tübingen.

40. [Owen], "Anatomy," in Brande and Cox, *Dictionary of Science*, vol. 1, 97. See also Owen, *Lectures on the Comparative Anatomy of Invertebrate Animals* (1843 ed.), 5–6.

41. [T. Huxley], "Owen and Rymer Jones." For a more recent and more favorable comparison of Owen with Siebold and Stannius, see Nordenskiöld, *History of Biology*, 417–18.

42. "Professor Owen's Comparative Anatomy," *London Review*, Apr. 28, 1866, 482.

43. [M. Foster], "Higher and Lower Animals," 381–82.

44. On Müller and his school, see Rothschuh, *Geschichte der Physiologie*, 112–49.

45. J. Carus, *Geschichte der Zoologie*, 638–40.

46. Van der Hoeven to Owen, Jan. 9, 1850, in Klaauw, "Scientific Correspondence," 338.

47. See Rupke, *Great Chain of History*, 24–26, 31–74.

48. Owen to Whewell, Oct. 31, 1837, Trinity College Library, Add. MS a. 210 (no. 54).

49. Owen, Hunterian lecture, Mar. 30, 1841, RCS, Owen Papers.

50. Owen to Silliman, Dec. 6, 1846, in Fisher, *Life of Silliman*, vol. 2, 172. See also Brown, *Benjamin Silliman.*

51. Ruse, *Darwinian Revolution*, 116.

52. Ospovat, *Development of Darwin's Theory*, 129.

53. Sloan, *Richard Owen*, 37–51; see also Sloan's "Whewell's Philosophy of Discovery."

54. A. Desmond, *Politics of Evolution*, passim. R. J. Richards takes issue with Desmond's social reductionism, in particular with respect to Green's position; see his *Meaning of Evolution*, 78 and passim.

55. "My duties in the Museum of the Rl. College of Surgeons, especially in preparing the Catalogue of Comparative Osteology, led me to test the validity of homologizing in a higher or more abstract degree than the 'special' route; and, with concomitant study of development

of skeletons, enforced an advance of thought beyond the law of adaptation or conditions of existence." Owen's autobiographical sketch of the road that led him to the theory of organic evolution, BM(NH), L, OC 90, vol. 3.

56. Powell and Owen, "Abstract of Professor Owen's View of Vertebral Structure and Its Archetype." The French translation of *On the Archetype* was carried out by Owen himself, "with the aid of a Parisian friend" in London, and completed, at least in first draft, as early as the summer of 1848; see Owen to Milne Edwards, July 10, 1848, American Philosophical Society.

57. Whewell, *Novum Organon Renovatum*, 351–52.

58. Owen, *On the Archetype*, table 3. This table was also included in [Owen], *Descriptive Catalogue of the Osteological Series*, vol. 1, xxxviii. A table with synonyms of the bones of the head of fishes was already printed in Owen, *Lectures on Comparative Anatomy of Vertebrate Animals*, 158–59.

59. Owen, *On the Archetype*, 7. See also the glossary appended to Owen's *Lectures on Comparative Anatomy of Invertebrate Animals*, 374 and 379.

60. Owen, *On the Archetype*, 7. Mivart depicted the *Draco volans* in his *Genesis of Species*, 178, and in his booklet *Common Frog*, 33. Owen featured it also in *On the Anatomy of Vertebrates*, vol. 1, 58.

61. Westwood, "Illustrations," 409. See also Blaisdell, "Natural Theology."

62. Owen, *On the Archetype*, 73.

63. Ibid., 73–77.

64. Owen, *On the Nature of Limbs*, 41.

65. Owen, *On the Archetype*, 133.

66. Ibid., 114.

67. Ibid., 168.

68. Ibid., 170.

69. Ibid., 81. See also Owen, *On the Anatomy of Vertebrates*, vol. 1, x, where Owen defines a vertebrate as "a clothed sum of segments."

70. Owen, *On the Archetype*, 165.

71. Ibid., 133; Owen, *On the Nature of Limbs*, 70.

72. H. Miller, *Footprints of the Creator*, 69.

73. [Broderip and Owen], "Generalizations of Comparative Anatomy," 80.

74. Owen, *On the Nature of Limbs*, 10.

75. Ibid., 40.

76. See, e.g., Agassiz, "Observations on the Blind Fish," 127–28.

77. See Owen, *Memoir on the Dodo*, 39.

78. Owen, *On the Archetype*, 171.

79. Ibid.

80. A. Desmond, *Politics of Evolution*; Ospovat, "Perfect Adaptation"; Ospovat, *Development of Darwin's Theory*; Rehbock, *Philosophical Naturalists*; E. Richards, "'Metaphorical Mystifications'"; Sloan, *Richard Owen*; Sloan, "Whewell's Philosophy of Discovery"; Jacyna, "Romantic Programme"; R. Richards, *Romantic Conception of Life*.

81. Rehbock, *Philosophical Naturalists*, 56.

82. Knox, *Great Artists*, 212. See also Knox's partly autobiographical "Philosophy of Zoology." See further Lonsdale, *Life and Writings of Knox*.

83. Notes by Knox to his translation of the "Lectures by M. de Blainville on Comparative Osteology," *Lancet*, no. 1 (1839–40): 222, 191.

84. Knox, *Great Artists*, 206.

85. In addition to A. Desmond, *Politics of Evolution*, see his "Robert E. Grant."

86. Robert E. Grant, "Lectures on Comparative Anatomy," *Lancet*, no. 2 (1833–34): 1034.

87. See Rehbock, *Philosophical Naturalists*, 59–60.

88. Barry, "On the Unity of Structure," 116.

89. Barry, "Further Observations," 348.

90. Ibid., 354.

91. Sharpey, "Anatomy and Physiology," 492. See also D. Taylor, "William Sharpey."

92. Sloan, *Richard Owen*, 14–37.

93. Owen to John Simon, n.d., in Simon's "Memoir" to Green, *Spiritual Philosophy*, vol. 1, xiv–xv.

94. Sloan, *Richard Owen*, 6–7.

95. Ibid., 39.

96. Rev. Owen, *Life*, vol. 1, 200.

97. Rehbock, *Philosophical Naturalists*, 68–69.

98. Forbes, "On the Morphology of the Reproductive System."

99. Ibid., 978.

100. R. Murchison, "Address," xxxiv.

101. G. Argyll, "Address," lxxvi.

102. Daubeny, "Address," lxiv, lxxii.

103. T. Huxley, "Owen's Position," 319.

104. The designation "morphological period" was introduced by J. Carus, *Geschichte der Zoologie*, 573–726. It was given wider currency by Merz, *European Scientific Thought*, 200–275. This and the following section are taken with few alterations from Rupke, "Richard Owen's Vertebrate Archetype."

105. An account of Owen's two-part report appeared in *Athenaeum*, Sept. 19, 1846, 968–69; Sept. 26, 1846, 1004–5.

106. *BAAS, Report 1846*, 339–40.

107. Owen, *Lectures on Comparative Anatomy of Vertebrate Animals*, 41 and passim.

108. Maclise, "Skeleton," 623.

109. Maclise, "On the Nomenclature," 300.

110. Owen, *On the Archetype*, 177.

111. Owen, *Memoir on the Pearly Nautilus*, 2. On different concepts of type, see Farber, "Type-Concept in Zoology."

112. See Haupt, "Homologieprinzip bei Richard Owen," 160; A. Desmond, *Politics of Evolution*, 342.

113. Maclise, *Comparative Osteology*, text to go with pls. 15, 16.

114. Goodsir, "Anatomy of *Amphioxus lanceolatus*." Owen's interest in *Amphioxus* as a simple vertebrate is apparent from Owen to Whewell, Oct. 30, 1846, Trinity College Library, Whewell Papers, O. 15. 48 (39).

115. [Owen], "Archetype," in Brande and Cox, *Dictionary of Science*, vol. 1, 146.

116. Owen, *On the Archetype*, 189.

117. Ibid., 76 ("He, however, becomes the true discoverer who establishes the truth: and the sign of the proof is the general acceptance."). See also Owen, *Odontography*, xlix.

118. Owen, *On the Archetype*, 176.

119. [Broderip and Owen], "Generalizations of Comparative Anatomy," 78–79.

120. Owen, *On the Archetype*, 72–80, 164–71, and passim.

121. [Owen], "Richard Owen, M.D., D.C.L., F.R.S., Superintendent of the Natural History Departments, British Museum," in Owen's autobiographical scrapbook, BM(NH), L, OC 24, 13.

122. [Broderip and Owen], "Generalizations of Comparative Anatomy," 82.

123. Ibid., 74.

124. Rev. Owen, *Life*, vo1.1, 130–39.

125. C. Carus, *England und Schottland*, 115–26. See also Neuburger, "C. G. Carus."

126. E.g., Singer, *Short History of Biology*, 219; Rudwick, *Meaning of Fossils*, 211. This view has never been seriously challenged. W. C. Williams, however, in his entry on Owen in the *Dictionary of Scientific Biography* (New York, 1974), vol. 10, 260–63, correctly mentions Goethe and Carus, in addition to Oken, as formative influences on Owen. See further Gladwin, "Influence of Oken."

127. Owen, "Oken, Lorenz" (1858), 498–503.

128. Owen, untitled manuscript history of his contributions to the theory of organic evolution, BM(NH), L, OC 90, vol. 3, 179–90, on p. 179.

129. Trevor-Roper, *Hermit of Peking*, 344.

130. Lovejoy, *Great Chain of Being*, 279–80.

131. A fairly recent edition, with commentary, of Goethe's morphological writings is by Dorothea Kuhn, in Goethe, *Sämtliche Werke*, vol. 24, *Naturkundliche Schriften II: Schriften zur Morphologie*. See also Dobel, *Lexikon der Goethe-Zitate*, 961–62; Haecker, *Goethes morphologische Arbeiten*.

132. From among the various Carus studies, see Kern, *Carl Gustav Carus*; Prause, *Carl Gustav Carus*; Genschorek, *Carl Gustav Carus*; Meffert, *Carl Gustav Carus*.

133. It was translated by R. T. Gore into English: C. Carus, *Introduction to Comparative Anatomy*.

134. C. Carus, *Von den Ur-Theilen*, viii–ix.

135. Ibid., tables 4–7.

136. [Silliman], "Bibliography," *American Journal of Science*, July 1849, 153.

137. Owen, *On the Anatomy of Vertebrates*, vol. 3, 788.

138. Lewes, *Life of Goethe*, 357.

139. C. Carus, *Von den Ur-Theilen*, xiv.

140. Examples range from Haupt, who called his study of Owen a contribution to the history of biological Platonism, to A. Desmond, *Politics of Evolution*, 216 and passim.

141. Discussion on Plato's theory of ideas can be found in most general philosophy texts. On the image of the cave in Plato's *Republic*, see Plato, *Collected Dialogues*, 747ff.

142. Owen, *On the Anatomy of Vertebrates*, vol. 3, 807.

143. See Gode-von Aesch, *Natural Science*, 15.

144. [Owen], "Lyell—on Life," 450–51.

145. Ruse, *Darwinian Revolution*, 122.

146. Owen, "Report on the Archetype," 339; Owen, *On the Archetype*, 171.

147. Owen, *On the Archetype*, 172.

148. Owen, *On the Nature of Limbs*, 2.

149. Ibid., 83–84; Owen, *Principes d'ostéologie comparée*, 12.

150. "I also dare flatter myself that some day these attempts will bring to the Platonic philosophy and will facilitate the understanding of the idea of which the more than humanly gifted philosopher sketched the first outline" (Owen, *Principes d'ostéologie comparée*, 427–28).

151. Owen to his sister Maria, Nov. 7, 1852, RCS, OC, MS Add. 262 (letter 387).

152. Acland, *Harveian Oration*, 12.

153. The connection was made by Singer, *Short History of Biology*, 212–19.

154. Schopenhauer cited in particular Owen's *Principes d'ostéologie comparée*; see *World as Will and Idea*, vol. 2, 91–92; vol. 3, 131–32.

155. Glanvill, *Scepsis Scientifica*, 22; Watts, *Logick*, 13–14; Samuel Johnson, *Dictionary of the English Language* (London, 1819), s.v. "Archetype."

156. Sloan, *Richard Owen*, 15–39; Sloan, "Darwin." The Schelling-influenced Jena school has not yet received the same scholarly attention as the Kantian Göttingen school; on the latter, see Lenoir, "Göttingen School"; Lenoir, *Strategy of Life*; Lenoir, "Morphotypes."

157. Müller, *Elements of Physiology*, vol. 1, 25; see also vol. 2, 1338–39.

158. Green, *Vital Dynamics*, xxiv–xxviii.

159. Ibid., xxvi.

160. "Goethe was the Spinoza of poetry. . . . There can be no doubt that Goethe entirely subscribed to the teaching of Spinoza." "For the concept of *Naturphilosophie* is of course in essence nothing more than that of Spinoza, pantheism." Heine, *Über Deutschland*, 110. See further MacFarland, *Coleridge*.

161. The alarm was raised in particular by the *Manchester Spectator*. Clippings from Dec. 8 and 22 occur in Owen's personal, interleaved, and annotated copy of his *On the Nature of Limbs*, BM(NH), L, OC 18. See also E. Richards, "Question of Property Rights," 165–67.

162. Sedgwick, *Discourse*, supplement to the appendix, 230.

163. Ibid., 248.

164. Ibid., 253.

165. Ibid., preface, cxcix.

166. Whewell, "Second Memoir on the Fundamental Antithesis of Philosophy."

167. E.g., Whewell, "On the Platonic Theory of Ideas."

168. Whewell, *History*, vol. 1, 341.

169. See e.g., Raine, "Thomas Taylor."

170. Yeo, "William Whewell." For more on Whewell, see Fisch, *William Whewell*; Fisch and Schaffer, *William Whewel.*

171. Conybeare to Owen, n.d., BM(NH), L, OC 8 (374–75). The subject matter makes it clear that the letter was written in reaction to Owen's "Report" or to his book *On the Archetype.*

172. [Duns], "Professor Owen's Works," 344.

173. Daubeny, "Address," lxiv. See also [McCosh], "Typical Forms."

174. Owen to Acland, Dec. 28, 1848, Bodleian Library, MS Acland, d. 64 (fols. 135–36).

175. A. Desmond, *Archetypes and Ancestors*, 202 and passim; A. Desmond, *Politics of Evolution*, 216 and passim.

176. Owen, Hunterian lectures, May 1837, in Sloan, *Richard Owen*, 191.

177. T. Huxley, "Owen's Position," 320.

178. T. Huxley, "On the Theory of the Vertebrate Skull," 584–85.

179. Ibid., 585.

180. Ibid., 571.

181. T. Huxley, "Explanatory Preface," 131.

182. Spencer, *Works*, vol. 21, 24–25.

183. [Spencer], "Ultimate Laws of Physiology," 336.

184. [Spencer], "Owen on the Homologies," 412.

185. Ibid., 415.

186. Ibid.

187. Ibid., 416. See also Spencer, *Principles of Biology*, in *Works*, vol. 3, 123: "Everything, then, goes to show that the segmental composition which characterises the apparatus of external relation in most *Vertebrata*, is not primordial or genetic, but functionally determined or adaptive. Our inference must be that the vertebrate animal is an aggregate of the second order, in which a relatively superficial segmentation has been produced by mechanical intercourse with the environment."

188. Lewes, *Life of Goethe*, 360.

189. Lewes, *History of Philosophy*, vol. 1, lxxx.

190. Ibid., lxxxiv.

191. Ibid., lxxxv.

192. Ibid., lxxxvi.

193. Ibid., lxxxvi–lxxxvii.

194. Ibid., lxxxviii.

195. Lewes, "Reign of Law," 108–9.

196. Darwin to Owen, Feb. 12, 1847, RCS, 257.h.5/7 (23).

197. Darwin to Owen, n.d., in Rev. Owen, *Life*, vol. 1, 209. According to Rev. Owen, the letter was written in January 1843. This seems too early by five years. See also Darwin to Owen, Feb. 12, 1847: "Those vertebrae are awfully difficult to understand"; RCS, 257.h.5/7 (23).

198. The copy is part of the Darwin Library, held in Cambridge University Library.

199. C. Darwin, *Origin of Species*, 378–79.

200. E. Russell, *Form and Function*, 247; Ospovat, *Development of Darwin's Theory*, 117.

201. See Stuart A. Kauffman, "Biological Homologies and Analogies," in Philip P. Wiener, *Dictionary of the History of Ideas* (New York: Scribner, 1973–74), vol. 1, 236. See also Jardine, "Concept of Homology in Biology."

202. *Medical Times and Gazette*, no. 1 (1863): 35.

203. Darwin to Gray, June 8, 1860, in F. Darwin and Seward, *More Letters of Darwin*, vol. 1, 153.

Chapter Five

1. Ospovat, "God and Natural Selection," 193.

2. Owen, *On the Anatomy of Vertebrates*, vol. 3, 789. In his autobiographical account of the steps which led to his belief in an origin of species by natural law, Owen stated: "The works 'On the Archetype and Homologies of the Vertebrate Skeleton' and 'On the Nature of Limbs' issued in an abandonment of the hypothesis of the origin of species by primary law or direct exercise of creative force limited in time, and in the proposition, based mainly on a recognition of the skeleton of all vertebrates consisting of a series of essentially similar segments, that such vertebrate species exemplified a continuous operation of a secondary cause or law, which had operated working not only successively but progressively." BM(NH), L, OC 90, vol. 3, 180.

3. Examples include Wendt, *In Search of Adam*; Glass, Temkin, and Straus, *Forerunners of Darwin*; Irvine, *Apes, Angels and Victorians*; Himmelfarb, *Darwin and the Darwinian Revolution* (for its time, Himmelfarb's book was one of the better studies); Beer, *Charles Darwin*; Hull, *Darwin and His Critics*. On the "Darwin industry," see Oldroyd, "How Did Darwin Arrive at His Theory?"

4. MacLeod, "Evolutionism and Richard Owen"; Ross, "Survey"; Ospovat, *Development of Darwin's Theory*; Dobson, "Life and Achievements"; A. Desmond, *Archetypes and Ancestors*;

A. Desmond, *Politics of Evolution*; E. Richards, "Question of Property Rights"; E. Richards, "Political Anatomy of Monsters"; Sloan, *Richard Owen*; Sloan, "Whewell's Philosophy of Discovery"; Camardi, "Richard Owen."

5. Owen, "Report on British Fossil Reptiles, Part II," 196.

6. Ibid., 202.

7. Owen, "On British Fossil Reptiles."

8. Murchison to Owen, Apr. 2, 1845, in Rev. Owen, *Life*, vol. 1, 254.

9. [Sedgwick], "Natural History of Creation."

10. Owen to Whewell, Feb. 14, [1845], in Brooke, "Owen, Whewell, and the *Vestiges*," 142; see also Rev. Owen, *Life*, vol. 1, 253. For correspondence with Sedgwick, see Rev. Owen, *Life*, vol. 1, 255; J. Clark and Hughes, *Adam Sedgwick*, vol. 2, 87.

11. Owen to "Author of 'Vestiges,'" n.d., in Rev. Owen, *Life*, vol. 1, 249–52.

12. Owen to Whewell, Feb. 24, [1845], in Brooke, in Brooke, "Owen, Whewell, and the *Vestiges*," 142.

13. Brooke, "Owen, Whewell, and the *Vestiges*," 134, 138.

14. A. Desmond, *Archetypes and Ancestors*, 31–37; Ruse, *Darwinian Revolution*, 124–25; E. Richards, "Question of Property Rights."

15. A. Desmond, *Archetypes and Ancestors*, 210–11. For a full account of the issue of the authorship of *Vestiges*, see Secord, *Victorian Sensation*, 364–400.

16. Owen to "Author of 'Vestiges,'" n.d., in Rev. Owen, *Life*, vol. 1, 249–50.

17. Rev. Owen, *Life*, vol. 1, 255.

18. Owen, "Report on the Extinct Mammals of Australia," 240.

19. See Browne, *Secular Ark*, 97–98. See also Kinch, "Geographical Distribution."

20. Pentland, "Fossil Bones of Wellington Valley."

21. *Literary Gazette*, 1846, 847.

22. Forbes presented his ideas in a report, "On the Distribution of Endemic Plants," to the BAAS. In 1846 he gave a course of lectures at the London Institution entitled "The Geographical and Geological Distribution of Organised Beings." In the same year, he delivered a lecture at the Royal Institution, "On the Question, Whence and When Came the Plants and Animals Now Inhabiting the British Isles and Seas?" in which he postulated a Pleistocene dispersal of plants from the Continent to the British Isles by means of icebergs and an earlier, Pliocene landmass connecting Ireland to Spain; *Athenaeum*, Mar. 7, 1846, 247–48. These ideas were more fully developed in his "On the Connexion between the Distribution of the Existing Fauna and Flora of the British Isles and the Geological Changes Which Have Affected Their Area." He brought this innovative bio- and paleogeographical work to bear on his theory of centers of creation, and in 1849, for example, he delivered a lecture at the Royal Institution entitled "Have New Species of Organised Beings Appeared since the Creation of Man?"; *Athenaeum*, Mar. 24, 1849, 304. For a discussion of Forbes's significance, see Rehbock, *Philosophical Naturalists*, passim.

23. The theory of the spontaneous origin of species is discussed in Rupke, "Neither Creation nor Evolution"; and Rupke, "The Origin of Species from Linnaeus to Darwin."

24. Forbes, "Abstract of the Theory of Specific Centres."

25. Prichard, *Physical History of Mankind*, vol. 1, 96–97. Lyell discussed the theory in his *Principles of Geology*, 3rd ed., vol. 3, 29–30, 92.

26. [Owen], "Broderip's *Zoological Recreations*," 124.

27. On the theory of the autogenous origin of species, see Rupke, "Neither Creation nor Evolution." A fuller monograph account is in preparation.

28. Owen to [John Chapman], early 1848, in Rev. Owen, *Life*, vol. 1, 310. This was the first of two letters, written about a fortnight apart.

29. A. Desmond also states that Owen's claim "might have been a bluff, an attempt at self-aggrandisement" (*Archetypes and Ancestors*, 34).

30. [Owen], "Species," in Brande and Cox, *Dictionary of Science*, vol. 3, 524.

31. See Winsor, *Starfish*, 44–72.

32. Anon., "Steenstrup on the Alternation of Generations," 34.

33. Owen, *On Parthenogenesis*, 5. See also a report of Owen's two introductory lectures, "Generation and Development of the Invertebrated Animals," in *Medical Times* 19 (1848–49): 467–68, 483–85.

34. Mivart, "Beginnings and End of Life," 370. This was repeated in [Mivart], "Century of Science," 400. See also Mivart, *Essays and Criticisms*, vol. 2, 384.

35. Owen to his sister Eliza, [May] 1849, in Rev. Owen, *Life*, vol. 1, 338. See also Henry Holland to Owen, June 12, 1849, CUL, Add. 5354 (fol. 32).

36. Owen to Murray, Dec. 29, 1856, RCS, Add. MS 262 (letter 462).

37. Anon., review of Owen and Siebold on parthenogenesis, *Midland Quarterly Journal of the Medical Sciences*, Jan. 1858, 1–19.

38. [Owen], "Metagenesis," in Brande and Cox, *Dictionary of Science*, vol. 2, 501.

39. Owen, "Description of a Microscopic Entozoon." See also Owen, "Entozoa." James Paget believed that he had discovered the new entozoon and given the first account of it, even though Owen gave it its name; however, he regarded Owen's personal friendship of greater value than public recognition of his priority, so he decided not to press his claim; see S. Paget, *Memoirs and Letters*, 55–56. See also Campbell, "History of Trichinosis."

40. Owen, "Address" (1858), lxxv.

41. A. Desmond, *Archetypes and Ancestors*, 37.

42. T. Huxley, "Upon Animal Individuality."

43. C. Lyell, *Geological Evidences of the Antiquity of Man*, 421.

44. [Duns], "Genesis and Science," 346–47; [Duns], "Professor Owen's Works," 324–25.

45. Owen, "On Metamorphosis and Metagenesis."

46. Q. in the Corner, "Impromptu on Hearing a Certain Discourse on 'Metamorphosis' and 'Metagenesis,'" *Lancet*, no. 1 (Mar. 22, 1851): 314.

47. Nägeli, *Individualität in der Natur*, 39.

48. Kölliker, *Ueber die Darwinsche Schöpfungstheorie*, 13. The translation into English is from Anon., "Criticism on: 'The Origin of Species,'" *Natural History Review*, 1864, 574.

49. A rare Englishman to refer to metagenesis as a mode of evolution, albeit in fiction, was Charles Kingsley, who cited the "transformation" of jellyfish and other creatures to illustrate how "water babies" may have come into being: Kingsley, *Water Babies*, 77–78.

50. Oken, *Elements of Physiophilosophy*, 186.

51. Owen, "Oken, Lorenz," 500.

52. [Wilberforce], "Darwin's *Origin of Species*," 258. See also H. Miller, *Footprints of the Creator*, 21–22; Hitchcock, *Religion of Geology*, 243: "In the works of Professor Lorenz Oken, of Zurich, we see fully developed the tendencies and results of this hypothesis of development by law, combined with the unintelligible idealism of Kant, Fichte, Schelling, etc. In his Physiophilosophy, translated by the Ray Society for the edification of sober, matter-of-fact Anglo-Saxons, we find a man, of strong mind and extensive knowledge, taking the most ridiculous

positions with the stoutest dogmatism, and the most imperturbable gravity, yet whose blasphemy is equalled only by their absurdity."

53. "Rules and Regulations of the Ray Society," BM(NH), L, MSS Ray (1844–47), fol. 4. See also Hume, *Learned Societies*, 289.

54. BM(NH), L, MSS Ray (1844–47), fol. 128.

55. Minutes of Council, Nov. 5, 1847, ibid., fol. 127.

56. Minutes of Council, Nov. 19, 1847, Dec. 17, 1847, ibid., fols. 132, 134. See also Thomas Bell to William Jardine, Dec. 20, 1847, ibid., fols. 130–31.

57. Owen, "Report on the Archetype," 274; Owen, *On the Archetype*, 106.

58. Owen, On the Nature of Limbs, 86.

59. Owen, "A Brief Summary of the Steps Leading to [My] Present Creed on the Origin of Species," BM(NH), L, OC 90, vol. 3, 179–90.

60. Curwen, *Journal of Gideon Mantell*, 232.

61. Clippings from the *Manchester Spectator*, Dec. 8 and Dec. 22, 1849, occur in Owen's personal, interleaved, and annotated copy; BM(NH), L, OC 18. See also E. Richards, "Question of Property Rights," 161–63.

62. *Manchester Spectator*; see n. 61 above.

63. Ibid.

64. Sedgwick, *Discourse*, ccxiv.

65. Rupke, "Neither Creation nor Evolution," 162–63.

66. [Broderip and Owen], "Generalizations of Comparative Anatomy," 50.

67. Owen, *Principal Forms of the Skeleton*, 228 (Philadelphia ed.), 263 (London ed.).

68. Owen, "On Dinornis (Part IV)," 15.

69. Ibid.

70. The point was made by Robert M. Young in *Darwin's Metaphor*, 43–44. The expression "survival of the fittest" itself was introduced by Spencer in 1852.

71. Mantell, "Geology of New Zealand," *Proceedings of the Geological Society*, Feb. 27, 1850, 342.

72. E.g., Owen's Royal Institution lecture "On the Geographical Distribution of Mammalia."

73. Forbes, "On the Connexion between the Distribution of the Existing Fauna and Flora of the British Isles."

74. Prichard, *Physical History of Mankind*.

75. Owen, "Address" (1858), xc.

76. Ibid., li.

77. Ibid., xcii.

78. Owen's positive reaction had been preceded by a negative one by Samuel Houghton; see Cohen, "Three Notes on the Reception of Darwin's Ideas on Natural Selection," 590.

79. Owen, *On the Classification and Geographical Distribution of the Mammalia*, appendix, 60.

80. E.g., Beer, *Charles Darwin*, 164.

81. [Owen], "Darwin on the Origin of Species," 503–4.

82. Ibid., 503.

83. Ibid., 521.

84. [Wilberforce], "Darwin's *Origin of Species*," 247.

85. E.g., Beer, *Charles Darwin*, 165. See also Irvine, *Apes, Angels and Victorians*, 5: "His chief qualifications for propounding on a scientific subject derived, like nearly everything else that was solid in his career, from the undergraduate remoteness of a first in mathematics."

86. Rev. Owen, *Life*, vol. 1, 356.

87. Owen, "Address" (1858), lxxvi.

88. C. Darwin, *Origin of Species*, 1st ed., 484.

89. Gillespie, *Charles Darwin*, 93.

90. Himmelfarb, *Darwin and the Darwinian Revolution*, 277 ("perhaps in subservience to the powers-that-be").

91. [Owen], "Darwin on the *Origin of Species*," 511.

92. Beer, *Darwin and Huxley*, 61.

93. W. Carpenter, "Darwin on the Origin of Species," 214.

94. L. Huxley, *Joseph Dalton Hooker*, vol. 2, 50.

95. C. Darwin, *Origin of Species*, 6th ed. (1872), "Historical Sketch," xxviii.

96. Conybeare to Owen, n.d., BM(NH), L, OC 8 (fols. 374–75).

97. Sedgwick to Owen, Mar. 28, 1860, in J. Clark and Hughes, *Adam Sedgwick*, vol. 2, 361.

98. Powell, *Essays*, 370.

99. A. Gray, *Darwiniana*, 73.

100. Dowager Duchess of Argyll, *George Douglas*, vol. 1, 581.

101. [Mivart], "Century of Science," 396. See also Mivart to Owen, Apr. 13, 1872, in Gruber, *Conscience in Conflict*, 20. Darwin himself, in a letter to J. S. Henslow, May 8, [1860], expressed the suspicion that Owen "goes as far as I do"; in Barlow, *Darwin and Henslow*, 203.

102. [Mivart], "Darwin's Descent of Man," 86.

103. See n. 3 above.

104. Sedgwick to Owen, n.d., in Rev. Owen, *Life*, vol. 2, 96.

105. *Athenaeum*, Apr. 7, 1860, 479.

106. F. T. C, "Palaeontology," *Fraser's Magazine* 62 (1860): 522.

107. Owen, "On the Characters of the Aye-aye."

108. Owen, *Monograph on the Aye-aye*, 61. Parts of the "Conclusion" were printed in the *American Journal of Science*, 2nd ser., 36 (1863): 296–99.

109. A. Gray, *Darwiniana*, 73, 84.

110. Owen, *Monograph on the Aye-aye*, 62–63.

111. Ibid., 64–65.

112. Ibid., 66.

113. Owen, *Memoir on the Dodo*, 39.

114. "Professor Owen's Comparative Anatomy," *London Review of Politics, Society, Literature, Art and Science* 12 (1866): 483.

115. "The Comparative Anatomy of Vertebrate Animals," *Popular Science Review* 5 (1866): 212.

116. "Professor Owen and the Darwinian Theory," *London Review* 12 (1866): 516.

117. C. Darwin, *Origin of Species*, 4th ed. (1866), "Historical Sketch," xviii.

118. MacLeod, "Evolutionism and Richard Owen," 278.

119. J. Moore, *Post-Darwinian Controversies*, 88; Ruse, *Darwinian Revolution*, 228.

120. E.g., Darwin to Lyell, Apr. 10, 1860; in K. Lyell, *Life of Charles Lyell*, vol. 2, 94. See also Hull, *Darwin and His Critics*, 171–75.

121. Owen, *On the Anatomy of Vertebrates*, vol. 3, 798.

122. [Mozley], "Argument of Design," 163.

123. [Owen], "Species," in Brande and Cox, *Dictionary of Science*, vol. 3, 526.

124. *American Journal of Science*, 2nd ser., 47 (1869): 33–67. Four fairly long footnotes were omitted.

125. ". . . that if species have changed step by step, one should find traces of these gradual modifications; that one should find between the palaeotherium and present-day species a number of intermediary forms, and that until now this has not happened" (G. Cuvier, *Ossemens fossiles*, 2nd ed., vol. 1, lvii).

126. Owen, *On the Anatomy of Vertebrates*, vol. 3, 793.

127. Ibid., 795. On the subject of polydactyl horses, see Gould, *Hen's Teeth*, 177–86.

128. Owen, *On the Anatomy of Vertebrates*, vol. 3, 795.

129. Ibid., 796.

130. Ibid., 807.

131. Bowler, *Eclipse of Darwinism*; also Bowler, *Non-Darwinian Revolution*. See further Ridley, "Coadaptation."

132. Owen, *Lectures on Comparative Anatomy of Invertebrate Animals*, 32, 79.

133. Owen, *On the Anatomy of Vertebrates*, vol. 3, 817–18.

134. Owen, "Spontaneous Generation," BM(NH), L, OC 90, vol. 3, 30, 33.

135. See Farley, *Spontaneous Generation Controversy*; also Latour, "Pasteur et Pouchet."

136. See L. Huxley, *Joseph Dalton Hooker*, vol. 2, 51.

137. W. Carpenter, *Introduction to the Study of the Foraminifera*, viii.

138. *Athenaeum*, Mar. 28, 1863, 419.

139. L. Huxley, *Joseph Dalton Hooker*, vol. 2, 50.

140. C. Darwin, "Doctrine of Heterogeny and Modification of Species," *Athenaeum*, Apr. 25, 1863, 554–55; [Owen], "Origin of Species," *Athenaeum*, May 2, 1863, 586–87; C. Darwin, "Origin of Species," *Athenaeum*, May 9, 1863, 617; see also W. Carpenter, "Dr. Carpenter and His Reviewer," *Athenaeum*, Apr. 4, 1863, 461.

141. T. Huxley, "On Some Organisms Living at Great Depths in the North Atlantic Ocean," in M. Foster and Lankester, *Scientific Memoirs of Huxley*, vol. 3, 330–39. See also Rehbock, "Huxley, Haeckel and the Oceanographers"; Rupke, "*Bathybius Haeckelii*."

142. [M. Foster], "Higher and Lower Animals."

143. Mivart, *On the Genesis of Species*, 2nd ed., 273. H. Charlton Bastian, who was professor of pathological anatomy at University College, London, drew attention to Owen and Mivart's theory of an innate capacity to speciate, siding with the heterogenists in advocating spontaneous generation: *Beginnings of Life*, vol. 2, 580–640.

144. Owen, "On the Anatomy of the American King-Crab," 501–2.

145. T. Huxley, "Further Evidence."

146. Owen, "Monograph of a Fossil Dinosaur (*Omosaurus armatus*, Owen) of the Kimmeridge Clay," originally published in 1875 and reprinted as part of Owen, *Monograph on the Fossil Reptilia*, 45–93. Owen, *Memoirs on the Extinct Wingless Birds*, vol. 1, 460–61.

147. Owen, "Monograph of the Fossil Reptilia, Part II (Pterosauria)"; Seeley, "Remarks." See also Padian, "Case of the Bat-Winged Pterosaur." In the early 1880s Owen used the issue of the origin of birds to address once more the general question of the origin of species, to emphasize geographic isolation as a cause of speciation, to repudiate Darwin and Wallace, and to express a measure of agreement with Lamarck: "On the Sternum of *Notornis*," 695.

148. Owen, "On the Influence of the Advent of a Higher Form of Life."

149. [T. Huxley], "Darwin on the Origin of Species," 569–70.

150. Owen to Groom-Napier, Mar. 19, 1878, RCS, Misc. MSS/Q(xxii); also in Owen Correspondence at the Wellcome Institute Library.

151. Owen to Burrows, Oct. 15, 1881, Temple University, Miscellaneous Collection, Pc5 (1040).

152. Owen, "Brief Summary of the Steps Leading to [My] Present Creed on the Origin of Species," BM(NH), L, OC, vol. 3, 179–90.

153. Owen to Walpole, Nov. 5, 1882, RCS, OC, file L(I), (k).

154. Hodge, "Darwin," 2. See also Olby, "Retrospect."

Chapter Six

1. [Sedgwick], "Natural History of Creation," 3.

2. Darwin, *Origin of Species*, 1st ed., 488.

3. C. Lyell, *Geological Evidences of the Antiquity of Man*, 471–506.

4. [Blake], "Professor Huxley," 546.

5. Rev. Owen, *Life*, vol. 1, 251.

6. Owen, "Synopsis of the Hunterian Lectures on the Comparative Anatomy and Physiology of the Vertebrated Animals with Warm Blood," 1848, lecture 24, BM(NH), L, OC 38. These lectures were repeated in 1854, and the last one, "On the Comparison of the Apes or Anthropoid Quadrumana with Man," was published in *Medical Times and Gazette.*

7. Owen, "Osteological Contributions" (1849). The paper was originally read on Feb. 22, 1848.

8. Blunt, *Ark in the Park*, 38–40.

9. Owen, "On the Anatomy of the Orang Outang"; Owen, "On the Osteology of the Chimpanzee." See also Bowler, *Theories of Human Evolution*; Herbert, "Place of Man."

10. Broderip to Owen, July 6, 1840, RCS, OC, 275/h.7 (U-7).

11. The watercolor accompanies a letter on a species of orangutan; BL, Add. MS 42,581 (fol. 236).

12. Owen, "Report on the Dissection of the Chimpanzee," Jan. 26, 1844, RCS, OC, 275/g.6.

13. A brief history of the early acquisition of gorilla specimens is given by Owen, *Memoir on the Gorilla*, 3–5. See also his "On the Gorilla (Troglodytes Gorilla, Sav.)" in *Proceedings of the Zoological Society of London* and "On the Gorilla" in *Proceedings of the Royal Institution*, both 1859.

14. Owen, "On the Anthropoid Apes."

15. Owen, "Osteological Contributions" (1849), 402–3.

16. Owen, "On the Anthropoid Apes, and Their Relations to Man," 26.

17. Owen, "Osteological Contributions" (1849), 413–14.

18. Owen, "Osteological Contributions, No. IV" (1853), 86. See also a report of this paper, read on Nov. 11, 1851, in the *Literary Gazette*, Nov. 15, 1851, 777–78. See also Owen, "Osteological Contributions, No. V" (1857). Although this paper was published later than no. 4 of the "Osteological Contributions," it had been read earlier, on Sept. 9, 1851, and also been reported in the *Literary Gazette*, Sept. 13, 1851, 630. The continuation of this paper was "Osteological Contributions, No. VII" (1862).

19. Owen, "On the Gorilla (Troglodytes Gorilla, Sav.)," 12. See also Owen's Hunterian lecture "On the Comparison of the Apes or Anthropoid Quadrumana with Man."

20. Owen, *Essays and Observations, by John Hunter*, 43.

21. Owen, "Osteological Contributions" (1849), 417.

22. Ibid., 416.

23. Ibid., 417. See also Owen, "On the Anthropoid Apes," 111. Owen stated this conclusion in virtually all of his papers on the osteology of manlike apes.

24. Owen, "Osteological Contributions" (1849), 417. From among the several publications in which Owen discussed "transmutation" and the "unity" of the human species, see, e.g., "On the Anthropoid Apes, and Their Relations to Man"; also "On the Psychical and Physical Characters of the Mincopies."

25. See, e.g., Owen, "On the Comparison of the Apes or Anthropoid Quadrumana with Man," 514. See also Owen's *Principal Forms of the Skeleton*, 262–63 (London ed.).

26. Bonaparte, "New Systematic Arrangement"; in this paper, which was read on Nov. 7, 1837, Bonaparte divided the placental mammals into two subclasses, one having a "cerebrum bi- (vel tri-) lobum," the other a "cerebrum unilobum" (248–49). For Cuvier, too, the nervous system had been of prime taxonomic significance.

27. Owen, "On the Characters, Principles of Division, and Primary Groups of the Class Mammalia." See also Owen's Rede lecture *On the Classification and Geographical Distribution of the Mammalia*, 43–51; and his Fullerian lecture "On the Cerebral System of Classification."

28. Owen, *On the Anatomy of Vertebrates*, vol. 3, 138.

29. Owen, "On the Characters, Principles of Division, and Primary Groups of the Class Mammalia," 19–20.

30. I. Geoffroy St-Hilaire, *Histoire naturelle générale*, vol. 2, 259–60.

31. Owen, "On the Characters, Principles of Division, and Primary Groups of the Class Mammalia," 33.

32. Owen to Sedgwick, May 11, 1859, CUL, Add 7652/11B, 4a. For Whewell's interest, see Whewell to Owen, Apr. 3, 1859, BM(NH), L, OC 26 (fols. 285–86).

33. Owen, "On the Characters, Principles of Division, and Primary Groups of the Class Mammalia," 20. See also "Notes for the First Rede Lecture 1859 by Sir Richard Owen K.C.B.," CUL, MS Add. 8063; this file contains the larger part of the rough draft of Owen's Rede lecture, with a number of the pages of the Linnean original inserted, showing the changes, both large and small, that Owen made to this original.

34. Owen, *On the Classification and Geographical Distribution of the Mammalia*, 51.

35. [Wilberforce], "Darwin's *Origin of Species*," 258.

36. Lucas, "Wilberforce and Huxley." See also Jensen, "Wilberforce-Huxley Debate"; Jensen, *Thomas Henry Huxley*; Livingstone, "The Bishop and the Bulldog."

37. This is especially the case in the older literature on Huxley and Darwin: e.g., H. Peterson, *Huxley*; Irvine, *Apes, Angels and Victorians*; Bibby, "Huxley"; Beer, *Charles Darwin*; R. Clark, *The Huxleys*. Also in the more recent literature, historians continue to cast Owen in the role of the "bad guy" playing opposite the "good" Darwinians; see, e.g., Ruse, *Darwinian Revolution*; Gregorio, *Huxley's Place*. The exemplary exception is A. Desmond, *Archetypes and Ancestors*.

38. Since the first edition of this book in 1994, the hippocampus controversy has been retold a number of times; see Cosans, "Anatomy, Metaphysics, and Values"; Gould, "A Seahorse for All Races"; Gross, *Brain, Vision, Memory*, 137–78; Lyons, *Thomas Henry Huxley*, 189–230; C. Smith, "Worlds in Collision"; Wilson, "Gorilla."

39. Beer, *Charles Darwin*, 165. A more detailed and sympathetic account is given by C. Smith, "Hippopotamus Test."

40. L. Huxley, *Life of Huxley*, vol. 1, 194; L. Huxley, *Joseph Dalton Hooker*, vol. 1, 522.

41. T. Huxley, "On Species and Races," 392.

42. L. Huxley, *Life of Huxley*, vol. 1, 194.

43. *Athenaeum*, July 7, 1860, 26.

44. T. Huxley, "On Species and Races," 392.

45. Mivart, "On the Appendicular Skeleton"; the complete version appeared in *Philosophical Transactions of the Royal Society* 157 (1867): 299–429.

46. T. Huxley, *Evidence as to Man's Place in Nature*, 105–6.

47. See, e.g., Owen, "On the Zoological Significance of the Brain and Limb Characters," 373.

48. T. Huxley, "On the Zoological Relations of Man," 71.

49. Ibid., 76.

50. Ibid., 79.

51. Ibid., 83.

52. Ibid., 75.

53. Marshall, "Brain of a Young Chimpanzee," 314.

54. Rolleston, "Brain of the Orang Outang," 215.

55. T. Huxley, "On the Brain of Ateles paniscus," 494.

56. Owen, "The Gorilla and the Negro," *Athenaeum*, Mar. 23, 1861, 395–96.

57. T. Huxley, "Man and the Apes," *Athenaeum*, Mar. 30, 1861, 433.

58. Owen, "The Gorilla and the Negro," *Athenaeum*, Apr. 6, 1861, 467.

59. T. Huxley, "Man and the Apes," *Athenaeum*, Apr. 13, 1861, 498.

60. Ibid.

61. Martin and Saller, *Lehrbuch der Anthropologie*, vol. 1, 432ff.

62. T. Huxley, "Man and the Apes," *Athenaeum*, Apr. 13, 1861, 498.

63. Owen, "On the Cerebral Characters of Man and the Ape," 458.

64. Schroeder van der Kolk and Vrolik, "L'encéphale de l'orang-outang." Vrolik's international reputation as an expert on simians was reflected in the invitation to contribute the major entry "Quadrumana" to Todd's *Cyclopaedia of Anatomy and Physiology*.

65. *Athenaeum*, Sept. 14, 1861, 348.

66. *Athenaeum*, Sept. 21, 1861, 378; Sept. 28, 1861, 415. See also *BAAS, Report 1861*, Transactions of the Sections, 146.

67. Flower, "On the Posterior Lobes," 187. A parallel dispute between Owen and Flower developed over whether or not there exists a proper corpus callosum in the brains of marsupials and monotremes; see Owen, "On Zoological Names of Characteristic Parts"; Flower, "Reply." Against Owen, Flower asserted that the whole of the corpus callosum, "only in somewhat reduced proportions," is present in the nonplacental mammals. Owen's view in this controversy was long regarded to have been mistaken, but Mivart later stated that Owen had in fact been substantially right; see [Mivart], "Century of Science," 389.

68. Owen, "On the Zoological Significance of the Cerebral and Pedal Characters of Man," 117.

69. *Medical Times and Gazette*, no. 2 (1862): 373.

70. Ibid.

71. Ibid.

72. *Athenaeum*, Oct. 11, 1862, 468.

73. Ibid.

74. Ibid.

75. Rolleston's Royal Institution talk, "On the Affinities and Differences between the Brain of Man and the Brains of Certain Animals," delivered on Jan. 24, 1862, was reprinted in W. Turner, *Scientific Papers by George Rolleston*, vol. 1, 24–52.

76. Rolleston, "On the Distinctive Characters of the Brain."

77. T. Huxley, "The Brain of Man and Apes," *Medical Times and Gazette*, no. 2 (1862): 449.

78. [T. Huxley(?)], "St. Hilaire on the Systematic Position of Man," *Natural History Review*, 1862, 2.

79. Schroeder van der Kolk and Vrolik, "L'encéphale de l'orang-outang," 111.

80. [T. Huxley(?)], "Schleiden's Essays," *Natural History Review*, 1862, 199.

81. Wagner to Owen, Nov. 30, 1860, BM(NH), L, OC 26 (fols. 12–13); Jan. 28, 1861, ibid. (fols. 15–16).

82. Wagner, "Ueber die Hirnfunktionen." This was a reprint of the article from the *Nachrichten von der Gesellschaft der Wissenschaften zu Göttingen*.

83. Huxley to Wagner, Aug. 6, 1861, Niedersächsische Staats- und Universitätsbibliothek Göttingen, Handschriftenabteilung, "R. Wagner, Band 5."

84. Huxley to Wagner, Sept. 17, 1861, ibid.

85. Huxley to Wagner, Oct. 5, 1861, ibid.

86. Bartholomew, "Huxley's Defence of Darwin"; Gregorio, *Huxley's Place*, 77–79 and passim; Gregorio, "Dinosaur Connection."

87. T. Huxley, *Evidence as to Man's Place in Nature*, 106.

88. Wagner to Huxley, Nov. 25, 1861, Imperial College, Huxley Papers, vol. 28 (letter 86).

89. Huxley to Wagner, Dec. 4, 1862 (see n. 83 above).

90. Wagner to Huxley, Jan. 4, 1863, Imperial College, Huxley Papers, vol. 28 (letter 88). See also Huxley to Wagner, Jan. 14, 1862 (the actual date was 1863) (see n. 83 above).

91. Wagner to Huxley, Jan. 4, 1863, Imperial College, Huxley Papers, vol. 28 (letter 88).

92. Owen, "The Brain of Man and Apes," *Medical Times and Gazette*, no. 2 (1862): 473–74.

93. C. Lyell, *Geological Evidences of the Antiquity of Man*, 476–93.

94. *Athenaeum*, Feb. 21, 1863, 262–63. See also subsequent letters by Rolleston, *Athenaeum*, Feb. 28, 1863, 297; by Lyell, *Athenaeum*, Mar. 7, 1863, 331–32; and by Owen, *Athenaeum*, Mar. 7, 1863, 332.

95. Bynum, "Lyell's *Antiquity of Man*," 154–59.

96. T. Huxley, *Evidence as to Man's Place in Nature*, 91.

97. Ibid., 118.

98. T. Huxley, "Resemblances and Differences," in C. Darwin, *Descent of Man*, 2nd ed. (1874), 309–18.

Chapter Seven

1. Irvine, *Apes, Angels and Victorians*, 108.

2. Kühl, *Beiträge*, 70.

3. Tiedemann, *Icones Cerebri Simiarum*, 14, 51.

4. Cruveilhier, *Anatomie descriptive*, vol. 4, 697.

5. Quoted in [Blake], "Professor Huxley," 563–64. Blake also cited Knox as having denounced the hippocampus controversy as a "silly dispute."

6. *Lancet*, no. 2 (1862): 487.

7. *Westminster Review* 79 (1863): 584.

8. [Blake], "Professor Huxley," 549.

9. Ibid., 551. See also Owen, *On the Anatomy of Vertebrates*, vol. 3, 59–60.

10. [Blake], "Professor Huxley," 557.

11. Ibid., 559.

12. Ibid., 563.

13. Owen, *Memoir on the Gorilla*, 50.

14. Ibid., 38.

15. Ibid., 39–40.

16. Owen, *On the Anatomy of Vertebrates*, vol. 2, 273: "To adduce beginnings of structures in one group which reach their full development in another, as invalidating their zoological application in such higher group is puerile; to reproduce the facts of such incipient and indicatory structures as new discoveries is ridiculous; to represent the statement of the zoological character of a higher group as a denial of the existence of homologous parts in a lower one is disgraceful." The wording of this passage is the strongest used by Owen in print on the hippocampus issue.

17. Owen, *Memoir on the Gorilla*, 39. Owen agreed with the views expressed in Tiedemann's "egalitarian" classic "On the Brain of the Negro." The German edition of 1837 was reprinted in 1984: *Hirn des Negers.*

18. Brande and Cox, *Dictionary of Science*, vol. 2, 127.

19. Owen, preface to Blake, *Zoology for Students*, xi.

20. *Westminster Review* 79 (1863): 584.

21. Portmann, *Biologische Fragmente.*

22. As early as his 1851 review "Lyell—on Life and Its Successive Development," Owen quoted from Sedgwick: "Man stands by himself the despotic lord of the living world; not so great in organic strength as many of the despots that went before him in Nature's chronicle, but raised far above them all by a higher development of the brain" (418).

23. Already John Clark had written: "During the whole of his [Owen's] life we believe that he never owned to a mistake" (*Old Friends*, 395).

24. P. P. King to Owen, Apr. 30, 1850, BM(NH), L, OC 16 (fols. 409–10).

25. MacLeay to Owen, Apr. 28, 1850, BM(NH), L, OC 18 (fols. 331–32).

26. Owen to Francis Baring, n.d., RCS, OC, Misc. MS, Q (xx).

27. Huxley to MacLeay, Nov. 9, 1851, in L. Huxley, *Life of Huxley*, vol. 1, 102.

28. M. Foster and Lankester, *Scientific Memoirs of Huxley*, passim.

29. E.g., T. Huxley, "On the Premolar Teeth of Diprotodon."

30. See A. Desmond, *Archetypes and Ancestors*, passim; Gregorio, *Huxley's Place*, passim; Winsor, *Starfish*, 117–18.

31. A. Desmond, *Archetypes and Ancestors*, 37–41.

32. [T. Huxley], "Vestiges of Creation."

33. M. Foster and Lancester, *Scientific Memoirs of Huxley*, supplementary vol., 19.

34. [T. Huxley], "Owen and Rymer Jones."

35. Gregorio, *Huxley's Place*, 187.

36. Owen to Acland, Oct. 9, 1862, Bodleian Library, MS Acland, d. 200 (fols. 14–17).

37. Eve and Creasey, *John Tyndall*, 157–59. See also Tyndall to Owen, June 13, 1871, BM(NH), L, OC 25 (fols. 259–60).

38. Note written on Owen to Tyndall, June 14, 1871, BM(NH), L, OC 21 (fols. 28–29).

39. Owen to Tyndall, June 14, 1871, BM(NH), L, OC 21 (fols. 28–29).

40. Huxley to MacLeay, Nov. 9, 1851, in L. Huxley, *Life of Huxley*, vol. 1, 101.

41. Acland, "Letters on 'Characters' of Man," Bodleian Library, MS Acland, d. 200.

42. Acland to Owen, Oct. 4, 1862, ibid. (fols. 2–12) (II).

43. Ibid. (fol. 12). In response, Owen to Acland, Oct. 9, 1862, ibid. (fols. 14–17).

44. Acland to Owen, Oct. 10, 1862, ibid. (fols. 18–21).

45. Owen to Acland, Oct. 13, 1862, ibid. (fols. 19–20).

46. Acland to Longley, Feb. 11, 1863, reproduced in Atlay, *Sir Henry Wentworth Acland*, 306–8. Part of Acland to Owen, Oct. 4, 1862, is reproduced on 304–6.

47. Phillips to Acland, Feb. 24, 1863, Bodleian Library, MS Acland, d. 200 (fol. 42).

48. "Men or Monkeys?" *British Medical Journal*, Oct. 18, 1862, 419.

49. [Leifchild], "British Association at Cambridge," 365.

50. [Egerton], "Monkeyana."

51. [Egerton], "Gorilla's Dilemma." Bibby ("Huxley") believes that the two poems expressed Owen's defeat at the hands of Huxley.

52. Kingsley, "Speech of Lord Dundreary."

53. Kingsley, *Water Babies*, 70.

54. Ibid., 153.

55. Ibid., 156.

56. Ibid., 156–57. For more on Victorian literature and the life sciences, see Chappie, *Science and Literature*, 58–98.

57. [Pycroft], *Sad Case*, 4. Bibby ("Huxley," 79) is mistaken to believe that this pamphlet was intended for "the lower orders" and reflected the fact that even "streetsweepers" knew that Huxley was right and Owen wrong. An early summary of various literary reactions to the Owen-Huxley clash was given by Blinderman, "Great Bone Case."

58. [Pycroft], *Sad Case*, 5–7.

59. Ibid., 7–8.

60. *Punch* 49 (Sept. 23, 1865): 113.

61. [Jenkins], *Lord Bantam*, vol. 2, 158.

62. Quoted in "Men or Monkeys?" *British Medical Journal*, Oct. 18, 1962, 419.

63. [Leifchild], "British Association at Cambridge," 366.

64. "Men or Monkeys?" *British Medical Journal*, Oct. 18, 1862, 419–20; Quatrefages de Bréau, *L'espèce humaine*, 1–20.

65. Gratiolet, "Note sur l'encéphale du gorille."

66. Leuret and Gratiolet, *Anatomie comparée*, vol. 2, 687. See also Gratiolet, *Mémoire*, 100–103.

67. Quoted by William Coleman, "Louis Pierre Gratiolet," in *Dictionary of National Bibliography*.

68. Gratiolet, "Mémoire sur la microcéphalie," 66. The bulk of this paper was translated into German by Wagner and incorporated in his "Ueber die Hirnbildung des Menschen." This in turn was translated into English and published, with minor omissions and a postscript: "Upon the Structure of the Brain."

69. Wagner, *Lehrbuch der vergleichenden Anatomie*, 45.

70. Ibid., 412.

71. Wagner, *Menschenschöpfung und Seelensubstanz*; Wagner, *Ueber Wissen und Glauben*. This episode has been described by Degen, "Vor hundert Jahren."

72. Vogt, *Köhlerglaube und Wissenschaft*, xxxiii.

73. See also Gregory, "Scientific versus Dialectical Materialism."

74. Wagner, *Der Kampf um die Seele*, 214–18.

75. Wagner, *Gespräche mit Gauss.*

76. Wagner, *Vorstudien*, vol. 2, 79.

77. The broad context of Sömmerring's neuroanatomical and neurophysiological work is discussed in Mann and Dumont, *Gehirn-Nerven-Seele.*

78. C. Carus, *Versuch einer Darstellung*, 299.

79. Ibid., 301–2. See also Feremutsch, "Grundzüge der Hirnanatomie."

80. C. Carus, *Symbolik der menschlichen Gestalt*; C. Carus, "Zur vergleichenden Symbolik."

81. Huschke, *Schaedel, Hirn und Seele*, 160 and passim.

82. Schroeder van der Kolk, *Voorlezing.* See Snelders, "Schroeder van der Kolk, Jacobus Ludovicus Conradus," in *Dictionary of Scientific Biography.*

83. Dana, "Higher Subdivisions," 66. See also his review of T. Huxley's *Evidence as to Man's Place in Nature*, *American Journal of Science*, 2nd ser., 35 (1863): 452.

84. Owen, *On the Anatomy of Vertebrates*, vol. 3, 819.

85. Ibid., 821–23.

86. An early instance of this is [Mivart], "Century of Science," 781.

87. Owen, "On the 'Vital Principle' and the 'Soul,'" RCS, OC, file L(I), (j).

88. Brodie, *Psychological Inquiries*, 103, 115, 167. See also Donovan, *Reply*; Hawkins, *Works of Brodie*, vol. 1, 117ff.

89. W. Lawrence, *Introduction*, 174.

90. Abernethy, *Physiological Lectures*, 52–53, 332–40. The quotation is Lawrence's paraphrase of Abernethy's accusation, in W. Lawrence, *Lectures on Physiology*, 1. See also Macilwain, *Memoirs of John Abernethy*, 181–96.

91. W. Lawrence, *Lectures on Physiology*, 5.

92. See, e.g., Goodfield-Toulmin, "Some Aspects."

93. Anon., Facts versus Fiction! vi.

94. For a detailed discussion of the Lawrence affair, see Bynum, "Time's Noblest Offspring."

95. W. Lawrence, *Lectures on Physiology*, 152.

96. Ibid., 152.

97. Wagner, "Ueber die Hirnbildung des Menschen," 76–78.

98. Owen, "Our Origin as a Species," 64.

99. Tyndall argued by analogy with magnetism that life may be immanent in matter. See also Barton, "John Tyndall, Pantheist." For the views of Carpenter, whose *Principles of Mental Physiology* appeared in 1874, see R. Smith, "Human Significance of Biology."

100. On the politics of human origins, see Livingstone, *Adam's Ancestors.*

101. Du Chaillu to Owen, Dec. 21, 1860, BM(NH), L, OC 10 (fols. 162–63).

102. Rev. Owen, *Life*, vol. 2, 123.

103. Du Chaillu, "Geographical Features," 111. The quotation is taken from the more detailed page proofs, RCS, Misc./MS/Q.

104. Du Chaillu, "Geographical Features," 112.

105. Gunther, *Century of Zoology*, 133.

106. [Brewster], "Du Chaillu's Explorations," 248.

107. Owen, "On a National Museum," 120.

108. E.g., Ballantyne, *Gorilla Hunters.* Also, Gosse, in his *Romance of Natural History* (1861), 257–59, contributed to the image of the gorilla as a "terrible" beast.

109. Du Chaillu, *Explorations and Adventures*, 70–71.

110. Blunt, *Ark in the Park*, 137.

111. *Athenaeum*, May 11, 1861, 623.

112. *Athenaeum*, May 18, 1861, 662–63; also May 25, 1861, 695; June 1, 1861, 728; June 8, 1861, 765–66; June 15, 1861, 798. The issue was discussed at the 1861 meeting of the BAAS at Manchester; *Athenaeum*, Sept. 28, 1861, 414. The issue of Du Chaillu's fieldwork is discussed by McCook, "Paul du Chaillu."

113. *Athenaeum*, Sept. 21, 1861, 372–73; Sept. 28, 1861, 408; Oct. 12, 1861, 479.

114. Waterton to George Ord, May 31, 1861, in Blackburn, *Charles Waterton*, 181. See also *Athenaeum*, Oct. 12, 1861, 478–79; Oct. 19, 1861, 509; Oct. 26, 1861, 543–44.

115. Du Chaillu, "The New Traveller's Tales," *Athenaeum*, May 25, 1861, 694–95.

116. *Athenaeum*, Sept. 21, 1861, 373; Oct. 5, 1861, 445–46. See also Owen, "On Some Objects of M. du Chaillu."

117. *Athenaeum*, Oct. 12, 1861, 478–79.

118. "Dr Barth on M. du Chaillu's Adventures," *Literary Gazette*, Aug. 3, 1861, 111–13; Aug. 10, 1861, 132–37. "M. du Chaillu and His Critics," *Critic*, June 8, 1861, 721–24.

119. Blunt, *Ark in the Park*, 145.

120. Reade, *Savage Africa*, 179–80.

121. See, e.g., Kingsley to Rolleston, Oct. 12, 1862, in Kingsley, *Kingsley*, vol. 2, 143.

122. Kingsley, *Water Babies*, 247–48.

123. [Brewster], "Du Chaillu's Explorations," 252.

124. [Cox], "Du Chaillu's Adventures."

125. Barry Butcher describes how in Melbourne, in the National Museum of Victoria, a group of gorillas was displayed to illustrate–in the words of Frederick McCoy–"how infinitely remote the creature is from humanity" ("Gorilla Warfare in Melbourne," 165).

126. [Chapman], "Equatorial Africa."

127. [T. Huxley(?)], "The Fauna of Equatorial Africa," *Natural History Review*, 1861, 290.

128. Huxley to Wagner, Oct. 5, 1861, Niedersächsische Staats- und Universitätsbibliothek Göttingen, Handschriftenabteilung, "R. Wagner, Band 5."

129. T. Huxley, *Evidence as to Man's Place in Nature*, 54.

130. [T. Huxley(?)], "The Habits of the Gorilla," *Natural History Review*, 1861, 337, 339.

131. Du Chaillu, *Journey to Ashango-Land*, ix–x.

132. Owen, "Descriptions of Three Skulls," 460.

133. Murchison to Owen, Oct. 1864, in Rev. Owen, *Life*, vol. 2, 147.

134. Gunther, *Century of Zoology*, 133.

135. Du Chaillu to Owen, Aug. 19, 1864, BM(NH), L, OC 10, (fols. 181–91).

136. For instance, Carl Gustav Carus was inspired by the photograph to write his *Zur vergleichenden Symbolik zwischen Menschen- und Affen-Skelett*. On Fenton's photographic services to the British Museum, see Hannavy, *Roger Fenton of Crimble Hall*, chap. 4. See also Mivart, *Man and Apes*.

137. *London Quarterly Review* 17 (1861–62): 109.

138. [Sala], "With Mr. Gorilla's Compliments," 489–90.

139. [Sala], "A New Temple to Nature," *Daily Telegraph*, Nov. 23, 1880; a clipping of the article is in BM(NH), L, OC 90, vol. 4 (fols. 65–67).

140. [Mivart], "Century of Science."

141. "Armorial Bearings of Sir Richard Owen," BM(NH), L, OC 69.

142. Owen to "Will," Aug. 30, 1851, RCS, OC, MS Add. 262 (letter 353). Another instance is Owen to William Clift, n.d., BM, Add. MS 39,955 (fol. 263): "Whilst our dear boy is with you you

have a dear and deep trust. Every seed now sown will spring up for good or for evil. His only happiness and ours depends upon his truly fearing and reverencing God and his Holy Word and Laws. Guide him to that by example. Let the short time that is allotted to us be spent, as far as we are able, in showing our gratitude to the Heavenly giver of our peace and competence and the blessings of this life, which, if we look around us, we cannot but see to have been abundantly bestowed."

143. See F. Turner, "Victorian Conflict; F. Turner, *Contesting Cultural Authority.*

144. See the 1994 edition of this book, 324–32 and 342–51.

145. The so-called three estates originally referred to the composition of Parliament (the Crown, the Lords, and the Commons or, in a somewhat-different division of political power, the Lords Spiritual, the Lords Temporal, and the Commons). The term "fourth estate" was used for a variety of extraparliamentary forces, such as the army, the mob, and, during the early part of the nineteenth century, the press. More recently, the notion of the three estates has been used to denote the social grouping of cultural authority in Victorian society (aristocracy, clergy, and judiciary), with science constituting the fourth estate.

146. Among Owen's contributions to the general and popular press were the following. [Owen, using the pseudonym "Silas Seer"], "Recollections and Reflections of Gideon Shaddoe, Esq." (1844 and 1845); [Owen], "Poisonous Serpents"; [Owen, using the pseudonym "Ennoo," written in Greek capitals, or "Zoologus"], "Visit to Selborne"; "Milton and Galileo"; "Fate of the 'Jardin d'acclimatation' during the Late Sieges of Paris"; "Serpent-Charming in Cairo"; "On Petroleum and Oil Wells."

147. See Rowse, *Controversial Colensos.*

148. Owen to G. Rorison, n.d., in Goulburn et al., *Replies,* 517–18.

149. Some of the commotion can be gauged from [Wilberforce], "Essays and Reviews"; [Fairbairn], "Recent Attacks"; [Pattison], "Bishop Colenso."

150. "Declaration of Students of the Natural and Physical Sciences," p. 3, CUL, Add. 5989 (a collection of letters and newspaper reactions to the Scientists' Declaration, with a manuscript history, copies of circulars, and the declaration itself).

151. See W. Brock and MacLeod, "Scientists' Declaration."

152. Argyll to Anon., Apr. 30, 1864, CUL, Add. 5989 (fol. 12); Murchison to Anon., May 25, 1864, ibid. (fol. 22); Tyndall to C. H. Berger, June 11, 1864, ibid. (fol. 24).

153. Owen to Herbert MacLeod, May 10, 1864, ibid. (fol. 17).

154. F. James, *Chemistry and Theology.*

155. H. Miller, *Footprints of the Creator,* 155.

156. Owen, *Instances of the Power of God,* 22–23.

157. Ibid., 35–49.

158. Ibid., 37.

159. Ibid., 55.

160. See Gunther, *Century of Zoology,* 201.

161. Ibid.

162. Owen, "On British Eocene Serpents."

163. Owen, "On Longevity," 218.

164. Ibid.

165. Ibid., 223.

166. Ibid., 233.

167. Ibid., 224.

168. Ibid., 219.

169. Ibid., 233.

170. Owen to J. Crampton, Mar. 23, 1872, BM(NH)/L, OC 55 (59); see also RCS, OC, file L(I)j (f). A further letter illustrating Owen's critical view of Old Testament natural history is Owen to Anthony Ashley Cooper, seventh Earl of Shaftesbury, Apr. 3, 1863, Temple University, OC, vol. 4 (fol. 576), stating that contrary to Leviticus 11:6, the hare is not a ruminant animal.

171. H[obson], *Longevity*, 13–14.

172. For the reaction to scientific naturalism in late Victorian England, see F. Turner, *Between Science and Religion.*

173. Owen to Smyth, Jan. 3, 1853, RCS, MS Add. 292 (no. 392). On the Victorian fascination with spiritualism, see Oppenheim, *Other World.*

174. Gladstone, *Impregnable Rock*, 56.

175. Owen to Gladstone, Jan. 5, 1884, BM, Add. MS 44,485 (fol. 32); see also Owen to Gladstone, Oct. 21, 1885, BM, Add. MS 44,492 (fol. 205); Dec. 7, 1885, BM, Add. MS 44,493 (fol. 188); Jan. 5, 1885, CUL, Add. 5354 (fol. 180).

176. Owen, "The Reign of Law," BM(NH), L, OC 59 (6), p. 46.

177. Comment written on J. P. Faunthorpe to Owen, June 19, 1876, BM(NH), L, OC 12 (fols. 184–86).

178. See, e.g., [Wace], "Scientific Lectures," 35–37. Also Hays, "London Lecturing Empire."

179. Owen, Lectures on the Comparative Anatomy of Vertebrate Animals, 2.

180. Owen to Acland, Mar. 23, 1855, Bodleian Library, MS Acland, d. 81 (fols. 101–2).

181. [Sala], "A New Temple to Nature," *Daily Telegraph*, Nov. 23, 1880; a clipping of the article is in BM(NH), L, OC 90, vol. 4 (fols. 65–67).

182. *Times*, May 2, 1881; a clipping of the review is in Owen's autobiographical scrapbook, BM(NH), L, OC 24.

183. Owen to Noble, May 23, 1885, Bodleian Library, MS Autogr., c. 17 (fols. 357–58).

184. *British Medical Journal*, no. 2 (1892): 597, 650, 753, 805, 859, 1185, 1247, 1301.

185. Flower, "Obituary Notices," xii.

186. *British Medical Journal*, no. 2 (1892): 1400. This was followed by a lengthy obituary (1411–15).

Appendix

1. E.g., Cave, "Glands of Owen"; Cave, "Richard Owen and the Discovery of the Parathyroid Glands"; Cave, "Muscles of Owen"; Kier and Smith, "Tongues, Tentacles and Trunks."

Bibliography

Manuscripts

American Philosophical Society, Philadelphia: Richard Owen.
Bodleian Library, Oxford: H. W. Acland.
British Geological Survey, Nottingham: Richard Owen.
British Museum Library, London (abbreviated BM): W. E. Gladstone, Richard Owen, Robert Peel.
British Museum (Natural History), London (abbreviated BM(NH)): Richard Owen.
Cambridge University Library (abbreviated CUL): Richard Owen, Adam Sedgwick.
Imperial College, London: T. H. Huxley.
Linnean Society, London: W. S. MacLeay.
Mitchell Library, Sydney: T. L. Mitchell, Richard Owen, Henry Parkes.
National Museum of Wales, Cardiff: H. T. De la Beche.
Niedersächsische Staats- und Universitätsbibliothek, Göttingen: Rudolph Wagner.
Royal Botanic Gardens, Kew: W. J. Hooker.
Royal College of Surgeons, London (abbreviated RCS): Richard Owen.
Royal Society, London (abbreviated RS): William Buckland, J. F. W. Herschel.
Temple University, Philadelphia: Richard Owen.
Trinity College Library, Cambridge: William Whewell.
Turnbull Library, Wellington: Gideon Mantell and Walter Mantell.
University College, London: Edwin Chadwick.
University Museum, Oxford: William Buckland.
Wellcome Institute, London: Richard Owen.

Printed Sources

Abernethy, John. *Physiological Lectures, Exhibiting a General View of Mr Hunter's Physiology, and of His Researches in Comparative Anatomy, Delivered before the College of Surgeons in the Year 1817*. London: Longman, Hurst, Rees, Orme and Brown, 1825.

Acland, Henry Wentworth. *Ground-work of Culture. Address Delivered in King's College, London at the Distribution of Prizes on October 2, 1883*. London: J. and A. Churchill, 1883.

______. *The Harveian Oration 1865*. London: Macmillan, 1865.

[______]. "Obituary. Sir Richard Owen, K.C.B." *British Medical Journal*, no. 2 (1892): 1411–15.

Agassiz, Louis. *Essay on Classification*. Edited by Edward Lurie. Cambridge, MA: Harvard University Press; Oxford: Oxford University Press, 1962.

______. "Observations on the Blind Fish of the Mammoth Cave." *American Journal of Science*, 2nd ser., 11 (1851): 127–28. Also in *Edinburgh New Philosophical Journal* 50 (1851): 254–56.

______. "On a New Classification of Fishes, and on the Geological Distribution of Fossil Fishes." *Proceedings of the Geological Society* 2 (1838): 99–102.

______. "On the Differences between Progressive, Embryonic, and Prophetic Types in the Succession of Organized Beings through the Whole Range of Geological Times." *Edinburgh New Philosophical Journal* 49 (1850): 160–65.

______. "On the Succession and Development of Organized Beings at the Surface of the Terrestrial Globe." *Edinburgh New Philosophical Journal* 33 (1842): 388–99.

______. *Recherches sur les poissons fossiles*. 5 vols. Neuchâtel: Petitpierre, 1833–43.

Alter, Peter. *Wissenschaft, Staat, Mäzene: Anfänge moderner Wissenschaftspolitik in Grossbritannien, 1850–1920*. Stuttgart: Klett-Cotta, 1982. Translated as *The Reluctant Patron: Science and the State in Britain, 1850–1920*. Oxford: Berg, 1987.

Amundson, Ron. "Richard Owen and Animal Form." In Richard Owen, *On the Nature of Limbs*, edited by Ron Amundson, xv–li. Chicago: University of Chicago Press, 2007.

______. "Typology Reconsidered: Two Doctrines on the History of Evolutionary Biology." *Biology and Philosophy* 13 (1998): 153–77.

Andrews, J. R. H. *The Southern Ark: Zoological Discovery in New Zealand, 1769–1900*. Honolulu: University of Hawaii Press, 1988.

Anon. "Du Chaillu's Explorations." *London Review* 17 (1861): 73–117.

______. *Facts versus Fiction! An Essay on the Functions of the Brain*. 2nd ed. London: Watson, 1840.

______. "The Fossil Reptiles of England." *Literary Gazette*, Aug. 14, 1841, 513–19.

______. *London Interiors: A Grand National Exhibition of the Religious, Regal and Civic Solemnities, Public Amusements, Scientific Meetings, and Commercial Scenes of the British Capital*. London: J. Mead, 1841–44.

______. "A Popular Course of Geology. Introduction." *Magazine of Popular Science, and Journal of the Useful Arts* 2 (1836): 1–13.

______. Review of Owen and Siebold on parthenogenesis. *Midland Quarterly Journal of the Medical Sciences*, Jan. 1858, 1–19.

______. "Steenstrup on the Alternation of Generations." *Medico-Chirugical Review*, July 1, 1846, 22–34.

Appel, Toby A. *The Cuvier-Geoffroy Debate: French Biology in the Decades before Darwin*. New York: Oxford University Press, 1987.

______. "Henri de Blainville and the Animal Series: A Nineteenth-Century Chain of Being." *Journal of the History of Biology* 13 (1980): 291–319.

Argyll, Dowager Duchess of, ed. *George Douglas, Eighth Duke of Argyll, K.G., K.T. (1823–1900): Autobiography and Memoirs*. 2 vols. London: John Murray, 1906.

Argyll, George Douglas, Eighth Duke of. "Address." *BAAS, Report 1855*, lxxiii–lxxxvi.

______. *The Philosophy of Belief or Law in Christian Theology*. London: John Murray, 1896.

______. *The Reign of Law*. 5th ed. London: Strahan, 1870.

______. "The Supernatural." *Edinburgh Review* 116 (1862): 378–97.

______. *The Unity of Nature*. London: Alexander Strahan, 1884.

Ashton, Rosemary. *The German Idea: Four English Writers and the Reception of German Thought, 1800–1860*. Cambridge: Cambridge University Press, 1980.

______. *G. H. Lewes: A Life*. Oxford: Oxford University Press, 1991.

Atlay, J. B. *Sir Henry Wentworth Acland, Bart.; K.C.B., F.R.S., Regius Professor of Medicine in the University of Oxford: A Memoir*. London: Smith, Elder, 1903.

Austin, Frances, and Bernard Jones. "The Clifts and the Arts." *Annals of the Royal College of Surgeons of England* 66 (1984): 63–70.

Baer, Karl Ernst von. *Ueber Entwickelungsgeschichte der Thiere*. 2 vols. Königsberg: Bornträger, 1828–37.

Bakewell, P. C. *Geology for Schools and Students: or, Former Worlds, Their Structure, Condition, and Inhabitants*. London: National Illustrated Library, 1854.

Balan, Bernard. *L'ordre et le temps: L'anatomie comparée et l'histoire des vivants au XIXe siècle*. Paris: J. Vrin, 1979.

Balfour, Francis M. *A Treatise on Comparative Embryology*. 2 vols. London: Macmillan, 1880–81.

Ball, Valentine, et al. "On the Provincial Museums of the United Kingdom." *BAAS, Report 1887*, 97–130; *BAAS Report, 1888*, 124–32.

Ballantyne, R. M. *The Gorilla Hunters: A Tale of the Wilds of Africa*. London: T. Nelson, 1861.

Barber, Lynn. *The Heyday of Natural History, 1820–1870*. London: Cape, 1980.

Barlow, Nora, ed. *Darwin and Henslow: The Growth of an Idea; Letters, 1831–1860*. London: Bentham-Moxon Trust; John Murray, 1967.

Barry, Martin. "Further Observations on the Unity of Structure in the Animal Kingdom, and on Congenital Anomalies, Including 'Hermaphrodites'; with Some Remarks on Embryology, as Facilitating Animal Nomenclature, Classification, and the Study of Comparative Anatomy." *Edinburgh New Philosophical Journal* 22 (1837): 345–64.

______. "On the Unity of Structure in the Animal Kingdom." *Edinburgh New Philosophical Journal* 22 (1837): 116–41.

Bartholomew, Michael. "Huxley's Defence of Darwin." *Annals of Science* 32 (1975): 525–35.

Barton, Ruth. "'Huxley, Lubbock, and Half a Dozen Others': Professionals and Gentlemen in the Formation of the X Club, 1851–1864." *Isis* 89 (1998): 410–44.

______. "'An Influential Set of Chaps': The X-Club and Royal Society Politics, 1864–85." *British Journal for the History of Science* 23 (1990): 53–81.

______. "John Tyndall, Pantheist: A Rereading of the Belfast Address." *Osiris*, 2nd ser., 3 (1987): 111–34.

Bastian, H. Charlton. *The Beginnings of Life: Being some Account of the Nature, Modes of Origin and Transformations of Lower Organisms*. 2 vols. London: Macmillan, 1872.

Becker, Bernard H. *Scientific London*. London: H. S. King, 1874.

Beer, Gavin de. *Charles Darwin: A Scientific Biography*. New York: Doubleday, 1964.

______, ed. *Charles Darwin and T. H. Huxley: Autobiographies*. Oxford: Oxford University Press, 1983.

Bell, Charles. *The Hand: Its Mechanism and Vital Endowments, as Evincing Design*. 6th ed. London: John Murray, 1860.

Bennett, George. *Gatherings of a Naturalist in Australasia*. London: van Voorst, 1860.

Benton, Michael. "Progressionism in the 1850s: Lyell, Owen, Mantell and the Elgin Fossil Reptile *Leptopleuron* (*Telerpeton*)." *Archives of Natural History* 11 (1982): 123–36.

Berman, Morris. *Social Change and Scientific Organization: The Royal Institution, 1799–1844*. London: Heinemann Educational, 1978.

Bibby, Cyril. "Huxley and the Reception of the 'Origin.'" *Victorian Studies* 3 (1959): 76–86.

Blackburn, Julia. *Charles Waterton, 1782–1865. Traveller and Conservationist.* London: Bodley Head, 1989.

Blainville, Henri Marie Ducrotay de. "Lectures on Comparative Osteology. Additionally Illustrated with Numerous Notes, Observations, and Drawings, by Robert Knox." *Lancet,* no. 1 (1839–40): 137–45, 185–92, 217–22, 297–307; no. 2 (1839–40): 209–20, 289–96, 321–30, 353–68, 385–91, 433–39, 465–70, 513–17, 545–48, 593–97, 625–28, 657–61, 689–93, 737–39, 767–71, 801–5, 833–36, 881–85, 913–16.

______. *Ostéographie ou description iconographique comparée du squelette et du système dentaire des cinq classes d'animaux vertébrés récents et fossiles pour servir de base à la zoologie et à la géologie.* 4 vols. Paris: Baillière, 1839–64.

Blaisdell, Muriel. "Natural Theology and Nature's Disguises." *Journal of the History of Biology* 15 (1982): 163–89.

[Blake, C. Carter]. "Professor Huxley on Man's Place in Nature." *Edinburgh Review* 117 (1863): 541–69.

______. *Zoology for Students: A Handbook, with a Preface by Richard Owen.* London: Daldy, Isbister, 1875.

Blinderman, Charles S. "The Great Bone Case." *Perspectives in Biology and Medicine* 14 (1970–71): 370–93.

Blumenbach, Johann Friedrich. *Handbuch der vergleichenden Anatomie.* Göttingen: Dieterich, 1824.

______. "Ueber das Schnabelthier (*Ornithorhynchus paradoxus*) ein neuentdecktes Geschlecht von Säugethieren des fünften Welttheils." *Magazin für den neuesten Zustand der Naturkunde* 2 (1800): 205–14.

Blunt, Wilfrid. *The Ark in the Park: The Zoo in the Nineteenth Century.* London: Hamilton; Tryon Gallery, 1976.

______. *In for a Penny: A Prospect of Kew Gardens; Their Flora, Fauna and Falballas.* London: Hamilton; Tryon Gallery, 1978.

Bompas, George C. *Life of Frank Buckland.* London: Nelson, n.d.

Bonaparte, C. L. "A New Systematic Arrangement of Vertebrated Animals." *Transactions of the Linnean Society* 18 (1841): 247–304.

Bourdier, Frank. "Geoffroy Saint-Hilaire versus Cuvier: The Campaign for Palaeontological Evolution (1825–1838)." In *Toward a History of Geology,* edited by Cecil J. Schneer, 36–61. Cambridge, MA: MIT Press, 1969.

Bowler, Peter J. *The Eclipse of Darwinism: Anti-Darwinian Evolution Theories in the Decades around 1900.* Baltimore, MD: Johns Hopkins University Press, 1983.

______. *Evolution: The History of an Idea.* Berkeley and Los Angeles: University of California Press, 2003.

______. *Fossils and Progress: Paleontology and the Idea of Progressive Evolution in the Nineteenth Century.* New York: Science History Publications, 1976.

______. *The Invention of Progress: The Victorians and the Past.* Oxford: Basil Blackwell, 1989.

______. *Life's Splendid Drama: Evolutionary Biology and the Reconstruction of Life's Ancestry, 1869–1940.* Chicago: University of Chicago Press, 1996.

______. *The Non-Darwinian Revolution: Reinterpreting a Historical Myth.* Baltimore, MD: Johns Hopkins University Press, 1988.

———. *Theories of Human Evolution: A Century of Debate, 1844–1944*. Baltimore, MD: Johns Hopkins University Press, 1986.

Branagan, David. "Richard Owen in the Antipodean Context (a Review)." *Journal and Proceedings, Royal Society of New South Wales* 125 (1992): 95–102.

Brande, W. T., and G. W. Cox, eds. *A Dictionary of Science, Literature and Art*. 3 vols. London: Longmans, 1865–67.

[Brewster, David]. "Du Chaillu's Explorations and Adventures." *North British Review* 35 (1862): 219–52.

[———]. "Life and Works of Baron Cuvier." *Edinburgh Review* 62 (1836): 265–96.

[———]. "Professor Faivre's Scientific Biography of Goethe." *North British Review* 38 (1863): 107–33.

Briggs, Asa. *The Age of Improvement, 1783–1867*. London: Longman, 1979.

Brock, Michael G., and Richard Brent. "The Oxford of Peel and Gladstone, 1800–1833. Note. The Oriel Noetics." In *History of the University of Oxford*, edited by Michael G. Brock and Mark C. Curthoys, vol. 6, 7–76. Oxford: Clarendon Press, 1997.

Brock, W. H., and R. M. MacLeod. "The Scientists' Declaration: Reflexions on Science and Belief in the Wake of *Essays and Reviews*, 1864–5." *British Journal for the History of Science* 9 (1976): 39–66.

[Broderip, William J.] "Agassiz on Fossil Fishes." *Quarterly Review* 55 (1836): 433–45.

[———]. "The Zoological Gardens—Regent's Park." *Quarterly Review* 56 (1836): 309–32.

———. *Zoological Recreations*. London: Henry Colburn, 1847.

[Broderip, William J., and Richard Owen]. "Generalizations of Comparative Anatomy." *Quarterly Review* 93 (1853): 46–83.

[———]. "Professor Owen—Progress of Comparative Anatomy." *Quarterly Review* 90 (1852): 362–413.

Brodie, Benjamin. *The Hunterian Oration, Delivered in the Theatre of the Royal College of Surgeons in London, on the 14th of February, 1837*. London: Longman, 1837.

———. *Psychological Inquiries*. London: Longman, 1854.

Brongniart, Adolphe. *Prodrome d'une histoire des végétaux fossiles*. Paris: Levrault, 1828.

Brooke, John Hedley. "Natural Theology and the Plurality of Worlds: Observations on the Brewster-Whewell Debate." *Annals of Science* 34 (1977): 221–86.

———. "The Natural Theology of the Geologists: Some Theological Strata." In *Images of the Earth: Essays in the History of the Environmental Sciences*, edited by L. J. Jordanova and R. S. Porter, 39–64. Chalfont St. Giles, Bucks.: British Society for the History of Science, 1979.

———. "Richard Owen, William Whewell, and the *Vestiges*." *British Journal for the History of Science* 35 (1977): 132–45.

Brougham, Henry Lord. *A Discourse of Natural Theology*. 4th ed. London: Charles Knight, 1835.

Brown, Chandos Michael. *Benjamin Silliman: A Life in the Young Republic*. Princeton, NJ: Princeton University Press, 1989.

Browne, Janet. *Charles Darwin: The Power of Place*. New York: Alfred A. Knopf, 2002.

———. *Charles Darwin: Voyaging*. Princeton, NJ: Princeton University Press, 1995.

———. "Natural Causes: 'Old Bones,' the Skeleton in the Cupboard of Evolutionary Science." *Times Literary Supplement*, Aug. 12, 1994, 3–4.

———. *The Secular Ark: Studies in the History of Biogeography*. New Haven, CT: Yale University Press, 1983.

Buckland, Francis T. *Curiosities of Natural History.* Popular ed. 4 vols. London: Richard Bentley and Son, 1891–93.

Buckland, William. *Geologie und Mineralogie in Beziehung zur natürlichen Theologie.* 2 vols. Translated and annotated by Louis Agassiz. Neuchâtel: Leibrock, 1838–39.

———. *Geology and Mineralogy Considered with Reference to Natural Theology.* 2 vols. London: William Pickering, 1836.

———. *Geology and Mineralogy Considered with Reference to Natural Theology.* 3rd ed. Edited by Frank Buckland, with additions by Richard Owen, John Phillips, and Robert Brown. 2 vols. London: George Routledge, 1858.

———. "Instructions for Conducting Geological Investigations, and Collecting Specimens." *American Journal of Science and Arts* 3 (1821): 249–51.

———. "Notice on the Megalosaurus, or Great Fossil Lizard of Stonesfield." *Transactions of the Geological Society* 1 (1824): 390–96.

———. "On the Fossil Remains of the Megatherium." *BAAS, Report 1831, 1832*, 104–7.

———. *Reliquiae Diluvianae; or, Observations on the Organic Remains Contained in Caves, Fissures, and Diluvial Gravel, and on other Geological Phenomena, Attesting the Action of an Universal Deluge.* London: John Murray, 1823.

———. *Vindiciae Geologicae; or the Connexion of Geology with Religion Explained.* Oxford: University Press, 1820.

Buffetaut, Eric. *A Short History of Vertebrate Palaeontology.* London: Croom Helm, 1987.

Burgess, G. H. O. *The Curious World of Frank Buckland.* London: John Baker, 1967.

Burrow, John W. "In the Iguanodon Diner." *London Review of Books*, Oct. 6, 1994, 13–14.

Butcher, Barry W. "Gorilla Warfare in Melbourne." In *Australian Science in the Making*, edited by R. W. Home, 153–69. Cambridge: Cambridge University Press, 1988.

Bynum, W. F. "The Anatomical Method, Natural Theology, and the Functions of the Brain." *Isis* 64 (1973): 445–68.

———. "Charles Lyell's *Antiquity of Man* and Its Critics." *Journal of the History of Biology* 17 (1984): 153–87.

———. "Time's Noblest Offspring: The Problem of Man in the British Natural Historical Sciences, 1800–1863." PhD diss., University of Cambridge, 1974.

C., F. T. "Palaeontology." *Fraser's Magazine* 62 (1860): 505–25.

Camardi, Giovanni. "Richard Owen, Morphology and Evolution." *Journal of the History of Biology* 34 (2001): 481–515.

Camerini, Jane. "The Power of Biography." *Isis* 88 (1997): 306–11.

Cameron, Ian. *To the Farthest Ends of the Earth: The History of the Royal Geographical Society, 1830–1980.* London: Macdonald and Jane's, 1980.

Campbell, William C. "History of Trichinosis: Paget, Owen and the Discovery of *Trichinella spiralis.*" *Bulletin of the History of Medicine* 53 (1979): 520–52.

Cannon, Susan Faye. *Science in Culture: The Early Victorian Period.* New York: Science History Publications, 1978.

Cantor, Geoffrey. *Michael Faraday: Sandemanian and Scientist; A Study of Science and Religion in the Nineteenth Century.* London: Macmillan, 1991.

Cardwell, Donald Stephen Lowell. *The Organisation of Science in England.* Rev. ed. London: Heinemann, 1972.

Carlyle, Thomas. *Oliver Cromwell's Letters and Speeches, with Elucidations.* London: Chapman and Hall, 1845.

———. *On Heroes, Hero-Worship and the Heroic in History.* In *The Works of Thomas Carlyle in Thirty Volumes*, vol. 5. London: Chapman and Hall, 1901.

Carpenter, J. Estlin. *Nature and Man: Essays Scientific and Philosophical by William B. Carpenter with an Introductory Memoir.* London: Kegan Paul, 1888.

[Carpenter, W. B.]. "Darwin on the Origin of Species." *National Review* 10 (1860): 188–214.

———. *Introduction to the Study of the Foraminifera.* With William K. Parker and T. Rupert Jones. London: Robert Hardwicke for the Ray Society, 1862.

[———?]. "Owen and Coote on the Homologies of the Vertebrate Skeleton." *British and Foreign Medico-Chirurgical Review* 4 (1849): 175–83.

[———?]. "Owen and Maclise on the Archetype Skeleton." *British and Foreign Medico-Chirurgical Review* 2 (1848): 107–21.

———. *Principles of Comparative Physiology.* 3rd ed. London: Churchill, 1851.

———. *Principles of Comparative Physiology.* 4th ed. London: Churchill; Savill and Edwards, 1854.

[———]. "Professor Owen on the Comparative Anatomy and Physiology of the Vertebrate Animals." *British and Foreign Medical Review* 23 (1847): 472–92.

Carus, Carl Gustav. *England und Schottland im Jahre 1844.* Berlin: Duncker, 1845.

———. *Goethe: Zu dessen näherem Verständnis.* Leipzig: Weichardt, 1843.

———. *An Introduction to the Comparative Anatomy of Animals.* 2 vols. London: Longman, 1827.

———. *Lehrbuch der Zootomie.* Leipzig: Fleischer, 1818.

———. *Symbolik der menschlichen Gestalt: Ein Handbuch zur Menschenkenntnis.* Leipzig: Brockhaus, 1853.

———. *Versuch einer Darstellung des Nervensystems und insbesondere des Gehirns nach ihrer Bedeutung, Entwicklung und Vollendung im thierischen Organismus.* Leipzig: Breitkopf und Härtel, 1814.

———. *Von den Ur-Theilen des Knochen- und Schalengerüstes.* Leipzig: Fleischer, 1828.

———. "Zur vergleichenden Symbolik zwischen Menschen- und Affen-Skelett." *Verhandlungen der Kaiserlichen Leopoldinisch-Carolinischen deutschen Akademie der Naturforscher* 20 (1862): 1–18.

Carus, J. Victor. *Geschichte der Zoologie bis auf Joh. Müller und Charl. Darwin.* Munich: Oldenbourg, 1872.

Casier, Edgard. *Les iguanodons de Bernissart.* Brussels: Institut Royal des Sciences Naturelles de Belgique, 1960.

Cave, A. J. E. "The Glands of Owen." *St. Bartholomew's Hospital Journal* 57 (1953): 131–33.

———. "The Muscles of Owen." *St. Bartholomew's Hospital Journal* 61 (1957): 138–40.

———. "Richard Owen and the Discovery of the Parathyroid Glands." In *Science, Medicine and History*, edited by E. Ashworth Underwood, 217–22. London: Oxford University Press, 1953.

[Chambers, Robert]. *Vestiges of the Natural History of Creation.* 11th ed. London: Churchill, 1860.

———. *Vestiges of the Natural History of Creation and Other Evolutionary Writings.* Edited by James A. Secord. Chicago: University of Chicago Press, 1994.

[Chapman, John]. "Equatorial Africa, and Its Inhabitants." *Westminster Review*, o.s., 76 (1861): 137–87.

Chappie, J. A. V. *Science and Literature in the Nineteenth Century.* London: Macmillan, 1986.

Clark, John Willis. *Old Friends at Cambridge and Elsewhere.* London: Macmillan, 1900.

Clark, John Willis, and Thomas McKenny Hughes. *The Life and Letters of the Reverend Adam Sedgwick.* 2 vols. Cambridge: Cambridge University Press, 1890.

Clark, Ronald William. *The Huxleys.* London: Heinemann, 1968.

Cohen, I. Bernard. "Three Notes on the Reception of Darwin's Ideas on Natural Selection." In *The Darwinian Heritage,* edited by David Kohn and Malcolm J. Kottler, 589–607. Princeton, NJ: Princeton University Press, 1985.

Coleman, William. *Georges Cuvier, Zoologist: A Study in the History of Evolution Theory.* Cambridge, MA: Harvard University Press, 1964.

Comte, Isidore Auguste Marie François Xavier. *Cours de philosophie positive.* 6 vols. Paris: Bachelier, 1830–42.

Conway Morris, Simon. *The Crucible of Creation: The Burgess Shale and the Rise of Animals.* Oxford: Oxford University Press, 1998.

______. *Life's Solution: Inevitable Humans in a Lonely Universe.* Cambridge: Cambridge University Press, 2003.

Cook, E. T., and A. Wedderburn, eds. *The Works of John Ruskin.* Vols. 18, 36. London: G. Allen, 1905, 1909.

[Cooper, Daniel]. "A Brief Sketch of the Rise and Progress of Microscopic Science." *Microscopic Journal and Structural Record* 1 (1841): 1–4.

Coote, Holmes. *The Homologies of the Human Skeleton.* London: Samuel Highley, 1849.

Cope, Zachary. *The Royal College of Surgeons of England: A History.* London: Anthony Blond, 1959.

Corsi, Pietro. *Science and Religion: Baden Powell and the Anglican Debate, 1800–1860.* Cambridge: Cambridge University Press, 1988.

Cosans, Christopher. "Anatomy, Metaphysics, and Values: The Ape Brain Debate Reconsidered." *Biology and Philosophy* 9 (1994): 129–65.

Cowell, F. R. *The Athenaeum Club and Social Life in London, 1824–1974.* London: Heinemann, 1975.

[Cox, G. W.]. "Du Chaillu's Adventures in Equatorial Africa." *Edinburgh Review* 114 (1861): 212–32.

Cruveilhier, J. *Anatomie descriptive.* Vol. 4. Paris: Bechet; Locquin, 1836.

Curwen, E. Cecil, ed. *The Journal of Gideon Mantell, Surgeon and Geologist.* London: Oxford University Press, 1940.

Cuvier, Frédéric. *Des dents des mammifères considérées comme caractères zoologiques.* Strasbourg: Levrault, 1825.

Cuvier, Georges. *Essay on the Theory of the Earth.* Translated by Robert Kerr. Edinburgh: William Blackwood, 1813.

______. *Leçons d'anatomie comparée.* 5 vols. Paris: Baudouin, 1799–1805.

______. *Leçons d'anatomie comparée.* 2nd ed. Paris: Crochard, 1835–37.

______. *Recherches sur les ossemens fossiles de quadrupèdes.* 4 vols. Paris: Deterville, 1812.

______. *Recherches sur les ossemens fossiles, ou l'on retablit les caractères de plusieurs animaux dont les révolutions du globe ont détruit les espèces.* 2nd ed. 5 vols. Paris: Dufour and d'Ocagne, 1821–24.

______. *Recherches sur les ossemens fossiles, ou l'on retablit les caractères de plusieurs animaux dont les révolutions du globe ont détruit les espèces.* 3rd ed. Paris: Dufour, 1825.

______. *Le règne animal distribué d'après son organisation, pour servir de base à l'histoire naturelle des animaux et d'introduction à l'anatomie comparée.* 4 vols. Paris: Deterville, 1817.

Dana, J. D. "On the Higher Subdivisions in the Classification of Mammals." *American Journal of Science,* 2nd ser., 35 (1863): 65–71.

Darwin, Charles Robert, ed. *The Descent of Man and Selection in Relation to Sex.* 2nd ed. London: Murray, 1874.

______. *On the Origin of Species by Means of Natural Selection; or, the Preservation of Favoured Races in the Struggle for Life.* London: John Murray, 1859.

______. *On the Origin of Species by Means of Natural Selection; or, the Preservation of Favoured Races in the Struggle for Life.* 4th ed. London: John Murray, 1866.

______. *On the Origin of Species by Means of Natural Selection; or, the Preservation of Favoured Races in the Struggle for Life.* 6th ed. London: John Murray, 1872.

Darwin, Charles Robert, and Alfred Wallace. "On the Tendency of Species to Form Varieties; and on the Perpetuation of Varieties and Species by Natural Means of Selection." *Journal of the Proceedings of the Linnean Society* (Zoology) 3 (1859): 45–62.

Darwin, Francis, ed. *The Life and Letters of Charles Darwin, Including an Autobiographical Chapter.* 2 vols. New York: D. Appleton, 1901.

Darwin, Francis, and A. C. Seward, eds. *More Letters of Charles Darwin.* 2 vols. New York: D. Appleton, 1903.

Daubeny, C. G. B. 1856. "Address." *BAAS, Report 1856*, xlviii–lxxiii.

______. *Fugitive Poems Connected with Natural History and Physical Science.* Oxford, 1869.

Dean, Dennis R. *Gideon Mantell and the Discovery of Dinosaurs.* Cambridge: Cambridge University Press, 1999.

Dear, Pauline Carpenter. "Richard Owen and the Invention of the Dinosaur." Henry and Ida Schuman Prize Paper. Unpublished MS, 1984.

Degen, Heinz. "Vor hundert Jahren: Die Naturforscherversammlung zu Göttingen und der Materialismusstreit." *Naturwissenschaftliche Rundschau* 7 (1954): 271–77.

De la Beche, H. T. *Researches in Theoretical Geology.* London: Charles Knight, 1834.

Desmond, Adrian. *Archetypes and Ancestors: Palaeontology in Victorian London, 1850–1875.* London: Blond and Briggs, 1982.

______. "Artisan Resistance and Evolution in Britain, 1819–1848." *Osiris* 3 (1987): 77–110.

______. "Designing the Dinosaur: Richard Owen's Response to Robert Edmond Grant." *Isis* 70 (1979): 224–34.

______. *Huxley: The Devil's Disciple.* London: Michael Joseph, 1994.

______. *Huxley: Evolution's High Priest.* London: Michael Joseph, 1997.

______. "Lamarckism and Democracy: Corporations, Corruption, Comparative Anatomy in the 1830s." In *History, Humanity, and Evolution*, edited by J. R. Moore, 99–130. Cambridge: Cambridge University Press, 1989.

______. *The Politics of Evolution: Morphology, Medicine, and Reform in Radical London.* Chicago: University of Chicago Press, 1989.

______. "Richard Owen's Reaction to Transmutation in the 1830s." *British Journal for the History of Science* 18 (1985): 25–50.

______. "Robert E. Grant: The Social Predicament of a Pre-Darwinian Transmutationist." *Journal of the History of Biology* 17 (1984): 189–223.

Desmond, Adrian, and James R. Moore. *Darwin.* London: Penguin, 1992.

Desmond, Ray. *The India Museum, 1801–1879.* London: H.M.S.O., 1982.

Dickens, Charles. *Our Mutual Friend.* London: Chapman and Hall, 1865.

Dobel, R., ed. *Lexikon der Goethe-Zitate.* Zurich: Artemis, 1968.

Dobson, Jessie. "An Account of the Life and Achievements of Richard Owen." British Museum (Natural History). Unpublished MS, 1981.

______. *Conservators of the Hunterian Museum.* London, n.d. (Reprinted from the *Annals of the Royal College of Surgeons of England.*)

______. "John Hunter's Museum." In *The Royal College of Surgeons of England*, by Zachary Cope, 274–306. London: Anthony Blond, 1959.

______. *William Clift.* London: William Heinemann Medical Books, 1954.

Donovan, C. *Reply to Sir B. Brodie's Attack on Phrenology.* London: H. Baillière, 1857.

Du Chaillu, P. B. *Explorations and Adventures in Equatorial Africa, 1856–9.* London: Murray, 1861.

______. "The Geographical Features and Natural History of a Hitherto Unexplored Region of Western Africa." *Proceedings of the Royal Geographical Society of London* 5 (1860–61): 108–12 (108–10 lecture; 110–12 discussion).

______. *A Journey to Ashango-Land: and Further Penetration into Equatorial Africa.* London: Murray, 1867.

[Duns, John]. "Fossil Footprints—Hitchcock." *North British Review* 32 (1860): 247–63.

[______]. "Genesis and Science." *North British Review* 27 (1857): 325–65.

[______]. "Professor Owen's Works." *North British Review* 28 (1858): 313–45.

[Eastlake, Elizabeth, and Harriet Grote]. "The British Museum." *Quarterly Review* 124 (1868): 147–79.

Ecker, Alexander. *Lorenz Oken: A Biographical Sketch.* Translated from the German by Alfred Tulk. London: Kegan Paul, etc., 1883.

[Egerton, Philip de Malpas Grey]. "The Gorilla's Dilemma." *Punch* 43 (1861): 164.

[______]. "Monkeyana." *Punch* 40 (1861): 206.

Ehrenberg, Christian Gottfried. *Die Infusionstierchen als vollkommene Organismen: Ein Blick in das tiefere organische Leben der Natur.* 2 vols. Leipzig: Voss, 1838.

Eiseley, Loren. *Darwin's Century.* New York: Doubleday, 1958.

Elwick, James. "Styles of Reasoning in Early to Mid-Victorian Life Research: Analysis, Synthesis and Palaetiology." *Journal of the History of Biology* 40 (2007): 35–69.

______. *Styles of Reasoning in the British Life Sciences: Shared Assumptions, 1820–1858.* London: Pickering and Chatto, 2007.

Erdl, Michael. "Untersuchungen über den Bau der Zähne bei den Wirbelthieren, insbesondere den Nagern." *Abhandlungen der mathematisch-physikalischen Classe der Königlich bayerischen Akademie der Wissenschaften* 3 (1837–43): 483–548.

Eve, A. S., and C. H. Creasy. *Life and Work of John Tyndall.* London: Macmillan, 1945.

[Fairbairn, Patrick]. "Recent Attacks on the Pentateuch." *North British Review* 38 (1863): 36–74.

Falconer, Hugh. "On the Disputed Affinity of the Mammalian Genus Plagiaulax, from the Purbeck Beds." *Quarterly Journal of the Geological Society* 18 (1862): 348–69.

Faraday, Michael. "On the Crystalline Polarity of Bismuth and Other Bodies, and Its Relation to the Magnetic Force." *Athenaeum*, Feb. 3, 1849, 120.

Farber, Paul L. "The Type-Concept in Zoology during the First Half of the Nineteenth Century." *Journal of the History of Biology* 9 (1976): 93–119.

Farley, John. *The Spontaneous Generation Controversy from Descartes to Oparin.* Baltimore, MD: Johns Hopkins University Press, 1977.

Feremutsch, Kurt. "Die Grundzüge der Hirnanatomie bei Carl Gustav Carus (1789–1869)." *Centaurus* 2 (1951–53): 52–85.

Finnegan, Diarmid A. "The Spatial Turn: Geographical Approaches in the History of Science." *Journal of the History of Biology* 41 (2008): 369–88.

Fisch, Menachem. *William Whewell, Philosopher of Science.* Oxford: Clarendon, 1991.

Fisch, Menachem, and Simon Schaffer, eds. *William Whewell: A Composite Portrait.* Oxford: Clarendon, 1991.

Fischer, Gotthelf. *Das Nationalmuseum der Naturgeschichte zu Paris.* 2 vols. Frankfurt am Main: Esslinger, 1802.

Fisher, George P. *Life of Benjamin Silliman, M.D., LL.D., Late Professor of Chemistry, Mineralogy, and Geology in Yale College.* 2 vols. New York: C. Scribner, 1866.

Fitton, William Henry. "On the Strata from Whence the Fossil Described in the Preceding Notice Was Obtained." *Zoological Journal* 3 (1828): 408–12.

Flower, W. H. *Essays on Museums and Other Subjects Connected with Natural History.* London: Macmillan, 1898.

[______]. *A General Guide to the British Museum (Natural History).* London: printed by order of the Trustees, 1887. (Owen asserted that he was the author of this *Guide*; on the title page of his private copy he wrote: "by Sir Richard Owen, K.C.B.")

______. "Obituary Notices of Fellows Deceased. Richard Owen." *Proceedings of the Royal Society* 55 (1894): i–xiv.

______. "On the Affinity and Probable Habits of the Extinct Australian Marsupial, *Thylacoleo carnifex*, Owen." *Quarterly Journal of the Geological Society* 24 (1868): 307–19.

______. "On the Posterior Lobes of the Cerebrum of the Quadrumana." *Philosophical Transactions of the Royal Society* 152 (1862): 185–201.

______. 1865. "Reply to Prof. Owen's Paper 'On Zoological Names of Characteristic Parts and Homological Interpretations of Their Modifications and Beginnings, Especially in Reference to Connecting Fibres of the Brain.'" *Proceedings of the Royal Society of London* 14 (1865): 134–39.

Forbes, Edward. "Abstract of the Theory of Specific Centres." In *Essays on the Spirit of the Inductive Philosophy*, edited by Baden Powell, 498–500. London: Longman, Brown, Green, and Longmans, 1855.

______. "On the Connexion between the Distribution of the Existing Fauna and Flora of the British Isles and the Geological Changes Which Have Affected Their Area." *Memoirs of the Geological Survey of England and Wales* 1 (1846): 336–432.

______. "On the Distribution of Endemic Plants." *BAAS, Report 1845*, Transactions of the Sections, 67–68.

______. "On the Morphology of the Reproductive System of Sertularian Zoophytes, and Its Analogy with That of the Flowering Plants." *BAAS, Report 1844*, Transactions of the Sections, 68–69. Also in *Athenaeum*, Oct. 26, 1844, 977–78.

______. 1848. "On the Question in Natural History, Have Genera, Like Species, Centres of Distribution?" *Edinburgh New Philosophical Journal* 45 (1848): 175–76.

[Ford, Richard]. "The British Museum." *Quarterly Review* 88 (1850): 136–72.

Forgan, Sophie. "Building the Museum: Knowledge, Conflict, and the Power of Place." *Isis* 96 (2005): 572–85.

[Foster, Michael]. "Higher and Lower Animals." *Quarterly Review* 127 (1869): 381–400.

Foster, Michael, and E. Ray Lankester, eds. *Scientific Memoirs of Huxley.* 4 vols. London: Macmillan, 1898–1902.

Foster, William C. *Sir Thomas Livingston Mitchell and His World, 1792–1855.* Sydney: Institution of Surveyors, NSW, 1985.

Fox, Caroline. *See* Pym, Horace N.

Fox, R. W. "Report on Some Observations on Subterranean Temperature." *BAAS, Report 1840*, 309–19.

Franke, Johannes, ed. *Joseph Victor von Scheffels sämtliche Werke*. Vol. 4. Leipzig: Hesse and Becker, 1916.

Froude, James Anthony. *Thomas Carlyle: A History of His Life in London, 1834–1881*. 2 vols. London: Longmans, Green, 1884.

Garland, Martha McMakin. *Cambridge before Darwin: The Ideal of a Liberal Education, 1800–1860*. Cambridge: Cambridge University Press, 1980.

Gegenbaur, Carl. *Grundzüge der vergleichenden Anatomie*. 2nd ed. Leipzig: Engelmann, 1870.

Geikie, Archibald. *Annals of the Royal Society Club: The Record of a London Dining-Club in the Eighteenth and Nineteenth Centuries*. London: Macmillan, 1917.

______. *Life of Sir Roderick I. Murchison Based on His Journals and Letters*. 2 vols. London: J. Murray, 1875.

Genschorek, Wolfgang. *Carl Gustav Carus: Arzt, Künstler, Naturforscher*. Leipzig: Hirzel, 1978.

Geoffroy Saint-Hilaire, Étienne. "Betrachtungen über die Eier des Ornithorhynchus, welche neue Beweise für die Frage über die Classification der Monotremen abgeben." *Archiv für Anatomie und Physiologie*, 1830, 119–26.

______. *Philosophie anatomique*. 2 vols. Paris: Rignoux, 1822.

Geoffroy Saint-Hilaire, Isidore. *Histoire naturelle générale des règnes organiques*. Vol. 2. Paris: Masson, 1859.

Gijzen, Agatha. *'s Rijks Museum van Natuurlijke Historie, 1820–1915*. Rotterdam: W. L. and J. Brusse, 1938.

Gilbert, Scott F. "Owen's Vertebral Archetype and Evolutionary Genetics—a Platonic Appreciation." *Perspectives in Biology and Medicine* 23 (1980): 475–88.

Gillespie, Neal C. *Charles Darwin and the Problem of Creation*. Chicago: University of Chicago Press, 1979.

______. "The Duke of Argyll, Evolutionary Anthropology, and the Art of Scientific Controversy." *Isis* 68 (1977): 40–54.

Gillispie, Charles C. *The Edge of Objectivity: An Essay in the History of Scientific Ideas*. Reprint, Princeton, NJ: Princeton University Press, 1970.

______. *Genesis and Geology: A Study in the Relations of Scientific Thought, Natural Theology, and Social Opinion in Great Britain, 1790–1850*. New York: Harper, 1959.

Girouard, Mark. *Alfred Waterhouse and the Natural History Museum*. London: British Museum (Natural History), 1981.

Gladstone, William Ewart. *The Impregnable Rock of Holy Scripture*. Rev. ed. London: Isbister, 1903.

______. "Proem to Genesis: A Plea for a Fair Trial." *Nineteenth Century* 19 (1886): 1–22.

Gladwin, W. J. "The Influence of the Anatomical Work of Oken upon British and French Comparative Anatomy in the Nineteenth Century." PhD diss., University of London, 1970.

Glanvill, Joseph. *Scepsis Scientifica: or, Confest Ignorance, the Way to Science; in an Essay of the Vanity of Dogmatising, and Confident Opinion. With a Reply to the Exceptions of the Learned Thomas Albius*. London, 1665.

Glass, Bentley, Owsei Temkin, and William L. Straus, eds. *Forerunners of Darwin, 1745–1859*. Baltimore, MD: Johns Hopkins Press, 1959.

Glick, Thomas F., ed. *The Comparative Reception of Darwinism*. Chicago: University of Chicago Press, 1988.

Gode-von Aesch, Alexander. 1941. *Natural Science in German Romanticism.* New York: Columbia University Press, 1941.

Goethe, Johann Wolfgang von. *Sämtliche Werke: Briefe, Tagebücher und Gespräche.* Vol. 24, *Naturkundliche Schriften II: Schriften zur Morphologie.* Edited by Dorothea Kuhn. Frankfurt am Main: Deutscher Klassiker, 1987.

Goodfield-Toulmin, Jane. "Some Aspects of English Physiology: 1780–1840." *Journal of the History of Biology* 2 (1969): 283–320.

Goodsir, John. "On the Anatomy of *Amphioxus lanceolatus*; Lancelet, Yarrell." *Transactions of the Edinburgh Royal Society* 15 (1844): 247–64.

______. *Testimonials in Favour of John Goodsir, Candidate for the Chair of Anatomy in the University of Edinburgh.* Edinburgh: W. MacPhail, 1846.

Gordon, E. O. *The Life and Correspondence of William Buckland.* London: J. Murray, 1894.

Gosse, Philip Henry. 1860. *The Romance of Natural History.* London: James Nisbet, 1860.

______. *The Romance of Natural History.* London: James Nisbet, 1861.

Goulburn, E. M., et al. *Replies to "Essays and Reviews."* 2nd ed. Oxford: J. H. and James Parker, 1862.

Gould, Stephen Jay. *Hen's Teeth and Horse's Toes.* New York: Norton, 1983.

______. *Ontogeny and Phylogeny.* Cambridge, MA: Belknap Press of Harvard University Press, 1977.

______. "A Seahorse for All Races." In *Leonardo's Mountain of Clams and the Diet of Worms: Essays on Natural History,* 119–140. New York: Harmony Books, 1998.

Grant, Robert E. "Lectures on Comparative Anatomy and Physiology." *Lancet,* no. 1 (1833–34): 89–99, 121–128, 152–59, 193–200.

______. *Outlines of Comparative Anatomy.* London: Hippolyte Bailliére, 1835–41.

Gratiolet, Louis Pierre. "Mémoire sur la microcéphalie considerée dans ses rapports avec la question des caractères du genre humain." *Mémoires de la Société d'anthropologie de Paris* 1 (1860): 61–67.

______. *Mémoire sur les plis cérébraux de l'homme et des primates.* 2 vols. Paris: Bertrand, 1854.

______. "Note sur l'encéphale du gorille (*Gorillagina,* I. Geof.-St.-H.)." *Comptes rendus hebdomadaire des séances de l'Académie des sciences* 50 (1860): 801–5.

Gray, Asa. *Darwiniana: Essays and Reviews Pertaining to Darwinism.* Reprint, Cambridge, MA: Belknap Press of Harvard University Press, 1963.

Gray, J. E. "Address." *BAAS, Report 1864,* Transactions of the Sections, 75–86.

Green, Joseph Henry. *Spiritual Philosophy: Founded on the Teaching of the Late Samuel Taylor Coleridge. Edited, with a Memoir of the Author's Life, by John Simon.* 2 vols. London: Macmillan, 1865.

______. *Vital Dynamics: The Hunterian Oration before the Royal College of Surgeons in London, 14th February 1840.* London: W. Pickering, 1840.

Greenaway, Frank, ed. *The Archives of the Royal Institution of Great Britain in Facsimile: Minutes of the Managers' Meetings, 1799–1900.* Vols. 8, 11. Ilkley: Scolar Press, 1975–76.

Gregorio, Mario A. di. 1982. "The Dinosaur Connection: A Reinterpretation of T. H. Huxley's Evolutionary View." *Journal of the History of Biology* 15 (1982): 379–418.

______. *T. H. Huxley's Place in Natural Science.* New Haven, CT: Yale University Press, 1984.

______. "A Wolf in Sheep's Clothing: Carl Gegenbaur, Ernst Haeckel, the Vertebral Theory of the Skull, and the Survival of Richard Owen." *Journal of the History of Biology* 28 (1995): 247–80.

Gregory, Frederick. "Scientific versus Dialectical Materialism: A Clash of Ideologies in Nineteenth-Century German Radicalism." *Isis* 68 (1977): 206–23.

Gross, Charles G. *Brain, Vision, Memory: Tales in the History of Neuroscience.* Cambridge, MA: MIT Press, 1998.

Grove, William Robert. "Address." *BAAS, Report 1866,* liii–lxxxii.

Gruber, Jacob W. *A Conscience in Conflict: The Life of St. George Jackson Mivart.* New York: published for Temple University Publications by Columbia University Press, 1960.

______. "Does the Platypus Lay Eggs? The History of an Event in Science." *Archives of Natural History* 18 (1991): 51–123.

______. "From Myth to Reality: The Case of the Moa." *Archives of Natural History* 14 (1987): 339–52.

______. "Owen, Sir Richard (1804–1892)." In *Oxford Dictionary of National Biography,* vol. 42, 245–54. Oxford: Oxford University Press, 2004.

______. Review of *Richard Owen: Victorian Naturalist,* by Nicolaas Rupke. *Albion* 27 (1995): 329–31.

Gruber, Jacob W., and John C. Thackray. *Richard Owen Commemoration.* London: Natural History Museum Publications, 1992.

Guedalla, Philip, ed. *Gladstone and Palmerston; Being the Correspondence of Lord Palmerston with Mr. Gladstone, 1851–1865.* London: Victor Gollancz, 1928.

Günther, A. C. L. "Address." *BAAS, Report 1880,* Transactions of the Sections, 591–98.

Gunther, A. E. *A Century of Zoology at the British Museum through the Lives of Two Keepers, 1815–1914.* London: Dawsons, 1975.

______. *The Founders of Science at the British Museum, 1753–1900.* Halesworth, Suffolk: Halesworth Press, 1980.

Haecker, Valentin. *Goethes morphologische Arbeiten und die neuere Forschung.* Jena: Fischer, 1927.

Haight, Gordon S., ed. 1978. *The George Eliot Letters.* Vol. 8. New Haven, CT: Yale University Press, 1978.

Hall, Brian K. Introduction to *Homology: The Hierarchical Basis of Comparative Anatomy,* edited by Brian K. Hall, 1–19. San Diego, CA: Academic Press, 1994.

______. Preface to Richard Owen, *On the Nature of Limbs,* edited by Ron Amundson, vii–xiv. Chicago: University of Chicago Press, 2007.

Hall, Marie Boas. *All Scientists Now: The Royal Society in the Nineteenth Century.* Cambridge: Cambridge University Press, 1984.

Hambury, H. J. "A Visit of Professor Carus of Dresden to the Royal College of Surgeons in 1844." *Annals of the Royal College of Surgeons* 18 (1956): 262–65.

Hannavy, John. *Roger Fenton of Crimble Hall.* London: Gordon Fraser Gallery, 1975.

[Harkness, Robert]. "Fossil Footprints." *Edinburgh Review* 110 (1859): 109–31.

Haupt, Hans. "Das Homologieprinzip bei Richard Owen: Ein Beitrag zur Geschichte des Platonismus in der Biologie." *Sudhoffs Archiv für Geschichte der Medizin und der Naturwissenschaften* 28 (1935): 143–228.

Hawkins, Charles, ed. *The Works of Sir Benjamin Collins Brodie, with an Autobiography.* 3 vols. London: Longmans, 1865.

Hays, J. N. "The London Lecturing Empire, 1800–50." In *Metropolis and Province,* edited by Ian Inkster and Jack Morrell, 91–119. London: Hutchinson, 1983.

Hegel, G. W. F. *Enzyklopädie der philosophischen Wissenschaften im Grundrisse (1830).* Vol. 2, *Die Naturphilosophie.* In *Georg Friedrich Wilhelm Hegel Werke,* edited by Eva Moldenhauer and Karl Markus Michel, vol. 9. Frankfurt am Main: Suhrkamp, 1970.

Heine, Heinrich. *Über Deutschland*. Part 1, *Zur Geschichte der Religion und Philosophie in Deutschland: Die romantische Schule*. Amsterdam: Schadd, 1870.

Herbert, Sandra. *Charles Darwin, Geologist*. Ithaca, NY: Cornell University Press, 2005.

———. "The Place of Man in the Development of Darwin's Theory of Transmutation. Part I, To July 1837." *Journal of the History of Biology* 7 (1974): 217–58. "Part 2." 10 (1977): 155–227.

Herschel, J. F. W., ed. *Admiralty Manual of Scientific Enquiry, with a New Introduction by David Knight*. London: Dawsons, 1974.

Himmelfarb, Gertrud. 1962. *Darwin and the Darwinian Revolution*. New York: Doubleday, 1962.

Hitchcock, Edward. "Description of the Foot Marks of Birds (Ornithichnites) of New Red Sandstone in Massachusetts." *American Journal of Science* 29 (1836): 307–40.

———. *The Religion of Geology and Its Connected Sciences*. London: David Bogue, 1851.

H[obson], W. F. *Longevity; or, Professor Owen and the Speaker's Commentary*. Oxford and London, 1872.

Hodge, M. J. S. "Darwin and the Laws of the Animate Part of the Terrestrial System (1835–37): On the Lyellian Origins of His Zoonomical Explanatory Program." *Studies in the History of Biology* 6 (1982–83): 1–106.

Holland, Julian. "Thomas Mitchell and the Origins of Australian Vertebrate Palaeontology." *Journal and Proceedings, Royal Society of New South Wales* 125 (1992): 103–6.

Hollander, Bernard. *Scientific Phrenology*. London: Grant Richards, 1902.

Home, Everard. "A Description of the Anatomy of *Ornithorhynchus paradoxus*." *Philosophical Transactions* 92 (1802): 67–84.

———. *Lectures on Comparative Anatomy; in Which Are Explained the Preparations in the Hunterian Collection*. 6 vols. London: G. and W. Nichol, 1814–28.

———. "Some Observations on the Head of the *Ornithorhynchus paradoxus*." *Philosophical Transactions* 90 (1800): 432–36.

Hooker, J. D. *The Botany of the Antarctic Voyage of H.M. Discovery Ships* Erebus *and* Terror. Vol. 2, *Flora Novae-Zelandiae*. London: Reeve Brothers, 1853–55.

———. *The Botany of the Antarctic Voyage of H.M. Discovery Ships* Erebus *and* Terror. Vol. 3, *Flora Tasmaniae*. London: Reeve Brothers, 1860.

Hopkins, William. 1852. "Anniversary Address to the Geological Society, 1852." *Quarterly Journal of the Geological Society* 8 (1852): xxi–lxxx.

Howard, Robert W. *The Dawnseekers: The First History of American Paleontology*. New York: Harcourt Brace Jovanovich, 1975.

Howarth, O. J. R. *The British Association for the Advancement of Science: A Retrospect, 1831–1931*. London: The Association, 1931.

Hull, David L. *Darwin and His Critics: The Reception of Darwin's Theory of Evolution by the Scientific Community*. Cambridge, MA: Harvard University Press, 1973.

———. *Science as Process: An Evolutionary Account of the Social and Conceptual Development of Science*. Chicago: University of Chicago Press, 1988.

Hume, A. *The Learned Societies and Printing Clubs of the United Kingdom*. London: Willis, 1853.

Hunter, John. *Essays and Observations on Natural History, Anatomy, Physiology, Psychology, and Geology. Being His Posthumous Papers on Those Subjects, Arranged and Revised, with Notes. By Richard Owen*. London: J. Van Voorst, 1861.

———. *Observations and Reflections on Geology: Intended to Serve as an Introduction to the Catalogue of His Collection of Extraneous Fossils*. London: Taylor and Francis, 1859.

Huschke, Emil. *Schaedel, Hirn und Seele des Menschen und der Thiere nach Alter, Geschlecht und Raçe*. Jena: Mauke, 1854.

Hutton, James. "Theory of the Earth; or an Investigation of the Laws Discernible in the Composition, Dissolution and Restoration of Land upon the Globe." *Transactions of the Royal Society of Edinburgh* 1 (1788): 209–304.

Huxley, Leonard. *Life and Letters of Sir Joseph Dalton Hooker.* 2 vols. London, 1918.

______. *Life and Letters of Thomas Henry Huxley.* 2 vols. New York: D. Appleton, 1901.

Huxley, T. H. "Anniversary Address to the Geological Society, Febr. 21, 1862" (1862). In *Scientific Memoirs,* vol. 2, 512–29.

______. "Darwin on the Origin of Species." *Westminster Review,* o.s., 73 (1860): 541–70.

______. *Evidence as to Man's Place in Nature.* London: Williams and Norgate, 1863.

______. "Explanatory Preface to the Catalogue of the Palaeontological Collection in the Museum of Practical Geology" (1865). In *Scientific Memoirs,* vol. 3, 125–79.

______. "Further Evidence of the Affinity between Dinosaurian Reptiles and Birds" (1870). In *Scientific Memoirs,* vol. 3, 465–86.

______. "Man and the Apes." *Athenaeum,* Mar. 30, 1861, 433; Apr. 13, 1861, 498.

______. "Note on the Resemblances and Differences in the Structure and the Development of the Brain in Man and Apes." In *The Descent of Man,* by Charles Darwin, 309–18. 2nd ed. London: Murray, 1874.

______. "On Certain Zoological Arguments Commonly Adduced in Favour of the Hypothesis of the Progressive Development of Animal Life in Time" (1855). In *Scientific Memoirs,* vol. 1, 300–304.

______. "On Natural History, as Knowledge, Discipline, and Power" (1856). In *Scientific Memoirs,* vol. 1, 305–14.

______. "On Some Organisms Living at Great Depths in the North Atlantic Ocean." *Quarterly Journal of the Microscopical Society,* n.s., 8 (1868): 203–12. Also in *Scientific Memoirs,* vol. 3, 330–39.

______. "On Species and Races, and Their Origin" (1860). In *Scientific Memoirs,* vol. 2, 388–94.

______. "On the Agamic Reproduction and Morphology of *Aphis*" (1858). In *Scientific Memoirs,* vol. 2, 26–80.

______. "On the Brain of Ateles paniscus." *Proceedings of the Zoological Society,* 1861, 247–60. Also in *Scientific Memoirs,* vol. 2, 493–508.

______. "On the Classification of the Dinosauria with Observations on the Dinosauria of the Trias" (1870). In *Scientific Memoirs,* vol. 3, 487–509.

______. "On the Common Plan of Animal Forms" (1854). In *Scientific Memoirs,* vol. 1, 281–83.

______. "On the Method of Palaeontology" (1856). In *Scientific Memoirs,* vol. 1, 432–44.

______. "On the Persistent Types of Animal Life" (1859). In *Scientific Memoirs,* vol. 2, 90–93.

______. "On the Premolar Teeth of Diprotodon, and on a New Species of That Genus." *Quarterly Journal of the Geological Society* 18 (1862): 422–27. Also in *Scientific Memoirs,* vol. 2, 538–45.

______. "On the Theory of the Vertebrate Skull" (1858). In *Scientific Memoirs,* vol. 1, 538–606.

______. "On the Zoological Relations of Man with the Lower Animals." *Natural History Review,* 1861, 67–84.

[______]. 1856. "Owen and Rymer Jones on Comparative Anatomy." *British and Foreign Medico-Chirurgical Review* 35: 1–27.

______. "Owen's Position in the History of Anatomical Science." In *The Life of Richard Owen,* by Rev. Owen, vol. 2, 273–332. London: John Murray, 1894.

______. *The Scientific Memoirs of Thomas Henry Huxley.* Edited by Michael Foster and E. Ray Lankester. 4 vols. London: Macmillan, 1898–1902.

———. "Upon Animal Individuality" (1852). In *Scientific Memoirs*, vol. 1, 146–51.

[———]. "The Vestiges of Creation." *British and Foreign Medico-Chirurgical Review* 26 (1854): 425–39.

Ingles, Jean M., and Frederick C. Sawyer. "A Catalogue of the Richard Owen Collection of Palaeontological and Zoological Drawings in the British Museum (Natural History)." *Bulletin of the British Museum (Natural History)*, Historical ser., 6 (1979): 109–97.

Irvine, William. *Apes, Angels and Victorians: A Joint Biography of Darwin and Huxley*. London: Readers Union, Weidenfeld and Nicolson, 1956.

Jacyna, L. S. "Immanence or Transcendence: Theories of Life and Organization in Britain, 1790–1835." *Isis* 74 (1983): 311–29.

———. "The Physiology of Mind, the Unity of Nature, and the Moral Order in Late Victorian Thought." *British Journal of the History of Science* 14 (1981): 109–32.

———. "The Romantic Programme and the Reception of Cell Theory in Britain." *Journal of the History of Biology* 17 (1984): 13–48.

James, Frank A. J. L., ed. *Chemistry and Theology in Mid-Victorian London: The Diary of Herbert McLeod, 1860–1870*. Microfiche ed. London: Mansell, 1987.

James, Kenneth W. "*Damned Nonsense!*"*—the Geological Career of the Third Earl of Enniskillen*. Belfast: Ulster Museum, 1986.

Jardine, Nicholas. "The Concept of Homology in Biology." *British Journal for the Philosophy of Science* 18 (1967): 125–39.

[Jenkins, John Edward]. *Lord Bantam*. 2 vols. London: Strahan, 1872.

Jensen, J. Vernon. "Return to the Wilberforce-Huxley Debate." *British Journal for the History of Science* 21 (1988): 161–79.

———. *Thomas Henry Huxley: Communicating for Science*. Newark: University of Delaware Press, 1991.

———. "The X Club: Fraternity of Victorian Scientists." *British Journal for the History of Science* 5 (1970): 63–72.

[Jones, John W.]. "British Museum." *Quarterly Review* 104 (1858): 201–24.

Jones, T. Rymer. *A General Outline of the Animal Kingdom, and Manual of Comparative Anatomy*. London: John Van Voorst, 1838–41.

———. "Osseous System." In *Cyclopaedia of Anatomy and Physiology*, edited by Robert E. Todd, vol. 3, 820–47. London: Sherwood, Gilbert and Piper, 1839–47.

Kaplan, Fred. *Thomas Carlyle: A Biography*. Cambridge: Cambridge University Press, 1983.

Keith, Arthur. "Abstract of Minutes of the Museum Committee, Royal College of Surgeons of England, from 1800–1907." Royal College of Surgeons. Unpublished MS, 1908.

Kern, Hans. *Carl Gustav Carus: Persönlichkeit und Werk*. Berlin: Widukind-Verlag, 1942.

Kier, William M., and Kathleen K. Smith. "Tongues, Tentacles and Trunks: The Biomechanics of Movement in Muscular-Hydrostats." *Zoological Journal of the Linnean Society* 83 (1985): 307–24.

Kilian, H. F. *Die Universitäten Deutschlands in medicinisch-naturwissenschaftlicher Hinsicht betrachtet*. Facsimile reprint of the 1828 ed. Amsterdam: Israel, 1966.

Kinch, Michael P. "Geographical Distribution and the Origin of Life: The Development of Early Nineteenth-Century British Explanations." *Journal of the History of Biology* 13 (1980): 91–119.

Kingsley, Charles. *Charles Kingsley: His Letters and Memories of His Life; Edited by His Wife*. 2 vols. London: King, 1877.

______. "Speech of Lord Dundreary in Section D, on Friday Last, on the Great Hippocampus Question." In *Charles Kingsley: His Letters and Memories of His Life; Edited by His Wife*, vol. 3, 145–48. London: Macmillan, 1901.

______. *The Water Babies: A Fairy Tale for a Land-Baby.* 1st ed. London: Macmillan, 1863.

Klaauw, C. J. van der. "The Scientific Correspondence between Professor Jan van der Hoeven and Professor Richard Owen." *Janus* 36 (1932): 327–51.

Knox, Robert. *Great Artists and Great Anatomists: A Biographical and Philosophical Study.* London: J. Van Voorst, 1852.

______. "Introduction to Inquiries into the Philosophy of Zoology." *Lancet*, no. 1 (1855): 625–27; no. 2 (1855): 24–26, 45–46, 68–71, 162–64, 186–88, 216–18.

______. "On Organic Harmonies: Anatomical Co-relations, and Methods of Zoology and Palaeontology." *Lancet*, no. 2 (1856): 245–47, 270–71, 297–300.

Kohlstedt, Sally Gregory. "Australian Museums of Natural History: Public Priorities and Scientific Initiatives in the 19th Century." *Historical Records of Australian Science* 5 (1983): 1–29.

Kohn, David, ed. *The Darwinian Heritage.* Princeton, NJ: Princeton University Press in association with Nova Pacifica, 1985.

Kölliker, Rudolf Albert von. *Ueber die Darwinsche Schöpfungstheorie: Ein am 13. Februar 1864 in der Phys. Med. Gesellschaft von Würzburg gehaltener Vortrag.* Leipzig: Engelmann, 1864.

Köstlin, Otto. *Der Bau des knöchernen Kopfes in den vier Klassen der Wirbelthiere.* Stuttgart: Schweizerbart, 1844.

Kuhl, Heinrich. *Beiträge zur Zoologie und vergleichenden Anatomie.* Frankfurt am Main: Hermann, 1820.

Kuhn, Dorothea. *Empirische und ideelle Wirklichkeit: Studien über Goethes Kritik des französischen Akademiestreites.* Graz: Böhlau, 1967.

Kuhn-Schnyder, Emil. *Lorenz Oken (1779–1851).* Zurich: Rohr, 1980.

Lang, J. D. "Account of the Discovery of Bone Caves in Wellington Valley, about 210 Miles West from Sydney in New Holland." *Edinburgh New Philosophical Journal*, Oct. 1830–Apr. 1831: 364–68.

Langer, Wolfhart. "Frühe Bilder aus der Vorzeit." *Fossilien* 7 (1990): 202–4.

Lankester, E. Ray. "On the Use of the Term Homology in Modern Zoology, and the Distinction between Homogenetic and Homoplastic Agreements." *Annals and Magazine of Natural History*, 4th ser., 6 (1870): 34–43.

Larson, Edward J. *Evolution: The Remarkable History of a Scientific Theory.* New York: Modern Library, 2004.

Latour, Bruno. "Pasteur et Pouchet: Hétérogenèse de l'histoire des sciences." In *Éléments d'histoire des sciences*, edited by Michel Serres, 423–45. Paris: Bordas, 1989.

Lawrence, Christopher. "The Power and the Glory: Humphry Davy and Romanticism." In *Romanticism and the Sciences*, edited by Andrew Cunningham and Nicholas Jardine, 213–27. Cambridge: Cambridge University Press, 1990.

Lawrence, P. "Charles Lyell versus the Theory of Central Heat: A Reappraisal of Lyell's Place in the History of Geology." *Journal of the History of Biology* 11 (1978): 101–28.

______. "Heaven and Earth—the Relation of the Nebular Hypothesis to Geology." In *Cosmology, History, and Theology*, edited by W. Yourgrau and A. D. Breck, 253–81. New York: Plenum Press, 1977.

Lawrence, William. *An Introduction to Comparative Anatomy and Physiology: Being the Two Introductory Lectures Delivered at the Royal College of Surgeons, on the 21st and 25th of March, 1816.* London: Callow, 1816.

______. *Lectures on Physiology, Zoology, and the Natural History of Man.* 6th ed. London: Edward Portwine; John Thomas Cox, 1834.

[Leifchild, John R.]. "The British Association at Cambridge." *London Quarterly Review* 19 (1863): 362–92.

Leland, C. G. *Gaudeamus! Humorous Poems Translated from the German of J. V. Scheffel and Others.* London: Trubner, 1872.

Lenoir, Timothy. "The Göttingen School and the Development of Transcendental Naturphilosophie in the Romantic Era." *Studies in the History of Biology* 5 (1981): 111–205.

______. "Morphotypes and the Historical-Genetic Method in Romantic Biology." In *Romanticism and the Sciences*, edited by Andrew Cunningham and Nicholas Jardine, 119–29. Cambridge: Cambridge University Press, 1990.

______. Review symposium "Imposing Owen." *Metascience* 7 (1995): 67–72.

______. *The Strategy of Life: Teleology and Mechanics in Nineteenth Century German Biology.* Dordrecht: Reidel, 1982.

Lenoir, Timothy, and Cheryl Lynn Ross. "The Naturalized History Museum." In *The Disunity of Science: Boundaries, Contexts, and Power*, edited by Peter Galison and David J. Stump, 370–97. Stanford, CA: Stanford University Press, 1996.

Leuret, François, and Louis Pierre Gratiolet. *Anatomie comparée du système nerveux consideré dans ses rapports avec l'intélligence.* 2 vols. Paris: Baillière, 1839–57.

Levere, Trevor H. *Poetry Realized in Nature: Samuel Taylor Coleridge and Early Nineteenth-Century Science.* Cambridge: Cambridge University Press, 1981.

______. Review symposium "Imposing Owen." *Metascience* 7 (1995): 56–58.

L[ewes], G. H. 1856. "Professor Owen and the Science of Life." *Fraser's Magazine* 53 (1856): 79–92.

Lewes, G. H. *The History of Philosophy from Thales to Comte.* 3rd ed. 2 vols. London: Longmans, Green, 1867.

[______]. "Life and Doctrine of Geoffroy St. Hilaire." *Westminster Review* 61 (1854): 160–90.

______. *The Life of Goethe.* 2nd ed. London: Smith, Elder, 1864.

[______]. "The Reign of Law." *Fortnightly Review*, n.s., 2 (1867): 96–111.

______. *Studies in Animal Life.* London: Smith, Elder, 1862.

[Lewis, G. G.]. "Military Defence of the Colonies." *Edinburgh Review* 115 (1862): 104–26.

Limoges, Camille. "The Development of the Muséum d'histoire naturelle of Paris, c. 1800–1914." In *The Organization of Science and Technology in France, 1808–1914*, edited by Robert Fox and G. Weisz, 211–40. Cambridge: Cambridge University Press, 1980.

______. "Owen as Strategist." *Science* 265 (1994): 1468–69.

Livingstone, David N. *Adam's Ancestors: Race, Religion, and the Politics of Human Origins.* Baltimore, MD: Johns Hopkins University Press, 2008.

______. "The Bishop and the Bulldog." In *Galileo Goes to Jail and Other Myths about Science and Religion*, edited by Keith Benson and Ronald L. Numbers. Cambridge, MA: Harvard University Press, in press.

______. *Putting Science in Its Place: Geographies of Scientific Knowledge.* Chicago: University of Chicago Press, 2003.

Logan, W. E. "On the Footprints Occurring in the Potsdam Sandstone of Canada." *Quarterly Journal of the Geological Society* 8 (1852): 199–213.

______. "On the Occurrence of a Track and Foot-Prints of an Animal in the Potsdam Sandstone of Lower Canada." *Quarterly Journal of the Geological Society* 7 (1851): 247–50.

Lonsdale, Henry. *A Sketch of the Life and Writings of Robert Knox.* London: Macmillan, 1870.

Lovejoy, A. O. *The Great Chain of Being: A Study in the History of an Idea.* New York: Harper, 1936.

Lucas, J. R. "Wilberforce and Huxley: A Legendary Encounter." *Historical Journal* 22 (1979): 313–30.

Lurie, Edward. *Louis Agassiz: A Life in Science.* Chicago: University of Chicago Press, 1960.

Lyell, Charles. "Anniversary Address to the Geological Society, 1850." *Quarterly Journal of the Geological Society* 6 (1850): xxxii–lxvi.

______. "Anniversary Address to the Geological Society, 1851." *Quarterly Journal of the Geological Society* 7 (1851): xxv–lxxvi.

______. *The Geological Evidences of the Antiquity of Man with Remarks on Theories of the Origin of Species by Variation.* London: John Murray, 1863.

______. *Principles of Geology, Being an Attempt to Explain the Former Changes of the Earth's Surface, by Reference to Causes Now in Operation.* 3 vols. London: John Murray, 1830–33.

______. *Principles of Geology, Being an Attempt to Explain the Former Changes of the Earth's Surface, by Reference to Causes Now in Operation.* 3rd ed. 3 vols. London: John Murray, 1834.

[______]. "Scientific Institutions." *Quarterly Review* 34 (1826): 153–79.

______. "The Theory of Successive Development in the Scale of Being Both Animal and Vegetable, from the Earliest Periods to Our Own Time, as Deduced from Palaeontological Evidence." *Edinburgh New Philosophical Journal* 51 (1851): 1–31.

______. "The Theory of the Successive Geological Development of Plants, from the Earliest Periods to Our Own Time, as Deduced from Palaeontological Evidence." *Edinburgh New Philosophical Journal* 51 (1851): 213–26.

Lyell, Charles, et al. *To the Right Hon. W. E. Gladstone, First Lord of the Treasury, etc. etc.* London, 1872.

Lyell, K. M., ed. *Life, Letters and Journals of Sir Charles Lyell, Bart.* 2 vols. London: John Murray, 1881.

Lyons, Sherrie Lynne. *Thomas Henry Huxley: The Evolution of a Scientist.* Amherst, NY: Prometheus Books, 1999.

Mabberley, D. J. *Jupiter Botanicus: Robert Brown of the British Museum.* Braunschweig: J. Cramer, 1985.

MacFarland, T. *Coleridge and the Pantheist Tradition.* Oxford: Clarendon Press, 1969.

Macilwain, George, ed. *Memoirs of John Abernethy, with a View of His Lectures, His Writings, and Character.* 3rd ed. London: Hatchard, 1856.

MacLeod, Roy M. "The Ayrton Incident: A Commentary on the Relations of Science and Government in England, 1870–1873." In *Science and Values: Patterns of Tradition and Change,* edited by Arnold Thackray and Everett Mendelsohn, 45–78. New York: Humanities Press, 1974.

______. "Evolutionism and Richard Owen, 1830–1868: An Episode in Darwin's Century." *Isis* 56 (1965): 259–80.

______. "Science and the Treasury." In *Patronage of Science in the Nineteenth Century,* edited by Gerard Turner, 115–72. Leiden: Noordhoff, 1976.

______. "Whigs and Savants: Reflections on the Reform Movement in the Royal Society, 1830–48." In *Metropolis and Province,* edited by Ian Inkster and Jack B. Morrell, 55–90. London: Hutchinson, 1983.

______. "The X-Club: A Social Network of Science in Late-Victorian England." *Notes and Records of the Royal Society* 24 (1970): 305–22.

Maclise, J. *Comparative Osteology: Being Morphological Studies to Demonstrate the Archetype Skeleton of Vertebrated Animals.* London: Taylor and Walton, 1847.

______. "On the Nomenclature of Anatomy (Addressed to Professors Owen and Grant)." *Lancet,* no. 1 (1846): 298–301.

______. "Skeleton." In *Cyclopaedia of Anatomy and Physiology,* edited by Robert B. Todd, vol. 4, pt. 1, 622–76. London: Sherwood, Gilbert and Piper, 1847–49.

Mann, Gunter, and Franz Dumont, eds. *Gehirn-Nerven-Seele: Anatomie und Physiologie im Umfeld S. Th. Soemmerings.* Stuttgart: Fischer, 1988.

Mantell, Gideon A. "A Few Notes on the Prices of Fossils." *London Geological Journal* 1 (1846): 13–17.

______. "The Geological Age of Reptiles." *Edinburgh New Philosophical Journal,* Apr.–Oct. 1831, 181–85.

______. *The Medals of Creation; or, First Lessons in Geology, and in the Study of Organic Remains.* 2 vols. London: H. G. Bohn, 1844.

______. "Notice of the Remains of the Dinornis and Other Birds, and of Fossils and Rock-Specimens, Recently Collected by Mr. Walter Mantell, in the Middle Island of New Zealand; with Additional Notes on the Northern Island. With a Note on Fossiliferous Deposits in the Middle Island of New Zealand. By Prof. E. Forbes." *Quarterly Journal of the Geological Society* 6 (1850): 319–43.

______. "Notice on the Iguanodon, a Newly Discovered Fossil Reptile, from the Sandstone of Tilgate Forest, in Sussex." *Philosophical Transactions* 115 (1824): 179–86.

______. "On the Fossil Remains of Birds Collected in Various Parts of New Zealand by Mr. Walter Mantell." *Quarterly Journal of the Geological Society* 4 (1848): 225–41.

[______]. *Thoughts on a Pebble; or a First Lesson in Geology.* 6th ed. London: Relfe and Fletcher, 1842.

______. *The Wonders of Geology.* 2 vols. London: Relfe and Fletcher, 1838.

Mantell, Walter. "Notice on the Co-existence of Man with the Dinornis in New Zealand." *Natural History Review,* n.s., 2 (1862): 343–45.

Marshall, J. 1861. "On the Brain of a Young Chimpanzee." *Natural History Review,* n.s., 1 (1861): 296–315.

Martin, Rudolf, and Karl Saller. *Lehrbuch der Anthropologie in systematischer Darstellung mit besonderer Berücksichtigung der anthropologischen Methoden.* Vol. 1. 3rd ed. Stuttgart: Fischer, 1957.

Matthew, H. C. G. *Gladstone, 1809–1874.* Oxford: Clarendon, 1986.

Mayr, Ernst. *The Growth of Biological Thought: Diversity, Evolution, and Inheritance.* Cambridge, MA: Belknap Press of Harvard University Press, 1982.

Mazzolini, Renato G. *Politisch-biologische Analogien im Frühwerk Rudolf Virchows.* Marburg: Basilisken-Presse, 1988.

McCook, Stuart. "'It May Be Truth, but It Is Not Evidence': Paul du Chaillu and the Legitimation of Evidence in the Field Sciences." *Osiris,* 2nd ser., 11 (1996): 177–97.

[McCosh, James]. "Typical Forms: Goethe, Professor Owen, Mr. Fairbairn." *North British Review* 15 (1851): 389–418.

McCosh, James, and George Dickie. *Typical Forms and Special Ends in Creation.* 2nd ed. Edinburgh: Thomas Constable, 1857.

Meckel, Johann Friedrich. *Ornithorhynchi Paradoxi Descriptio Anatomica.* Leipzig: Fleischer, 1826.

———. *System der vergleichenden Anatomie.* Vol. 2, pt. 2. Halle: Renger, 1825.

———. "Ueber die Brustdrüse des Ornithorhynchus." *Archiv für Anatomie und Physiologie,* 1827, 23–27.

Meffert, Ekkehard. *Carl Gustav Carus: Sein Leben—seine Anschauung von der Erde.* Stuttgart: Verlag Freies Geistesleben, 1986.

Merz, J. T. *A History of European Scientific Thought in the Nineteenth Century.* 2 vols. Edinburgh: William Blackwood and Sons, 1896.

Meyer, Hermann von. "On the *Archaeopteryx lithographica,* from the Lithographic Slate of Solnhofen." *Annals and Magazine of Natural History* 9 (1862): 366–70.

Miller, Edward. *Prince of Librarians: The Life and Times of Antonio Panizzi of the British Museum.* Athens, OH: Ohio University Press, 1967.

———. *That Noble Cabinet: A History of the British Museum.* Athens, OH: Ohio University Press, 1974.

Miller, Hugh. *Footprints of the Creator: or, the Asterolepis of Stromness.* London: Johnstone and Hunter, 1849.

———. *The Old Red Sandstone; or, New Walks in an Old Field.* Edinburgh: J. Johnstone, 1841.

———. *The Testimony of the Rocks.* Edinburgh: T. Constable; Shepherd and Elliot, 1857.

Mitchell, Thomas L. "An Account of the Limestone Caves at Wellington Valley, and of the Situation, near One of Them, Where Fossil Bones Have Been Found." *Proceedings of the Geological Society* 1 (1834): 321–22.

———. *Three Expeditions into the Interior of Eastern Australia; with Descriptions of the Recently Explored Region of Australia Felix, and of the Present Colony of New South Wales.* 2nd ed. 2 vols. London: T. and W. Boone, 1839.

Mivart, St. George Jackson. "The Beginnings and End of Life." *Quarterly Review* 170 (1890): 370–93.

[———]. "A Century of Science." *Living Age* 205 (1895): 771–87. Also in *Quarterly Review* 180 (1895): 381–405.

———. *The Common Frog.* London: Macmillan, 1874.

[———]. "Darwin's Descent of Man." *Quarterly Review* 131 (1871): 47–90.

———. *Essays and Criticisms.* 2 vols. London: Osgood, McIlvaine, 1892.

———. "Likeness; or Philosophical Anatomy." *Contemporary Review* 26 (1875): 938–57.

———. *Man and Apes: An Exposition of Structural Resemblances and Differences Bearing upon Questions of Affinity and Origin.* London: Robert Hardwicke, 1873.

———. "On the Appendicular Skeleton of the Primates." *Proceedings of the Royal Society* 15 (1866–67): 320–21.

———. *On the Genesis of Species.* 2nd ed. London: Macmillan, 1871.

———. "On the Use of the Term 'Homology.'" *Annals and Magazine of Natural History* 6 (1870): 113–21.

———. "Sir Richard Owen's Hypothesis." *Natural Science* 2 (1893): 18–23.

———. "The Vertebrate Skeleton." *Nature,* Aug. 11, 1870, 291–92.

Moore, Carlisle. "Carlyle and Goethe as Scientist." In *Carlyle and His Contemporaries: Essays in Honor of Charles Richard Sanders,* edited by John Clubbe, 21–34. Durham, NC: Duke University Press, 1976.

Moore, J. R. "1859 and All That: Remaking the Story of Evolution-and-Religion." In *Charles Darwin, 1809–1882: A Centennial Commemorative,* edited by R. G. Chapman and C. T. Duval, 166–94. Wellington: Nova Pacifica, 1982.

______, ed. *History, Humanity and Evolution: Essays for John C. Greene.* Cambridge: Cambridge University Press, 1989.

______. *The Post-Darwinian Controversies: A Study of the Protestant Struggle to Come to Terms with Darwin in Great Britain and America.* Cambridge: Cambridge University Press, 1979.

Morrell, Jack B., and Arnold Thackray. *Gentlemen of Science: Early Years of the British Association for the Advancement of Science.* Oxford: Clarendon Press, 1981.

Moxon, R. K. "Richard Owen (1804–1892), Naturalist and Anti-Darwinist." *New England Journal of Medicine* 267 (1962): 35–37.

Moyal, Ann Mozley. *"A Bright and Savage Land": Scientists in Colonial Australia.* Sydney: Collins, 1986.

______. *Scientists in Nineteenth Century Australia: A Documentary History.* Melbourne: Cassell Australia, 1976.

______. "Sir Richard Owen and His Influence on Australian Zoological and Palaeontological Science." *Records of the Australian Academy of Sciences* 3 (1975): 41–56.

[Mozley, J. B.]. "The Argument of Design." *Quarterly Review* 127 (1869): 134–76.

______. "The Oxford Commission." *Quarterly Review* 93 (1853): 152–238.

Müller, Johannes. *Elements of Physiology.* Translated from the German by William Baly. 2 vols. London: Taylor and Walton, 1842.

Murchison, Charles, ed. *Palaeontological Memoirs and Notes of the Late Hugh Falconer, AM., M.D., with a Biographical Sketch of the Author.* 2 vols. London: Robert Hardwicke, 1868.

Murchison, Roderick Impey. 1846. "Address." *BAAS, Report 1846,* xxvii–xliii.

Murray, David. *Museums, Their History and Their Use. With a Bibliography and a List of Museums in the United Kingdom.* 3 vols. Glasgow: James Mac Lehose and Sons, 1904.

Nägeli, Carl Wilhelm von. *Die Individualität in der Natur mit vorzüglicher Berücksichtigung des Pflanzenreiches.* Zurich: Meyer and Zeller, 1856.

Nasmyth, Alexander. *Researches on the Development, Structure, and Diseases of the Teeth.* London: J. Churchill, 1839.

Negus, Victor. *History of the Trustees of the Hunterian Collection.* Edinburgh: E. and S. Livingstone, 1966.

Neuburger, Max. "C. G. Carus on the State of Medicine in Britain in 1844." In *Science, Medicine and History,* edited by E. A. Underwood, vol. 2, 262–73. London: Oxford University Press, 1953.

Neumark, Victoria. "Naturally the Man for the Job." *Times Educational Supplement,* Mar. 18, 1994.

Newland, Elisabeth D. "Dr George Bennett and Sir Richard Owen: A Case Study of the Colonization of Early Australian Science." In *International Science and National Scientific Identity: Australia between Britain and America,* edited by R. W. Home and Sally Gregory Kohlstedt, 55–74. Dordrecht: Kluwer, 1991.

Nordenskiöld, Erik. *The History of Biology: A Survey.* New York: Tudor Publishing, 1946.

Nyhart, Lynn K. *Biology Takes Form: Animal Morphology and the German Universities, 1800–1900.* Chicago: University of Chicago Press, 1995.

Oken, Lorenz. *Abriss der Naturphilosophie.* Göttingen: Vandenhoek und Ruprecht, 1805.

______. *Elements of Physiophilosophy.* Translated from the German by Alfred Tulk. London: Ray Society, 1847.

______. *Lehrbuch der Naturgeschichte.* 3 vols. Leipzig: Reclam, 1813–26.

______. *Lehrbuch der Naturphilosophie.* 2nd ed. Jena: Frommann, 1831.

———. *Rede über das Zahlengesetz in den Wirbeln des Menschen.* Munich: Lindauer, 1828.

———. "Ueber das Zahlengesetz in den Wirbeln des Menschen." *Isis von Oken* 22 (1829): 306–12.

———. *Über die Bedeutung der Schädelknochen.* Jena: Göpferdt, 1807.

———. *Die Zeugung.* Bamberg: Goebhardt, 1805.

Olby, Robert. "A Retrospect on the Historiography of the Life Sciences." In *The Light of Nature,* edited by J. D. North and J. J. Roche, 95–109. Dordrecht: Nijhoff, 1985.

Oldroyd, David R. "How Did Darwin Arrive at His Theory? The Secondary Literature to 1982." *History of Science* 22 (1984): 325–74.

Oppenheim, Janet. *The Other World: Spiritualism and Psychical Research in England, 1850–1914.* Cambridge: Cambridge University Press, 1985.

Ospovat, Dov. *The Development of Darwin's Theory: Natural History, Natural Theology, and Natural Selection, 1838–1859.* Cambridge: Cambridge University Press, 1981.

———. "God and Natural Selection: The Darwinian Idea of Design." *Journal of the History of Biology* 13 (1980): 169–94.

———. "The Influence of Karl Ernst von Baer's Embryology, 1828–1859: A Reappraisal in Light of Richard Owen's and William B. Carpenter's 'Palaeontological Application of "Von Baer's Law."'" *Journal of the History of Biology* 9 (1970): 1–28.

———. "Perfect Adaptation and Teleological Explanation: Approaches to the Problem of the History of Life in the Mid-Nineteenth Century." *Studies in the History of Biology* 2 (1978): 33–56.

Ostrom, John. "The Meaning of *Archaeopteryx.*" In *The Beginnings of Birds,* edited by M. K. Hecht et al., 161–76. Eichstätt: Freunde des Jura-Museums Eichstätt, Willibaldsburg, 1985.

Outram, Dorinda. *Georges Cuvier: Vocation, Science and Authority in Post-revolutionary France.* Manchester: Manchester University Press, 1984.

Owen, Richard. "Account of a *Thylacinus,* the Great Dog-Headed Opossum, One of the Rarest and Largest of the Marsupiate Family of Animals." *BAAS, Report 1841,* Transactions of the Sections, 70–71.

———. "An Account of the Dissection of the Parts concerned in the Aneurism for the Cure of Which Dr. Stevens Tied the Internal Iliac Artery at Santa Cruz, in the Year 1812." *Transactions of the Medico-Chirurgical Society* 16 (1830): 219–35.

———. "Acrita." In *Cyclopaedia of Anatomy and Physiology,* edited by Robert B. Todd, vol. 1, 47–49. London: Sherwood, Gilbert, and Piper, 1835–36.

———. "Additional Evidence Proving the Australian Pachyderm Described in a Former Number of the 'Annals' to Be a Dinotherium, with Remarks on the Nature and Affinities of That Genus." *Annals and Magazine of Natural History* 11 (1843): 329–32.

———. "Address." *BAAS, Report 1858,* xlix–cx.

———. "Address." In *St. Mary's Hospital Medical School: Addresses on Medical Education,* 5–10. London: Robert Hardwicke, 1868.

———. Address to the fellows of the Royal Society. London: privately printed, 1873.

[———]. "Anatomy." In *Dictionary of Science, Literature, and Art,* edited by W. T. Brande and G. W. Cox, vol. 1, 97. London: Longmans, 1865.

[———]. "Ancient Animals in South America." *Edinburgh Review* 155 (1882): 186–204.

———. "The Brain of Man and Apes." *Medical Times and Gazette,* no. 2 (1862): 473–74.

[———]. "Broderip's *Zoological Recreations.*" *Quarterly Review* 82 (1847): 119–42.

———. "Cephalopoda." In *Cyclopaedia of Anatomy and Physiology,* edited by Robert B. Todd, vol. 1, 517–62. London: Sherwood, Gilbert, and Piper, 1835–36.

______. "Considérations sur le plan organique et le mode de developpement des animaux." *Annales des sciences naturelles; zoologie et biologie animale*, série 3 (Zoologie), 2 (1844): 162–88.

______. "Contributions to the Natural History of the Anthropoid Apes, no. VIII, On the External Characters of the Gorilla (*Troglodytes gorilla*, Sav.)." *Transactions of the Zoological Society* 5, pt. 4 (1865): 243–84.

______. *A Cuvierian Principle in Palaeontology Tested by Evidence of an Extinct Leonine Marsupial* (Thylacoleo carnifex). London: Taylor and Francis, 1871.

[______]. "Darwin on the *Origin of Species*." *Edinburgh Review* 111 (1860): 487–532.

______. *Derivative Hypothesis of Life and Species, Being the Concluding Chapter of the Anatomy of Vertebrates.* London, 1868.

______. "Description of a Fossil Molar Tooth of a Mastodon Discovered by Count Strzlecki in Australia." *Annals and Magazine of Natural History* 14 (1844): 268–71.

______. "Description of a Microscopic Entozoon Infesting the Muscles of the Human Body." *Transactions of the Zoological Society* 1 (1835): 315–24.

______. 1840. "A Description of a Specimen of the *Plesiosaurus Macrocephalus*, Conybeare, in the Collection of Viscount Cole, M.P., D.C.L., F.G.S., etc." *Transactions of the Geological Society*, 2nd ser., 5 (1840): 515–35.

______. "Description of a Tooth and Part of the Skeleton of the *Glyptodon*." *Transactions of the Geological Society*, 2nd ser., 6 (1841): 81–106.

______. "Description of Part of the Skeleton and Teeth of Five Species of the Genus *Labyrinthodon*." *Transactions of the Geological Society*, 2nd ser., 6 (1842): 515–44.

______. "Description of the Cavern of Bruniquel, and Its Organic Contents, Part I. Human Remains." *Philosophical Transactions* 159 (1869): 517–33; "Part II. Equine remains." *Philosophical Transactions* 159 (1869): 535–38.

______. "Description of the Impressions and Foot-Prints of the Protichnites from the Potsdam Sandstone of Canada." *Quarterly Journal of the Geological Society of London* 8 (1852): 214–25.

______. "Description of the Impressions on the Potsdam Sandstone, Discovered by Mr. Logan in Lower Canada." *Quarterly Journal of the Geological Society* 7 (1851): 250–52.

______. *Description of the Skeleton of an Extinct Gigantic Sloth,* Mylodon robustus, Owen, *with Observations on the Osteology, Natural Affinities, and Probable Habits of the Megatherioid Quadrupeds in General.* London: R. and J. E. Taylor, 1842.

______. "Descriptions of Three Skulls of Western Equatorial Africans—Fan, Ashira, and Fernand Vaz—with Some Admeasurements of the Rest of the Collection of Skulls, Transmitted to the British Museum from the Fernand Vaz, by P. B. du Chaillu." In *A Journey to Ashango-Land*, by P. B. du Chaillu, 439–60. London: John Murray, 1867.

[______]. *Descriptive and Illustrated Catalogue of the Fossil Organic Remains of Mammalia and Aves.* London: R. and J. E. Taylor, 1845.

______. *Descriptive and Illustrated Catalogue of the Fossil Reptilia of South Africa in the Collection of the British Museum.* London: [British Museum], 1876.

______. "The Earliest Discovered Evidence of Extinct Struthious Birds in New Zealand." *Geological Magazine*, 1873, 478.

______. "Entozoa." In *Cyclopaedia of Anatomy and Physiology*, edited by Robert B. Todd, vol. 2, 111–44. London: Sherwood, Gilbert, and Piper, 1836–39.

______. *Essays and Observations on Natural History, Anatomy, Physiology, Psychology, and Geology, by John Hunter, F.R.S.; Being His Posthumous Papers on Those Subjects, Arranged and Revised,*

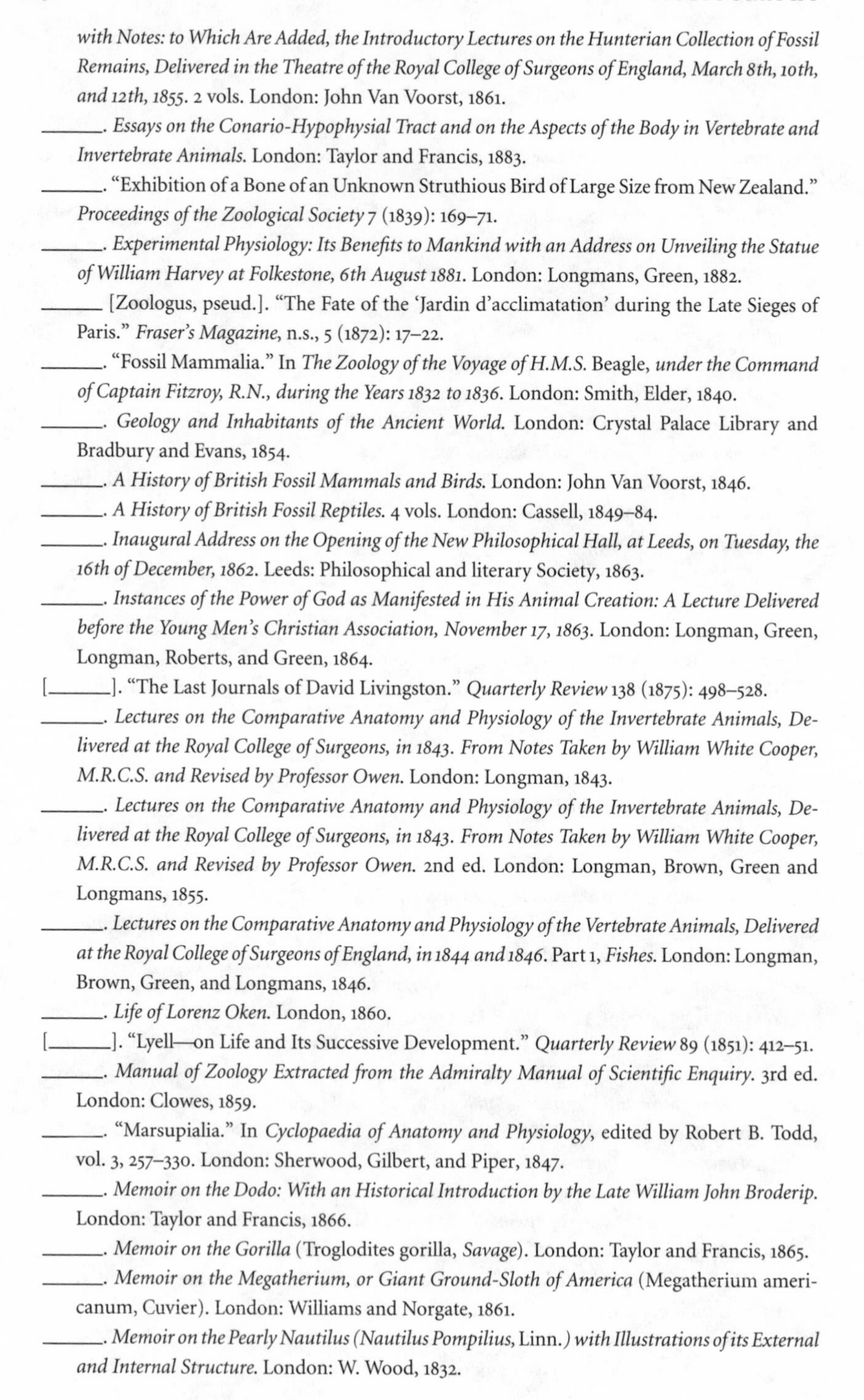

with Notes: to Which Are Added, the Introductory Lectures on the Hunterian Collection of Fossil Remains, Delivered in the Theatre of the Royal College of Surgeons of England, March 8th, 10th, and 12th, 1855. 2 vols. London: John Van Voorst, 1861.

———. *Essays on the Conario-Hypophysial Tract and on the Aspects of the Body in Vertebrate and Invertebrate Animals.* London: Taylor and Francis, 1883.

———. "Exhibition of a Bone of an Unknown Struthious Bird of Large Size from New Zealand." *Proceedings of the Zoological Society* 7 (1839): 169–71.

———. *Experimental Physiology: Its Benefits to Mankind with an Address on Unveiling the Statue of William Harvey at Folkestone, 6th August 1881.* London: Longmans, Green, 1882.

——— [Zoologus, pseud.]. "The Fate of the 'Jardin d'acclimatation' during the Late Sieges of Paris." *Fraser's Magazine,* n.s., 5 (1872): 17–22.

———. "Fossil Mammalia." In *The Zoology of the Voyage of H.M.S.* Beagle, *under the Command of Captain Fitzroy, R.N., during the Years 1832 to 1836.* London: Smith, Elder, 1840.

———. *Geology and Inhabitants of the Ancient World.* London: Crystal Palace Library and Bradbury and Evans, 1854.

———. *A History of British Fossil Mammals and Birds.* London: John Van Voorst, 1846.

———. *A History of British Fossil Reptiles.* 4 vols. London: Cassell, 1849–84.

———. *Inaugural Address on the Opening of the New Philosophical Hall, at Leeds, on Tuesday, the 16th of December, 1862.* Leeds: Philosophical and literary Society, 1863.

———. *Instances of the Power of God as Manifested in His Animal Creation: A Lecture Delivered before the Young Men's Christian Association, November 17, 1863.* London: Longman, Green, Longman, Roberts, and Green, 1864.

[———]. "The Last Journals of David Livingston." *Quarterly Review* 138 (1875): 498–528.

———. *Lectures on the Comparative Anatomy and Physiology of the Invertebrate Animals, Delivered at the Royal College of Surgeons, in 1843. From Notes Taken by William White Cooper, M.R.C.S. and Revised by Professor Owen.* London: Longman, 1843.

———. *Lectures on the Comparative Anatomy and Physiology of the Invertebrate Animals, Delivered at the Royal College of Surgeons, in 1843. From Notes Taken by William White Cooper, M.R.C.S. and Revised by Professor Owen.* 2nd ed. London: Longman, Brown, Green and Longmans, 1855.

———. *Lectures on the Comparative Anatomy and Physiology of the Vertebrate Animals, Delivered at the Royal College of Surgeons of England, in 1844 and 1846.* Part 1, *Fishes.* London: Longman, Brown, Green, and Longmans, 1846.

———. *Life of Lorenz Oken.* London, 1860.

[———]. "Lyell—on Life and Its Successive Development." *Quarterly Review* 89 (1851): 412–51.

———. *Manual of Zoology Extracted from the Admiralty Manual of Scientific Enquiry.* 3rd ed. London: Clowes, 1859.

———. "Marsupialia." In *Cyclopaedia of Anatomy and Physiology,* edited by Robert B. Todd, vol. 3, 257–330. London: Sherwood, Gilbert, and Piper, 1847.

———. *Memoir on the Dodo: With an Historical Introduction by the Late William John Broderip.* London: Taylor and Francis, 1866.

———. *Memoir on the Gorilla* (Troglodites gorilla, *Savage*). London: Taylor and Francis, 1865.

———. *Memoir on the Megatherium, or Giant Ground-Sloth of America* (Megatherium americanum, Cuvier). London: Williams and Norgate, 1861.

———. *Memoir on the Pearly Nautilus (Nautilus Pompilius,* Linn.*) with Illustrations of its External and Internal Structure.* London: W. Wood, 1832.

______. *Memoirs on the Extinct Wingless Birds of New Zealand; with an Appendix on Those of England, Australia, Newfoundland, Mauritius, and Rodriguez.* 2 vols. London: J. Van Voorst, 1879.

______. "Milton and Galileo." *Fraser's Magazine* 79 (1869): 678–84.

______. "Monograph of the Fossil Reptilia of the Liassic Formations, Part II (Pterosauria)." *Palaeontographical Society* 23 (1870): 41–81.

______. *Monograph on the Aye-aye* (Chiromys madagascariensis, *Cuvier*). London: Taylor and Francis, 1863.

______. *A Monograph on the Fossil Reptilia of the Mesozoic Formations.* London: printed for the Palaeontographical Society, 1874–89.

______. "Monotremata." In *The Cyclopaedia of Anatomy and Physiology*, edited by Robert B. Todd, vol. 3, 366–407. London: Sherwood, Gilbert, and Piper, 1847.

[______]. "Mr. Cumming's *Hunter's Life in South Africa.*" *Quarterly Review* 88 (1850): 1–41.

______. "Note on the Glyptodon." In *Buenos Ayres*, by Woodbine Parish, 178(b)–(e). London: J. Murray, 1838.

______. "Notice of a Fragment of the Femur of a Gigantic Bird of New Zealand." *Transactions of the Zoological Society* 3 (1842): 29–32.

______. "Notices of Some Fossil Mammalia of South America." *BAAS, Report 1846*, Transactions of the Sections, 65–67.

______. "Observations on Mr. Strickland's Article on the Structural Relations of Organized Beings." *Philosophical Magazine* 28 (1846): 525–27.

______. *Odontography; or, a Treatise on the Comparative Anatomy of the Teeth; Their Physiological Relations, Mode of Development, and Microscopic Structure, in the Vertebrate Animals.* 2 vols. London: Hippolyte Baillière, 1840–45.

______. "Oken, Lorenz." In *Encyclopaedia Britannica*, vol. 16, 498–503. 8th ed. Edinburgh: Adam and Charles Black, 1858. 9th ed. 1884, 749–52.

______. "Oken, Lorenz." In *Encyclopaedia Britannica*, vol. 16, 749–52. 9th ed. Edinburgh: Adam and Charles Black, 1884.

______. "On a National Museum of Natural History." *Athenaeum*, July 27, 1861, 118–20; Aug. 3, 1861: 153–55; Aug. 10, 1861: 187–89.

______. "On a New Genus (*Dimorphodon*) of Pterodactyle, with Remarks on the Geological Distribution of Flying Reptiles." *BAAS, Report 1858*, Transactions of the Sections, 97–98.

______. "On an Extinct Genus of Struthious Bird from New Zealand." *Proceedings of the Zoological Society* 10 (1843): 8–10.

______. "On British Eocene Serpents and the Serpent of the Bible." *Edinburgh New Philosophical Journal* 49 (1850): 239–42. Also in the *American Journal of Science*, 2nd ser., 11 (1851): 281–83.

______. "On British Fossil Reptiles." *Edinburgh New Philosophical Journal* 33 (1842): 65–88.

______. "On Dinornis (Part IV): Containing the Restoration of the Feet of That Genus and of *Palapteryx*, with a Description of the Sternum in *Palapteryx* and *Aptornis.*" *Transactions of the Zoological Society* 4 (1850): 1–20.

______. "On *Dinornis Novae-Zealandiae.*" *Proceedings of the Zoological Society* 11 (1843): 144–46.

______. "On Longevity." *Fraser's Magazine*, n.s., 5 (1872): 218–33.

______. "On *Macrauchenia patachonica.*" *Geological Magazine* 2 (1865): 520–23.

______. "On Marsupiata." *BAAS, Report 1838*, Transactions of the Sections, 105.

______. "On Metamorphosis and Metagenesis." *Proceedings of the Royal Institution* 1 (1851): 9–16. Also in *Medical Times and Gazette*, n.s., 23 (1851): 663–66.

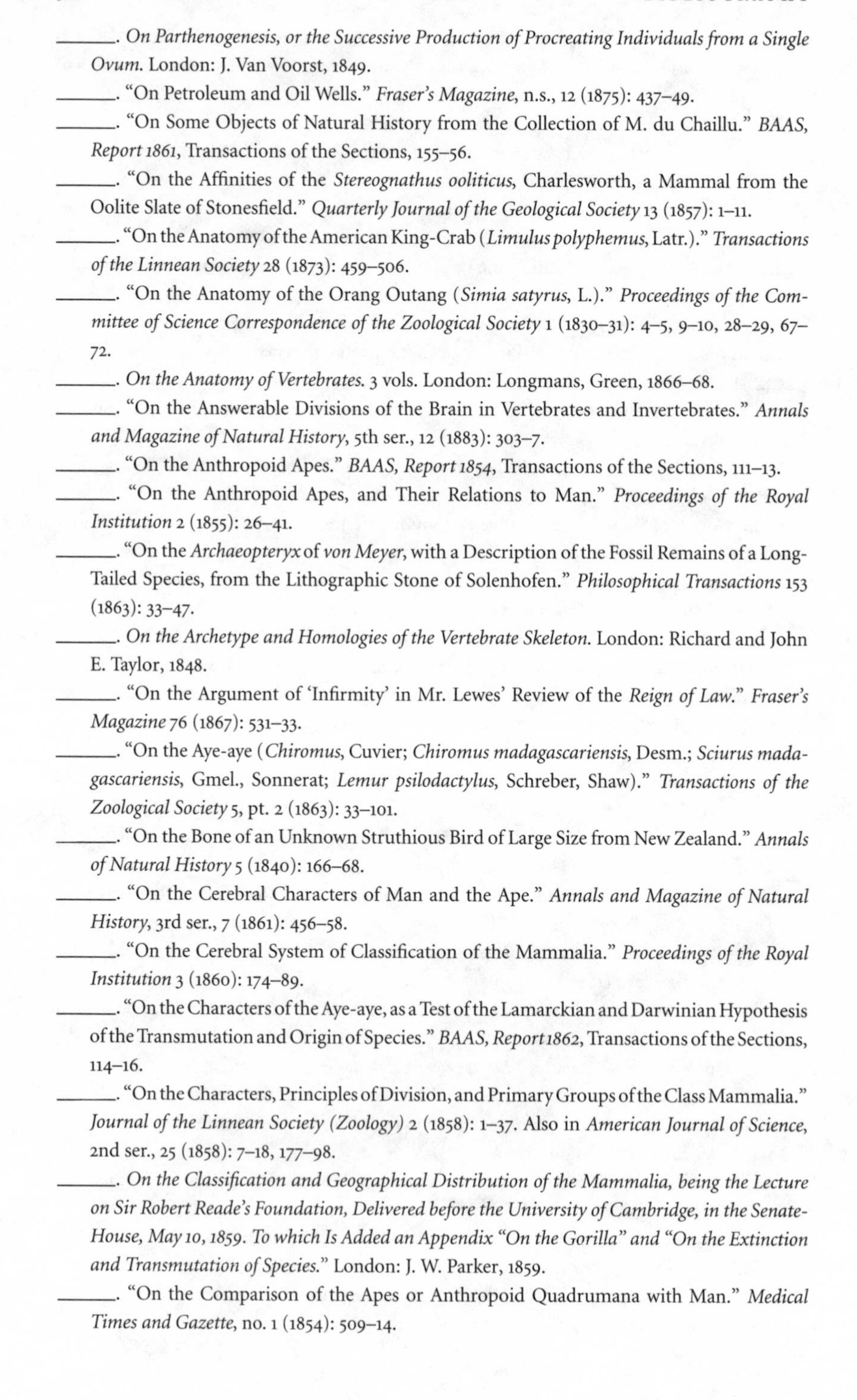

———. *On Parthenogenesis, or the Successive Production of Procreating Individuals from a Single Ovum.* London: J. Van Voorst, 1849.

———. "On Petroleum and Oil Wells." *Fraser's Magazine,* n.s., 12 (1875): 437–49.

———. "On Some Objects of Natural History from the Collection of M. du Chaillu." *BAAS, Report 1861,* Transactions of the Sections, 155–56.

———. "On the Affinities of the *Stereognathus ooliticus,* Charlesworth, a Mammal from the Oolite Slate of Stonesfield." *Quarterly Journal of the Geological Society* 13 (1857): 1–11.

———. "On the Anatomy of the American King-Crab (*Limulus polyphemus,* Latr.)." *Transactions of the Linnean Society* 28 (1873): 459–506.

———. "On the Anatomy of the Orang Outang (*Simia satyrus,* L.)." *Proceedings of the Committee of Science Correspondence of the Zoological Society* 1 (1830–31): 4–5, 9–10, 28–29, 67–72.

———. *On the Anatomy of Vertebrates.* 3 vols. London: Longmans, Green, 1866–68.

———. "On the Answerable Divisions of the Brain in Vertebrates and Invertebrates." *Annals and Magazine of Natural History,* 5th ser., 12 (1883): 303–7.

———. "On the Anthropoid Apes." *BAAS, Report 1854,* Transactions of the Sections, 111–13.

———. "On the Anthropoid Apes, and Their Relations to Man." *Proceedings of the Royal Institution* 2 (1855): 26–41.

———. "On the *Archaeopteryx* of *von Meyer,* with a Description of the Fossil Remains of a Long-Tailed Species, from the Lithographic Stone of Solenhofen." *Philosophical Transactions* 153 (1863): 33–47.

———. *On the Archetype and Homologies of the Vertebrate Skeleton.* London: Richard and John E. Taylor, 1848.

———. "On the Argument of 'Infirmity' in Mr. Lewes' Review of the *Reign of Law.*" *Fraser's Magazine* 76 (1867): 531–33.

———. "On the Aye-aye (*Chiromus,* Cuvier; *Chiromus madagascariensis,* Desm.; *Sciurus madagascariensis,* Gmel., Sonnerat; *Lemur psilodactylus,* Schreber, Shaw)." *Transactions of the Zoological Society* 5, pt. 2 (1863): 33–101.

———. "On the Bone of an Unknown Struthious Bird of Large Size from New Zealand." *Annals of Natural History* 5 (1840): 166–68.

———. "On the Cerebral Characters of Man and the Ape." *Annals and Magazine of Natural History,* 3rd ser., 7 (1861): 456–58.

———. "On the Cerebral System of Classification of the Mammalia." *Proceedings of the Royal Institution* 3 (1860): 174–89.

———. "On the Characters of the Aye-aye, as a Test of the Lamarckian and Darwinian Hypothesis of the Transmutation and Origin of Species." *BAAS, Report 1862,* Transactions of the Sections, 114–16.

———. "On the Characters, Principles of Division, and Primary Groups of the Class Mammalia." *Journal of the Linnean Society (Zoology)* 2 (1858): 1–37. Also in *American Journal of Science,* 2nd ser., 25 (1858): 7–18, 177–98.

———. *On the Classification and Geographical Distribution of the Mammalia, being the Lecture on Sir Robert Reade's Foundation, Delivered before the University of Cambridge, in the Senate-House, May 10, 1859. To which Is Added an Appendix "On the Gorilla" and "On the Extinction and Transmutation of Species."* London: J. W. Parker, 1859.

———. "On the Comparison of the Apes or Anthropoid Quadrumana with Man." *Medical Times and Gazette,* no. 1 (1854): 509–14.

______. "On the Dicynodont Reptilia, with a Description of Some Fossil Remains Brought by H.R.H. Prince Alfred from South Africa, November 1860." *Philosophical Transactions* 152 (1862): 455–67.

______. "On the Discovery of the Remains of a Mastodontoid Pachyderm in Australia." *Annals and Magazine of Natural History* 11 (1843): 7–12.

______. *On the Extent and Aims of a National Museum of Natural History.* London: Saunders, Otley, 1862.

______. "On the Extinct Animals of the Colonies of Great Britain." *Proceedings of the Royal Colonial Institute* 10 (1878–79): 267–97.

______. "On the Fossil Mammals of Australia. Part I. Description of a Mutilated Skull of a Large Marsupial Carnivore (*Thylacoleo carnifex*, Owen), from a Calcareous Conglomerate Stratum, Eighty Miles S.W. of Melbourne, Victoria." *Philosophical Transactions* 149 (1859): 309–22.

______. "On the Fossil Mammals of Australia. Part IV. Dentition and Mandible of *Thylacoleo carnifex*, with Remarks on the Argument for Its Herbivority." *Philosophical Transactions* 161 (1871): 213–66.

______. "On the Geographical Distribution of Extinct Mammalia." *Athenaeum*, Feb. 14, 1846, 178–79.

______. "On the Gorilla." *Proceedings of the Royal Institution* 3 (1859): 10–30.

______. "On the Gorilla (Troglodytes Gorilla, Sav.)." *Proceedings of the Zoological Society* 27 (1859): 1–23.

______. "On the Homologies and Notation of the Dental System in Mammalia." *BAAS, Report 1848*, Transactions of the Sections, 91–93.

______. "On the Homology of the Conario-Hypophysial Tract, or of the So-Called 'Pineal' and 'Pituitary Glands.'" *BAAS, Report 1881*, Transactions of Sections, 719–20.

______. "On the Homology of the Conario-Hypophysial Tract, or the So-Called Pineal and Pituitary Glands." *Journal of the Linnean Society (Zoology)* 16 (1882): 131–49.

______. "On the Influence of the Advent of a Higher Form of Life in Modifying the Structure of an Older and Lower Form." *Quarterly Journal of the Geological Society* 34 (1878): 421–30.

______. "On the Jaws of the *Thylacotherium Prevostii* (Valenciennes) from Stonesfield." *Proceedings of the Geological Society* 3 (1838): 5–9. Also in *Magazine of Natural History* 3 (1839): 201–9.

______. "On the Megatherium" (Croonian Lecture, May 8, 1851). *Abstracts of the Papers Communicated to the Royal Society* 6 (1850–54): 57–65.

______. *On the Nature of Limbs.* London: J. Van Voorst, 1849.

______. *On the Nature of Limbs: A Discourse.* Edited by Ron Amundson. Chicago: University of Chicago Press, 2007.

______. "On the Osteology of the Chimpanzee and Orang Outang." *Transactions of the Zoological Society* 1 (1835): 343–80.

______. "On the *Phascolotherium.*" *Proceedings of the Geological Society* 3 (1838): 17–21.

______. "On the Psychical and Physical Characters of the Mincopies, or Natives of the Andaman Islands, and on the Relations Thereby Indicated to Other Races of Mankind." *BAAS, Report 1861*, 241–9.

______. "On the Sternum of *Notornis* and on Sternal Characters." *Proceedings of the Zoological Society*, 1882, 689–97.

______. "On the Structure and Homology of the Cephalic Tentacles in the Pearly Nautilus." *Annals and Magazine of Natural History* 12 (1843): 305–11.

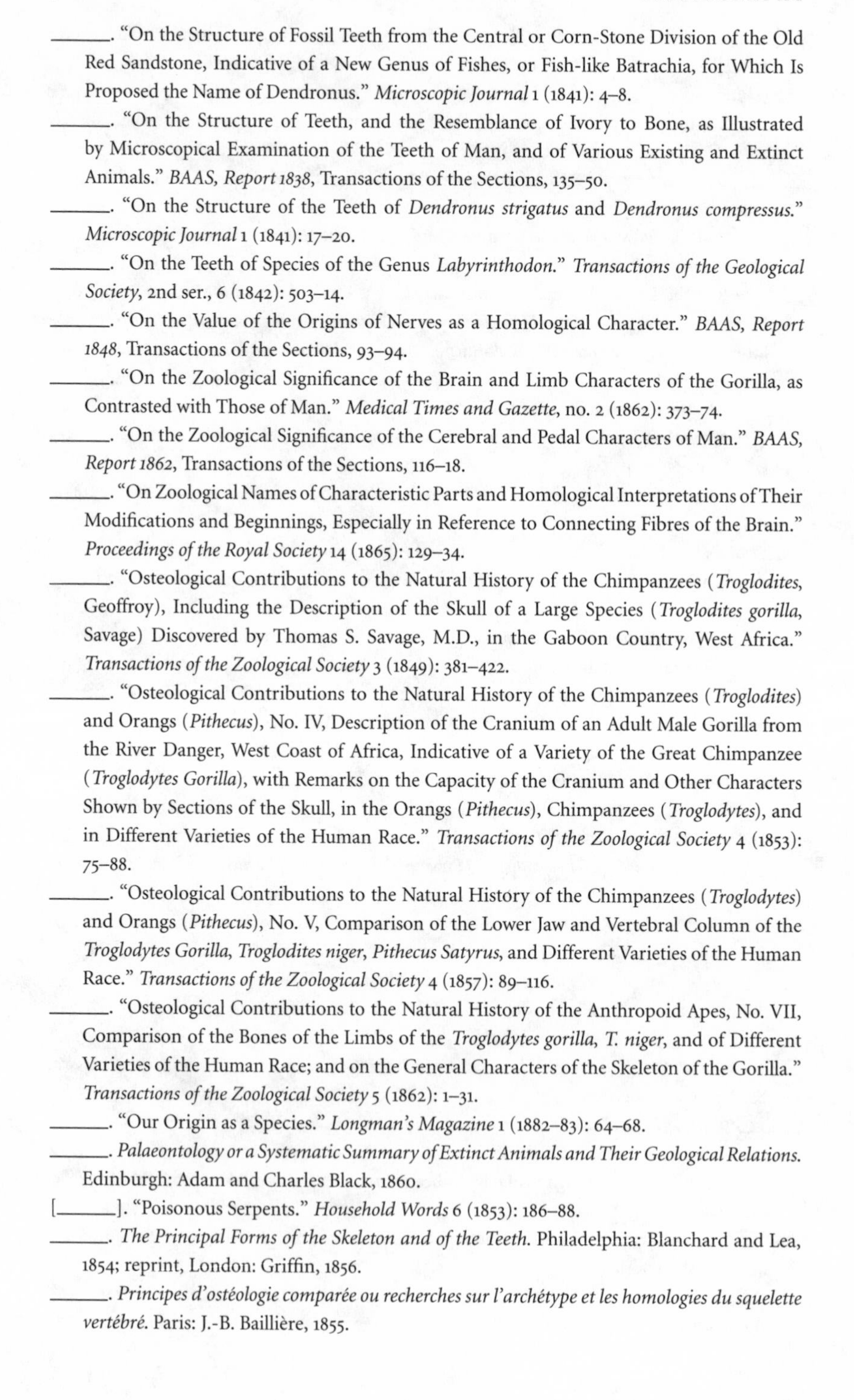

———. "On the Structure of Fossil Teeth from the Central or Corn-Stone Division of the Old Red Sandstone, Indicative of a New Genus of Fishes, or Fish-like Batrachia, for Which Is Proposed the Name of Dendronus." *Microscopic Journal* 1 (1841): 4–8.

———. "On the Structure of Teeth, and the Resemblance of Ivory to Bone, as Illustrated by Microscopical Examination of the Teeth of Man, and of Various Existing and Extinct Animals." *BAAS, Report 1838*, Transactions of the Sections, 135–50.

———. "On the Structure of the Teeth of *Dendronus strigatus* and *Dendronus compressus*." *Microscopic Journal* 1 (1841): 17–20.

———. "On the Teeth of Species of the Genus *Labyrinthodon*." *Transactions of the Geological Society*, 2nd ser., 6 (1842): 503–14.

———. "On the Value of the Origins of Nerves as a Homological Character." *BAAS, Report 1848*, Transactions of the Sections, 93–94.

———. "On the Zoological Significance of the Brain and Limb Characters of the Gorilla, as Contrasted with Those of Man." *Medical Times and Gazette*, no. 2 (1862): 373–74.

———. "On the Zoological Significance of the Cerebral and Pedal Characters of Man." *BAAS, Report 1862*, Transactions of the Sections, 116–18.

———. "On Zoological Names of Characteristic Parts and Homological Interpretations of Their Modifications and Beginnings, Especially in Reference to Connecting Fibres of the Brain." *Proceedings of the Royal Society* 14 (1865): 129–34.

———. "Osteological Contributions to the Natural History of the Chimpanzees (*Troglodites*, Geoffroy), Including the Description of the Skull of a Large Species (*Troglodites gorilla*, Savage) Discovered by Thomas S. Savage, M.D., in the Gaboon Country, West Africa." *Transactions of the Zoological Society* 3 (1849): 381–422.

———. "Osteological Contributions to the Natural History of the Chimpanzees (*Troglodites*) and Orangs (*Pithecus*), No. IV, Description of the Cranium of an Adult Male Gorilla from the River Danger, West Coast of Africa, Indicative of a Variety of the Great Chimpanzee (*Troglodytes Gorilla*), with Remarks on the Capacity of the Cranium and Other Characters Shown by Sections of the Skull, in the Orangs (*Pithecus*), Chimpanzees (*Troglodytes*), and in Different Varieties of the Human Race." *Transactions of the Zoological Society* 4 (1853): 75–88.

———. "Osteological Contributions to the Natural History of the Chimpanzees (*Troglodytes*) and Orangs (*Pithecus*), No. V, Comparison of the Lower Jaw and Vertebral Column of the *Troglodytes Gorilla, Troglodites niger, Pithecus Satyrus*, and Different Varieties of the Human Race." *Transactions of the Zoological Society* 4 (1857): 89–116.

———. "Osteological Contributions to the Natural History of the Anthropoid Apes, No. VII, Comparison of the Bones of the Limbs of the *Troglodytes gorilla, T. niger*, and of Different Varieties of the Human Race; and on the General Characters of the Skeleton of the Gorilla." *Transactions of the Zoological Society* 5 (1862): 1–31.

———. "Our Origin as a Species." *Longman's Magazine* 1 (1882–83): 64–68.

———. *Palaeontology or a Systematic Summary of Extinct Animals and Their Geological Relations.* Edinburgh: Adam and Charles Black, 1860.

[———]. "Poisonous Serpents." *Household Words* 6 (1853): 186–88.

———. *The Principal Forms of the Skeleton and of the Teeth.* Philadelphia: Blanchard and Lea, 1854; reprint, London: Griffin, 1856.

———. *Principes d'ostéologie comparée ou recherches sur l'archétype et les homologies du squelette vertébré.* Paris: J.-B. Baillière, 1855.

______. "Raw Materials from the Animal Kingdom." In *Lectures on the Results of the Great Exhibition of 1851, Delivered before the Society of Arts, Manufacturers, Commerce, at the Suggestion of H.R.H. Prince Albert*, 77–131. London: David Bogue, 1852.

______ [Silas Seer, pseud.]. "Recollections and Reflections of Gideon Shaddoe, Esq." *Hood's Magazinemic Miscellany* 2 (1844): 442–50.

______ [Silas Seer, pseud.]. "Recollections and Reflections of Gideon Shaddoe, Esq." *Hood's Magazinemic Miscellany* 3 (1845): 294–303.

______. "Remarks on the Entozoa." *Transactions of the Zoological Society* 1 (1835): 387–94.

______. "Remarks on the 'Observations sur l'Ornithorhynque' par M. Jules Verraux." *Annals and Magazine of Natural History*, 2nd ser., 2 (1848): 317–22.

______. "Reply to Some Observations of Prof. Wagner on the Genus *Mylodon*." *Annals and Magazine of Natural History* 16 (1845): 100–102.

______. "Report on British Fossil Reptiles." *BAAS, Report 1839*, 43–126.

______. "Report on British Fossil Reptiles, Part II." *BAAS, Report 1841*, 60–204.

______. "Report on the Archetype and Homologies of the Vertebrate Skeleton." *BAAS, Report 1846*, 169–340.

______. "Report on the British Fossil Mammalia. Part I." *BAAS, Report 1842*, 54–74.

______. "Report on the British Fossil Mammalia. Part II." *BAAS, Report 1843*, 208–41.

______. "Report on the Extinct Mammals of Australia, with Descriptions of Certain Fossils Indicative of the Former Existence in That Continent of Large Marsupial Representatives of the Order *Pachydermata*." *BAAS, Report 1844*, 223–40.

______. *Report on the State of Lancaster*. London: W. Clowes and Sons, 1845.

______. *Researches on the Fossil Remains of the Extinct Mammals of Australia; with a Notice of the Extinct Marsupials of England*. 2 vols. London: J. Erxleben, 1877.

______. "Serpent Charming in Cairo." *Blackwood's Edinburgh Magazine* 111 (1872): 169–75.

[______]. "Sir Emerson Tennent's *Ceylon*." *Edinburgh Review* 110 (1859): 343–75.

______. "Sketch of Hunter's Scientific Character and Works." In *Leicester Square*, by Tom Taylor, 420–33. London: Bickers and Son, 1874.

______. "Summary of the Succession in Time and Geographical Distribution of Recent and Fossil Mammalia" (Apr. 12, 1859). *Proceedings of the Royal Institution of Great Britain* 3 (1858–62): 109–16.

______. "Teleology of the Skeleton of Fishes." *Edinburgh New Philosophical Journal* 42 (1847): 216–27.

______ [Ennoo, pseud.]. "A Visit to Selborne." *Blackwood's Edinburgh Magazine* 80 (1856): 175–83.

______. "Wayside Gatherings and Their Teachings." *Gentleman's Magazine*, n.s., 4 (1867): 41–51.

______. "Zoology." In *A Manual of Scientific Enquiry; Prepared for the Use of Her Majesty's Navy and Adapted for Travellers in General*, edited by J. F. W. Herschel, 343–99. London: J. Murray, 1849.

______. *The Zoology of the Voyage of H.M.S. Beagle, under the Command of Captain Fitzroy, R.N., during the Years 1832 to 1836. Edited and Superintended by Charles Darwin, Esq., M.A. F.R.S. Sec. G.S. Naturalist to the Expedition*. Part 1, *Fossil Mammalia*. London: Smith, Elder, 1840.

Owen, Richard, et al. "Periodical Phaenomena of Animals and Plants." *BAAS, Report 1845*, 321–36.

Owen, Rev. Richard. *The Life of Richard Owen by His Grandson, with the Scientific Portions Revised by C. Davies Sherborn*. 2 vols. London: John Murray, 1894.

Padian, Kevin. "The Case of the Bat-Winged Pterosaur: Typological Taxonomy and the Influence of Pictorial Representation on Scientific Perception." In *Dinosaurs Past and Present*, edited by S. J. Czerkas and E. C. Olsen, vol. 2, 65–81. Seattle: Natural History Museum of Los Angeles County in association with University of Washington Press, 1987.

———. "Form versus Function: The Evolution of a Dialectic." In *Functional Morphology in Vertebrate Paleontology*, edited by J. J. Thomason, 264–77. Cambridge: Cambridge University Press, 1995.

———. "A Missing Hunterian Lecture on Vertebrae by Richard Owen, 1837." *Journal of the History of Biology* 28 (1995): 333–68.

———. "The Rehabilitation of Sir Richard Owen." *BioScience* 47 (1997): 446–53.

———. "Richard Owen's Quadrophenia: The Pull of Opposing Forces in Victorian Cosmogony." In Richard Owen, *On the Nature of Limbs*, edited by Ron Amundson, liii–xci. Chicago: University of Chicago Press, 2007.

Paget, Stephen, ed. *Memoirs and Letters of Sir James Paget*. 3rd ed. London: Longmans, 1903.

Panchen, Alec L. "Richard Owen and the Concept of Homology." In *Homology: The Hierarchical Basis of Comparative Anatomy*, edited by Brian K. Hall, 21–62. San Diego: Academic Press, 1994.

Pantin, C. F. A. *Science and Education*. Cardiff: University of Wales Press, 1963.

Parish, Woodbine. "An Account of the Discovery of Portions of Three Skeletons of the Megatherium in the Province of Buenos Ayres in South America." *Proceedings of the Geological Society* 1 (1834): 403–4.

———. *Buenos Ayres, and the Provinces of the Rio de la Plata: Their Present State, Trade, and Debt; with Some Account from Original Documents of the Progress of Geographical Discovery in Those Parts of South America during the Last Sixty Years*. London: J. Murray, 1838.

Parker, Charles Stuart, ed. *Sir Robert Peel, from His Private Papers*. 2nd ed. 3 vols. London: John Murray, 1899.

[Parker, John W.]. "William John Broderip: In Memoriam." *Fraser's Magazine* 59 (1859): 485–88.

Parkinson, James. *Organic Remains of a Former World*. Vol. 3. London: Sherwood, Neely and Jones, 1811.

Patterson, Colin. "Archetypes and Ancestors." *Nature* 368 (1994): 375–76.

[Pattison, Mark]. "Bishop Colenso and the Pentateuch." *Westminster Review*, n.s., 23 (1863): 57–76.

Peacock, A. 1982. "The Relationship between the Soul and the Brain." In *Historical Aspects of the Neurosciences*, edited by F. Clifford Rose and W. F. Bynum, 83–98. New York: Raven Press, 1982.

Pentland, Joseph. "On the Fossil Bones of Wellington Valley, New Holland, or New South Wales." *Edinburgh New Philosophical Journal*, Oct. 1831–Apr. 1832, 301–8.

Peterson, Houston. *Huxley: Prophet of Science*. London: Longmans, Green, 1932.

Peterson, M. J. *The Medical Profession in Mid-Victorian London*. Berkeley and Los Angeles: University of California Press, 1978.

Pevsner, Nikolaus. *The Buildings of England, London: The Cities of London and Westminster*. Harmondsworth: Penguin, 1962.

Pfannenstiel, Max, and Rudolph Zaunick. "Lorenz Oken und J. W. von Goethe." *Sudhoffs Archiv für Geschichte der Medizin und der Naturwissenschaften* 33 (1941): 113–73.

Phillips, John. *A Guide to Geology*. London: Longman, Rees, Orme, Brown, Green, and Longman, 1834.

______. *Treatise on Geology.* Vol. 1. London: Longman, Rees, Orme, Brown, Green, and Longman, 1837.

Plato. *The Collected Dialogues of Plato.* Edited by E. Hamilton and Huntington Cairns. Princeton, NJ: Princeton University Press, 1973.

Player, Ann. "Julian Tenison Woods, Richard Owen and Ancient Australia." *Journal and Proceedings, Royal Society of New South Wales* 125 (1992): 107–10.

Playfair, Lyon. "The Study of Abstract Science Essential to the Progress of Industry." In *Records of the School of Mines and of Science Applied to the Arts,* vol. 1, no. 1, 23–48. London: Longman, Brown, Green and Longmans, 1852.

Porter, Roy. "Bones, Stones and Buckland." *Nature* 306 (1983): 813.

Portmann, Adolf. *Biologische Fragmente zu einer Lehre vom Menschen.* 2nd ed. Basel: Schwabe, 1951.

Powell, Baden. *Essays on the Spirit of the Inductive Philosophy, the Unity of Worlds, and the Philosophy of Creation.* London: Longman, Brown, Green, and Longmans, 1855.

Powell, Baden, and Richard Owen. "Abstract of Professor Owen's View of the Vertebral Structure and Its Archetype." In *Essays on the Spirit of the Inductive Philosophy,* by Baden Powell, 490–95. London: Longman, Brown, Green, and Longmans, 1855.

Prause, Marianne. *Carl Gustav Carus: Leben und Werk.* Berlin: Deutscher Verlag für Kunstwissenschaft, 1968.

Prichard, James Cowles. *Researches into the Physical History of Mankind.* Vol. 1. 4th ed. London: Houlston and Stoneman, 1851.

[Pycroft, George]. *A Report: A Sad Case, Recently Tried before the Lord Mayor, Owen versus Huxley, in Which Will Be Found Fully Given the Merits of the Great Recent Bone Case.* London, 1863.

Pym, Horace N., ed. *Memories of Old Friends, Being Abstracts from the Journals and Letters of Caroline Fox of Penjerrick, Cornwall from 1835 to 1871.* 2 vols. London: Smith Elder, 1882.

Quatrefages de Bréau, Jean-Louis-Armand de. *L'espèce humaine.* 2nd ed. Paris: Baillière, 1877.

Raine, Kathleen. "Thomas Taylor, Plato and the English Romantic Movement." *British Journal of Aesthetics* 8 (1968): 99–123.

Reade, W. Winwood. *Savage Africa: Being the Narrative of a Tour in Equatorial, Southwestern, and Northwestern Africa; with Notes on the Habits of the Gorilla; on the Existence of Unicorns and Tailed Men; on the Slave-Trade; on the Origin, Character, and Capabilities of the Negro, and on the Future Civilization of Western Africa.* New York: Harper and Brothers, 1864.

Rehbock, Philip F. "Huxley, Haeckel and the Oceanographers: The Case of *Bathybius Haeckelii.*" *Isis* 66 (1975): 504–33.

______. *The Philosophical Naturalists: Themes in Early Nineteenth-Century British Biology.* Madison: University of Wisconsin Press, 1983.

______. "Transcendental Anatomy." In *Romanticism and the Sciences,* edited by Andrew Cunningham and Nicholas Jardine, 144–60. Cambridge: Cambridge University Press, 1990.

Reid, Stuart J. *A Sketch of the Life and Times of the Rev. Sydney Smith.* 2nd ed. London: Sampson Low, 1884.

Reid, T. Wemyss. *The Life, Letters, and Friendships of Richard Monckton Milnes, First Lord Houghton.* 2nd ed. 2 vols. London: Cassell, 1890.

______. 1899. *Memoirs and Correspondence of Lyon Playfair.* London: Cassell.

Richards, Evelleen. "'Metaphorical Mystifications': The Romantic Gestation of Nature in British Biology." In *Romanticism and the Sciences,* edited by Andrew Cunningham and Nicholas Jardine, 130–43. Cambridge: Cambridge University Press, 1990.

———. "A Political Anatomy of Monsters, Hopeful and Otherwise: Teratogeny, Transcendentalism, and Evolutionary Theorizing." *Isis* 85 (1994): 377–411.

———. "A Question of Property Rights: Richard Owen's Evolutionism Reassessed." *British Journal for the History of Science* 20 (1987): 129–71.

Richards, Robert J. *The Meaning of Evolution: The Morphological Construction and Ideological Reconstruction of Darwin's Theory.* Chicago: University of Chicago Press, 1992.

———. *The Romantic Conception of Life: Science and Philosophy in the Age of Goethe.* Chicago: University of Chicago Press, 2002.

Ridley, Mark. "Coadaptation and the Inadequacy of Natural Selection." *British Journal for the History of Science* 15 (1982): 45–68.

Ritvo, Harriet. *The Animal Estate: The English and Other Creatures in the Victorian Age.* Cambridge, MA: Harvard University Press, 1987.

———. *The Platypus and the Mermaid and Other Figments of the Classifying Imagination.* Cambridge, MA: Harvard University Press, 1997.

———. Review of *Richard Owen: Victorian Naturalist*, by Nicolaas Rupke. *Victorian Studies* 39 (1995): 298–300.

Roget, Peter Mark. *Animal and Vegetable Physiology Considered with Reference to Natural Theology.* 2 vols. London: W. Pickering, 1834.

Rolleston, George. "On the Affinities of the Brain of the Orang Outang." *Natural History Review*, 1861, 201–17.

———. "On the Distinctive Characters of the Brain in Man and in the Anthropomorphous Apes." *Medical Times and Gazette*, Oct. 18, 1862, 418–20.

Ross, Dale Lloyd. "A Survey of Some Aspects of the Life and Work of Sir Richard Owen, K.C.B., together with a Working Handlist of the Owen Papers at the Royal College of Surgeons of England." PhD diss., University of London, 1972.

Rosse, William Parsons, Lord. "Address." *Proceedings of the Royal Society* 6 (1850–54): 102–13.

Rothschuh, K. E. *Geschichte der Physiologie.* Berlin: Springer, 1953.

Rowse, A. L. *The Controversial Colensos.* Redruth, Cornwall: Dyllansow Truran, 1989.

Rudwick, M. J. S. *The Meaning of Fossils: Episodes in the History of Palaeontology.* London: Macdonald, 1972.

———. *Scenes from Deep Time: Early Pictorial Representations of the Prehistoric World.* Chicago: University of Chicago Press, 1992.

Rupke, Nicolaas, A. "*Bathybius Haeckelii* and the Psychology of Scientific Discovery." *Studies in the History and Philosophy of Science* 7 (1976): 53–62.

———. "Caves, Fossils and the History of the Earth." In *Romanticism and the Sciences*, edited by Andrew Cunningham and Nicholas Jardine, 241–62. Cambridge: Cambridge University Press, 1990.

———. "Darwin's Choice." In *Biology and Ideology*, edited by Denis Alexander and Ronald L. Numbers. Forthcoming.

———. "'The End of History' in the Early Picturing of Geological Time." *History of Science* 36 (1998): 61–90.

———. *The Great Chain of History: William Buckland and the English School of Geology (1814–1849).* Oxford: Clarendon Press, 1983.

———. "Metonymies of Empire: Visual Representations of Prehistoric Times." In *Non-verbal Communication in Science prior to 1900*, edited by Renato Mazzolini, 513–28. Florence: Olschki, 1993.

______. "Neither Creation nor Evolution: The Third Way in Mid–nineteenth Century Thinking about the Origin of Species." *Annals of the History and Theory of Biology* 10 (2005): 143–72.

______. "The Origin of Species from Linnaeus to Darwin." In *Aurora Torealis: Studies in the History of Science and Ideas in Honor of Tore Frängsmyr*, edited by Marco Beretta, Karl Grandin, and Svante Lindqvist, 73–87. Sagamore Beach, MA: Science History Publications, 2008.

______. "Oxford's Scientific Awakening and the Role of Geology." In *The History of the University of Oxford*, edited by Michael G. Brock and Marck C. Curthoys, vol. 6, 543–62. Oxford: Clarendon Press, 1997.

______. "Richard Owen: Evolution ohne Darwin." In *Die Rezeption von Evolutionstheorien im 19. Jahrhundert*, edited by Eve-Marie Engels, 214–24. Frankfurt am Main: Suhrkamp Verlag, 1995.

______. "Richard Owen's Hunterian Lectures on Comparative Anatomy and Physiology, 1837–55." *Medical History* 29 (1985): 237–58.

______. "Richard Owen's Vertebrate Archetype." *Isis* 84 (1993): 231–51.

______. *Richard Owen: Victorian Naturalist.* New Haven, CT: Yale University Press, 1994.

______. "The Road to Albertopolis: Richard Owen (1804–92) and the Founding of the British Museum of Natural History." In *Science, Politics and the Public Good*, edited by Nicolaas A. Rupke, 63–89. Basingstoke: Macmillan, 1988.

Ruse, Michael. *The Darwinian Revolution: Science Red in Tooth and Claw.* Chicago: University of Chicago Press, 1979.

______. *Monad to Man: The Concept of Progress in Evolutionary Biology.* Cambridge, MA: Harvard University Press, 1996.

Ruskin, John. *Sesame and Lilies* (1865). In *The Works of John Ruskin*, edited by E. T. Cook and A. Wedderburn, vol. 18. London: Allen, 1905.

Russell, Colin A. *Science and Social Change.* London: Macmillan, 1983.

Russell, E. S. *Form and Function: A Contribution to the History of Animal Morphology.* London: J. Murray, 1916.

Ryals, Clyde de L., and Kenneth J. Fielding. *The Collected Letters of Thomas and Jane Welsh Carlyle.* Vol. 15. Durham, NC: Duke University Press, 1987.

[Sala, G. A. H.]. "With Mr. Gorilla's Compliments." *Temple Bar* 3 (1861): 482–91.

Sanderson, Michael. *The Universities in the Nineteenth Century.* London: Routledge and Kegan Paul, 1975.

Sarjeant, William A. S., and Justin B. Delair. "An Irish Naturalist in Cuvier's Laboratory." *Bulletin of the British Museum (Natural History)*, Historical ser., 6 (1980): 245–319.

Schopenhauer, Arthur. *The World as Will and Idea.* 3 vols. London: Routledge and K. Paul, 1948.

Schroeder van der Kolk, J. L. C. *Eene Voorlezing over het Verschil tusschen Doode Natuurkrachten, Levenskrachten en Ziel.* Utrecht: Van der Post, 1835.

Schroeder van der Kolk, J. L. C., and W. Vrolik. "Note sur l'encéphale de l'orang-outang." *Natural History Review*, 1862, 111–17.

Sclater, P. L. "On Certain Principles to Be Observed in the Establishment of a National Museum of Natural History." *BAAS, Report 1870*, Transactions of the Sections, 123–28.

Secord, James A. "Extraordinary Experiment: Electricity and the Creation of Life in Victorian England." In *The Uses of Experiment*, edited by D. Gooding, T. Pinch, and Simon Schaffer, 337–83. Cambridge: Cambridge University Press, 1988.

———. *Victorian Sensation: The Extraordinary Publication, Reception, and Secret Authorship of "Vestiges of the Natural History of Creation."* Chicago: University of Chicago Press, 2000.

Sedgwick, Adam. "Anniversary Address to the Geological Society, Febr. 19, 1830." *Proceedings of the Geological Society* 1 (1834): 187–212.

———. "Anniversary Address to the Geological Society, Febr. 18, 1831." *Proceedings of the Geological Society* 1 (1834): 281–316.

———. *A Discourse on the Studies of the University of Cambridge.* 5th ed. Cambridge: John Deighton, 1850.

[———]. "Natural History of Creation." *Edinburgh Review* 82 (1845): 1–85.

Seeley, H. G. "Remarks on Professor Owen's Monograph on *Dimorphodon.*" *Annals and Magazine of Natural History*, 4th ser., 6 (1870): 129–52.

Sharpey, William. "Anatomy and Physiology: Introductory Lectures." *Lancet*, no. 1 (1840–41): 73–78, 142–47, 281–85, 425–28, 489–93.

Sheets-Pyenson, Susan. "Cathedrals of Science: The Development of Colonial Natural History Museums during the Late Nineteenth Century." *History of Science* 25 (1986): 279–300.

———. *Cathedrals of Science: The Development of Colonial Natural History Museums during the Late Nineteenth Century.* Kingston, Ontario: McGill-Queen's University Press, 1988.

———. "Horse Race: John William Dawson, Charles Lyell, and the Competition over the Edinburgh Natural History Chair in 1854–1855." *Annals of Science* 49 (1992): 461–78.

Siebold, C. T. E. von. 1856. *Über wahre Parthenogenesis bei Schmetterlingen und Bienen. Ein Beitrag zur Fortpflanzungsgeschichte der Thiere.* Leipzig: Engelmann, 1856.

Siebold, C. T. E. von, and H. Stannius. *Lehrbuch der vergleichenden Anatomie der wirbellosen Thiere.* 2 vols. Berlin: Veit, 1846–48.

Silliman, Benjamin. *A Visit to Europe in 1851.* 2 vols. New York: G. P. Putnam, 1853 (both vols. also give 1854 as date of publication).

Simcox, Edwin W. *Homer's Iliad Translated into English Hexameters.* London: Jackson, Walford and Hodder, 1865.

Simpson, George Gaylord. *Discoverers of the Lost World: An Account of Some of Those Who Brought Back to Life South American Mammals Long Buried in the Abyss of Time.* New Haven, CT: Yale University Press, 1984.

Singer, Charles. *A Short History of Biology: A General Introduction to the Study of Living Things.* Oxford: Clarendon Press, 1931.

Sloan, Phillip R. "Darwin, Vital Matter, and the Transformism of Species." *Journal of the History of Biology* 19 (1986): 369–445.

———, ed. *Richard Owen: The Hunterian Lectures in Comparative Anatomy, May–June 1837.* With an introductory essay and commentary. London: Natural History Museum, 1992.

———. "Whewell's Philosophy of Discovery and the Archetype of the Vertebrate Skeleton: The Role of German Philosophy of Science in Richard Owen's Biology." *Annals of Science* 60 (2003): 39–61.

Smith, Christopher U. M. "The Hippopotamus Test: A Controversy in Nineteenth-Century Brain Science." *Cogito* (supplement to the *Italian Journal of Neurological Sciences*) 1 (1992): 69–74.

———. "Owen and Huxley: Unfinished Business." *Endeavour* 22 (1998): 110–13.

———. "Worlds in Collision: Owen and Huxley on the Brain." *Science in Context* 10 (1997): 343–65.

Smith, Roger. "The Human Significance of Biology: Carpenter, Darwin and the *Vera Causa*." In *Nature and the Victorian Imagination*, edited by U. C. Knoepflmacher and G. B. Tennyson, 216–30. Berkeley and Los Angeles: University of California Press, 1977.

Sömmerring, Samuel Thomas. *Ueber das Organ der Seele*. Königsberg: Nicolovius, 1796.

Spary, Emma C. *Utopia's Garden: French Natural History from Old Regime to Revolution*. Chicago: University of Chicago Press, 2000.

[Spencer, Herbert]. "Owen on the Homologies of the Vertebrate Skeleton." *British and Foreign Medico-Chirurgical Review* 44 (1858): 400–416.

[______]. "The Ultimate Laws of Physiology." *National Review* 5 (1857): 332–55.

______. *The Works of Herbert Spencer*. 21 vols. Osnabrück: Zeller, 1966.

Spix, J. B. *Cephalogenesis sive Capitis Ossei Structura, Formatio et Significatio per Omnes Animalium Classes, Familias, Genera ac Aetates Digesta, atque Tabulis Illustrata, Legesque Simul Psychologiae, Cranioscopiae et Physionomiae Inde Derivatae*. Munich: Hübschmann, 1815.

Stafford, Robert A. "Geological Surveys, Mineral Discoveries, and British Expansion, 1835–71." *Journal of Imperial and Commonwealth History* 12 (1984): 5–32.

______. *Scientist of Empire: Sir Roderick Murchison, Scientific Exploration and Victorian Imperialism*. Cambridge: Cambridge University Press, 1989.

Stamp, Gavin, and Colin Amery. *Victorian Buildings of London, 1837–1887*. London: Architectural Press, 1980.

Stearn, William T. *The Natural History Museum at South Kensington*. London: Heinemann, 1981.

Steenstrup, J. J. S. *On the Alternation of Generations; or, the Propagation and Development of Animals through Alternate Generations: a Peculiar Form of Fostering the Young in the Lower Classes of Animals*. London: Ray Society, 1845.

Strahan, Ronald, et al. *Rare and Curious Specimens: An Illustrated History of the Australian Museum, 1827–1979*. Sydney: Australian Museum, 1979.

Strickland, Hugh Edwin. "On the Structural Relations of Organized Beings." *Philosophical Magazine* 28 (1846): 354–64.

Taylor, D. W. "William Sharpey." *Medical History* 15 (1971): 421–59.

[Taylor, Tom]. "The Clubs of London." *National Review* 4 (1857): 295–334.

______. *Leicester Square*. London: Bickers and Son, 1874.

Temple, Frederick, et al. *Essays and Reviews*. 10th ed. London: Longman, Green, Longman, and Roberts, 1862.

Tennyson, Hallam. *Alfred Lord Tennyson: A Memoir*. 2 vols. London: Macmillan, 1897.

[Thynne, Robert]. "The Explorers of Australia." *Edinburgh Review* 116 (1862): 1–46.

Tiedemann, Friedrich. *Das Hirn des Negers mit dem des Europäers und Orang-Outang verglichen*. Facsimile with an introduction by Hans-Konrad Schmutz. Marburg: Basilisken-Presse, 1984.

______. "Hirn des Orang-Outangs mit dem des Menschen verglichen." *Zeitschrift für Physiologie* 2 (1826): 17–28.

______. *Icones Cerebri Simiarum et Quorundam Mammalium Rariorum*. Heidelberg: Mohr et Winter, 1821.

______. "On the Brain of the Negro, Compared with That of the European and the Orang-outang." *Philosophical Transactions*, no. 2 (1836): 497–527.

Timbs, John. *Club Life of London with Anecdotes of the Clubs, Coffee-Houses and Taverns of the Metropolis during the 17th, 18th, and 19th Centuries*. 2 vols. London: Richard Bentley, 1866.

______. *The Yearbook of Facts in Science and Art*. London: Simpkin, Marshall, 1852.

Todd, Robert B., ed. *The Cyclopaedia of Anatomy and Physiology.* 5 vols. London: Sherwood, Gilbert, and Piper, 1835–59.

Todhunter, I. *William Whewell, D.D., Master of Trinity College, Cambridge: An Account of His Writings with Selections from His Literary and Scientific Correspondence.* 2 vols. London: Macmillan, 1876.

Torrens, Hugh. "When Did the Dinosaur Get Its Name?" *New Scientist,* Apr. 4, 1992, 40–44.

Tree, Isabella. *The Ruling Passion of John Gould: A Biography of the Bird Man.* London: Barrie and Jenkins, 1991.

Trevelyan, George O. *Life and Letters of Lord Macaulay.* Popular ed. London: Harper, 1901.

Trevor-Roper, Hugh. *Hermit of Peking: The Hidden Life of Sir Edmund Backhouse.* Harmondsworth, Middlesex: Penguin, 1978.

Triqueti, H. J. F. de. 1861. *Les trois musées de Londres: Le British Museum, la National Gallery, le South Kensington Museum.* Paris: chez l'auteur, 1861.

Tuckwell, W. *Reminiscences of Oxford.* 2nd ed. London: Smith, Elder, 1907.

Turner, Frank M. *Between Science and Religion: The Reaction to Scientific Naturalism in Late Victorian England.* New Haven, CT: Yale University Press, 1974.

———. *Contesting Cultural Authority: Essays in Victorian Intellectual Life.* Cambridge: Cambridge University Press, 1993.

———. "The Victorian Conflict between Science and Religion: A Professional Dimension." *Isis* 69 (1978): 356–76.

———. "Victorian Scientific Naturalism and Thomas Carlyle." *Victorian Studies* 18 (1974–75): 325–43.

Turner, Gerard, ed. *Patronage of Science in the Nineteenth Century.* Leiden: Noordhoff, 1976.

Turner, William, ed. *The Anatomical Memoirs of John Goodsir: With a Biographical Memoir by Henry Lonsdale.* 2 vols. Edinburgh: Adam and Charles Black, 1868.

———, ed. *Scientific Papers and Addresses by George Rolleston: With a Biographical Sketch by Edward B. Tylor.* 2 vols. Oxford: Clarendon Press, 1884.

Ulrich, John McAllister. "Thomas Carlyle, Richard Owen, and the Paleontological Articulation of the Past." *Journal of Victorian Culture* 11 (2006): 30–58.

Vidler, Edward A. "Notable Naturalists. II. Professor Owen." *Victorian Naturalist* 45 (1928): 74–77.

Visser, R. P. W. *The Zoological Work of Petrus Camper (1722–1789).* Amsterdam: Rodopi, 1985.

Vogt, Carl. *Köhlerglaube und Wissenschaft: Eine Streitschrift gegen Hofrath Rudolph Wagner in Göttingen.* 4th ed. Giessen: Ricker, 1856.

———. *Vorlesungen über den Menschen, seine Stellung in der Schöpfung und in der Geschichte der Erde.* 2 vols. Giessen: Ricker, 1863.

Vorzimmer, Peter J. *Charles Darwin: The Years of Controversy; The "Origin of Species" and Its Critics, 1859–1882.* Philadelphia: Temple University Press, 1970.

Vrolik, Willem. "Quadrumana." In *Cyclopaedia of Anatomy and Physiology,* edited by Robert B. Todd, vol. 4, no. 1, 194–221. London: Sherwood, Gilbert, and Piper, 1847–49.

———. *Recherches d'anatomie comparée sur le chimpansé.* Amsterdam: Müller, 1841.

[Wace, Henry]. "Scientific Lectures—Their Use and Abuse." *Quarterly Review* 145 (1878): 35–61.

[Waddy, Frederick]. *Cartoon Portraits and Biographical Sketches of Men of the Day: The Drawings by Frederick Waddy.* 2nd ed. London: Tinsley Brothers, 1874.

Wagner, Rudolph. *Gespräche mit Carl Friedrich Gauss in den letzten Monaten seines Lebens.* Edited by Heinrich Rubner. Göttingen: Vandenhoeck und Ruprecht, 1975.

______. *Der Kampf um die Seele vom Standpunkt der Wissenschaft.* Göttingen: Dieterich, 1857.

______. *Lehrbuch der vergleichenden Anatomie.* Leipzig: Voss, 1834–35.

______. *Lehrbuch der Zootomie: Anatomische Charakteristik der Thierklassen.* 2 vols. (vol. 2 by Heinrich Prey and Rudolph Leuckart). Leipzig: Voss, 1843–47.

______. *Menschenschöpfung und Seelensubstanz.* Göttingen: Wigand, 1854.

______. "Ueber die Hirnbildung des Menschen und der Quadrumanen und deren Verhältniss zur zoologischen Systematik, mit besonderer Rücksicht auf die Ansichten von Owen, Huxley und Gratiolet." *Archiv für Naturgeschichte* 27 (1861): 63–80.

______. "Ueber die Hirnfunctionen mit besonderer Beziehung zur allgemeinen Zoologie." *Archiv für Naturgeschichte* 27 (1861): 171–80.

______. *Ueber Wissen und Glauben mit besonderer Beziehung zur Zukunft der Seelen.* Göttingen: Wigand, 1854.

______. "Upon the Structure of the Brain in Man and Monkeys, and Its Bearing upon Classification, with Special Reference to the Views of Owen, Huxley and Gratiolet." *American Journal of Science,* 2nd ser., 34 (1862): 188–99.

______. *Vorstudien zu einer wissenschaftlichen Morphologie und Physiologie des menschlichen Gehirns als Seelenorgan.* 2 vols. Göttingen: Dieterich, 1860–62.

Ward, Peter Douglas. *In Search of Nautilus: Three Centuries of Scientific Adventures in the Deep Pacific to Capture a Prehistoric—Living—Fossil.* New York: Simon and Schuster, 1988.

Watts, Isaac. *Logick: or, the Right Use of Reason in the Enquiry after Truth, with a Variety of Rules to Guard against Error, in the Affairs of Religion and Human Life, as well as in the Sciences.* London: J. Clark and R. Hett, E. Matthews, and R. Ford, 1725.

Weismann, August. *Essays upon Heredity and Kindred Biological Problems.* Oxford: Clarendon Press, 1889.

Wells, Martin. "Legend of the Living Fossil." *New Scientist,* Oct. 23, 1986, 36–41.

Wendt, Herbert. *In Search of Adam: The Story of Man's Quest for the Truth about His Earliest Ancestors.* Boston: Houghton Mifflin, 1956.

Westwood, J. O. "Illustrations of the Relationships Existing amongst Natural Objects, Usually Termed Affinity and Analogy, Selected from the Class of Insects." *Transactions of the Linnean Society of London* 18 (1841): 409–21.

Wetzels, Walter D. "Johann Wilhelm Ritter: Romantic Physics in Germany." In *Romanticism and the Sciences,* edited by Andrew Cunningham and Nicholas Jardine, 199–212. Cambridge: Cambridge University Press, 1990.

Whewell, William. Anniversary address to the Geological Society, 16 Febr. 1838. *Proceedings of the Geological Society* 2 (1838): 624–49.

______. "Comte and Positivism." *Macmillan's Magazine* 13 (1866): 353–63.

______. *History of the Inductive Sciences, from the Earliest to the Present Time.* 3 vols. London: John W. Parker, 1857.

______. *Indications of the Creator, Extracts Bearing upon Theology, from the History and Philosophy of the Inductive Sciences.* London: John W. Parker, 1845.

______. *Novum Organon Renovatum: Being the Second Part of the Philosophy of the Inductive Sciences.* 3rd ed. London: John W. Parker and Son, 1858.

______. "On Plato's Survey of the Sciences." *Transactions of the Cambridge Philosophical Society* 9 (1856): 582–89.

______. "On the Platonic Theory of Ideas." *Transactions of the Cambridge Philosophical Society* 9 (1856): 94–104.

______. *The Philosophy of the Inductive Sciences, Founded upon Their History.* 2 vols. London: John W. Parker, 1840.

______. "Second Memoir on the Fundamental Antithesis of Philosophy." *Transactions of the Cambridge Philosophical Society* 8 (1849): 614–16.

[Wilberforce, Samuel]. "Darwin's *Origin of Species.*" *Quarterly Review* 108 (1860): 225–64.

[______]. "Essays and Reviews." *Quarterly Review* 109 (1861): 248–305.

Wilson, Leonard G. "The Gorilla and the Question of Human Origins: The Brain Controversy." *Journal of the History of Medicine and Allied Sciences* 51 (1996): 184–207.

Winsor, Mary P. *Reading the Shape of Nature: Comparative Zoology at the Agassiz Museum.* Chicago: University of Chicago Press, 1991.

______. *Starfish, Jellyfish, and the Order of Life.* New Haven, CT: Yale University Press, 1976.

Woodward, Horace B. *The History of the Geological Society of London.* London: Longmans, Green, 1908.

[Woodward, John]. *Brief Instructions for Making Observations in All Parts of the World: as also for Collecting, Preserving, and Sending over Natural Things. Being an Attempt to Settle an Universal Correspondence for the Advancement of Knowledge both Natural and Civil.* London: Richard Wilkin, 1696.

Yanni, Carla. *Nature's Museums: Victorian Science and the Architecture of Display.* Baltimore, MD: Johns Hopkins University Press, 1999.

Yeo, Richard. "The Principle of Plenitude and Natural Theology in Nineteenth-Century Britain." *British Journal for the History of Science* 19 (1986): 263–82.

______. "William Whewell, Natural Theology and the Philosophy of Science in Mid Nineteenth-Century Britain." *Annals of Science* 36 (1979): 493–516.

Young, Robert M. *Darwin's Metaphor: Nature's Place in Victorian Culture.* Cambridge: Cambridge University Press, 1985.

______. *Mind, Brain and Adaptation in the Nineteenth Century: Cerebral Localization and Its Biological Context from Gall to Ferrier.* Oxford: Clarendon Press, 1970.

Index